ADVANCED SPATIAL STATISTICS

ADVANCED STUDIES IN THEORETICAL AND APPLIED ECONOMETRICS
VOLUME 12

For a complete list of volumes in this series see final page of this volume.

Advanced Spatial Statistics

Special Topics in the Exploration of Quantitative Spatial Data Series

by

Daniel A. Griffith

(University of New York, Buffalo)

1988 **KLUWER ACADEMIC PUBLISHERS**
DORDRECHT / BOSTON / LANCASTER

Distributors

for the United States and Canada: Kluwer Academic Publishers, 101 Philip Drive, Norwell, MA 02061, USA
for the UK and Ireland: Kluwer Academic Publishers, MTP Press Limited, Falcon House, Queen Square, Lancaster LA1 1RN, UK
for all other countries: Kluwer Academic Publishers Group, Distribution Center, P.O. Box 322, 3300 AH Dordrecht, The Netherlands

Library of Congress Cataloging in Publication Data

Griffith, Daniel A.
Advanced spatial statistics.

(Advanced studies in theoretical and applied econometrics ; 12)
1. Spatial analysis (Statistics) I. Title.
II. Series: Advanced studies in theoretical and applied econometrics ; v. 12.
QA278.2.G74 1988 519.5 87-26259

ISBN 90-247-3627-7
ISBN 90-247-2622-0 (series)

PRINTED IN THE NETHERLANDS

In memory of Mary Jane Griffith
whose untimely and tragic death
has had such a profound impact upon my life

Preface

In recent years there has been a growing interest in and concern for the development of a sound spatial statistical body of theory. This work has been undertaken by geographers, statisticians, regional scientists, econometricians, and others (e.g., sociologists). It has led to the publication of a number of books, including Cliff and Ord's *Spatial Processes* (1981), Bartlett's *The Statistical Analysis of Spatial Pattern* (1975), Ripley's *Spatial Statistics* (1981), Paelinck and Klaassen's *Spatial Econometrics* (1979), Ahuja and Schachter's *Pattern Models* (1983), and Upton and Fingleton's *Spatial Data Analysis by Example* (1985). The first of these books presents a useful introduction to the topic of spatial autocorrelation, focusing on autocorrelation indices and their sampling distributions. The second of these books is quite brief, but nevertheless furnishes an eloquent introduction to the relationship between spatial autoregressive and two-dimensional spectral models. Ripley's book virtually ignores autoregressive and trend surface modelling, and focuses almost solely on point pattern analysis. Paelinck and Klaassen's book closely follows an econometric textbook format, and as a result overlooks much of the important material necessary for successful spatial data analysis. It almost exclusively addresses distance and gravity models, with some treatment of autoregressive modelling. *Pattern Models* supplements Cliff and Ord's book, which in combination provide a good introduction to spatial data analysis. Its basic limitation is a preoccupation with the geometry of planar patterns, and hence is very narrow in scope. Finally, Upton and Fingleton's book pays considerable attention to the spatial autoregressive model and accompanying spatial autocorrelation issues. It seems to be a natural sequel for Cliff and Ord's book.

As is indicated by this wide range of books, the subject of concern here has been discussed in different ways by members of a variety of disciplines. I will refer to it here as spatial statistics; but, it often is labelled spatial econometrics. The distinction actually is one of approach taken, and hence emphasis. But it has led to considerable controversy, as is indicated by the exchange between Haining and Anselin (1986). Haining has argued that the research agenda that needs to be pursued in order to develop a body of spatial statistical theory should include the themes of boundary value effects, development of autocorrelation tests for residuals from spatial models, covariance structures for geographical data, methods for spatial interpolation, and model applications. Griffith (1987) strongly agrees with this viewpoint. Furthermore, Haining's beliefs are mirrored to a large degree in this book. In contrast, Anselin has proposed the themes of (a) approaches to model validation in spatial econometrics, (b) robust techniques, soft spatial econometrics, and qualitative modelling, and (c) the use of non-parametric statistics and heuristics. The main problem, as he sees it, is that the spatial statistics and spatial econometrics literatures tend to be rather self-contained, with little cross-referencing occurring between them.

The purpose of this book is two-fold. First, it attempts to bring together recent works of a somewhat advanced nature that are concerned with spatial data analysis, highlighting much of my published and unpublished research findings uncovered during the time interval between my doctoral dissertation in geography (1978) and my master's thesis in statistics (1985). The importance of such a book lies in both the existing literature gap, alluded to by both Haining (1986) and Griffith (1987), and a need to disseminate research findings about advanced spatial statistics in a more palatable form. The second goal of this book is to help nurture a dialogue between spatial statisticians and spatial econometricians, in an attempt to address, at least in part, the major problem identified by Anselin (1986).

As mentioned previously, the materials in this book have been compiled largely from my research findings over the past decade. These materials first were organized and revised into a graduate level course that has been taught at SUNY/Buffalo. Next, they were modified to some degree after my completion of Keith Ord's spatial processes course at The Pennsylvania State University. And, they were finalized during my two-weeks visit to The Netherlands Economic Institute during the Spring of 1986.

I am indebted to Diane Griffith, who typed the original manuscript and all of its horrendous equations with their formidable notation, to Greg Theisen for preparing the cartographic work, and to Jean-Pierre Ancot for his invitation to write this book for the 'Advanced Studies in Theoretical and Applied Econometrics' series.

DANIEL A. GRIFFITH

BUFFALO, NEW YORK
JANUARY 20, 1987

Anselin, L., 1986, Some further notes on Spatial models and regional science, *Journal of Regional Science*, Vol. 26: 799–802.

Griffith, D., 1987, Toward a theory of spatial statistics: another step forward, *Geographical Analysis*, Vol. 19: 69–82.

Haining, R., 1986, Spatial models and regional science: a comment on Anselin's paper and research directions, *Journal of Regional Science*, Vol. 26: 793–798.

Contents

CHAPTER 1

Introduction to Spatial Statistics and Data Handling

Traditionally science has been concerned, in part, with the study of structure amongst variables. With the development of statistical concepts, such as the runs statistic, and the advent of time-series analysis, more recent attention has been turned to the study of structure amongst observations of a single variable. Within this context the notion of autocorrelation has received considerable treatment (see Bhat, 1974; Crowder, 1976; Silvey, 1961), with the special case of spatial autocorrelation receiving extensive consideration (see, for example, Cliff and Ord, 1981).

Autocorrelation refers to the pairwise correlation of univariate observations; correlation retains its classical meaning of association, whereas 'auto-' means self. The term spatial describes the manner in which this self-correlation arises. Moreover, the autocorrelation in question is attributable to the configurational arrangement of observations on a two-dimensional surface. This particular geometric property has been the focus of many investigations in the realm of spatial statistics (i.e., statistics dealing with realizations of random variables arranged over a two-dimensional surface). This notion is at odds with classical statistical analysis, which assumes that such an ordering does not exist, and hence observations always are pairwise independent (see Gould, 1970).

Spatial autocorrelation has two primary connotations as a statistical property of spatial data sets (Haining, 1980; Haining, Griffith and Bennett, 1983). It has been viewed most often as a property of the arrangement of n areal units upon a planar surface. Suppose there exists a set of n values of some variable

X, say $\{x_i; i = 1, 2, \ldots, n\}$. One observes a spatial distribution in which each of these n values is affiliated with one of the n areal units. One question that may be asked at this point has to do with the structure of an observed set of values, and may be phrased as follows:

> if the set of variable values $\{x_i\}$ had been allocated in a random fashion to the set of n areal units appearing on the two-dimensional map, would this observed spatial distribution be a likely outcome of such an allocation procedure?

In order to answer this particular question, several features of this problem need to be recognized. First, it is necessary to define a statistic that can discriminate between different arrangements of the set $\{x_i\}$ on a two-dimensional surface. Second, the different arrangements that are possible need to be identified, in order to construct the appropriate sampling distribution.

A first step in such an identification is the counting of these possible arrangements. There exist n values and n areal units. Once a value has been allocated to an areal unit, it is not available for allocation to any of the remaining areal units. To begin with then, there are n choices for the first areal unit. After selection of an x_i value, there remain $(n - 1)$ choices for the second areal unit. After selection of one of the remaining available x_i values for this areal unit, there remain $(n - 2)$ choices for the third areal unit. This selection procedure continues until all x_i's have been allocated, which is equivalent to each areal unit having had an x_i value allocated to it. Since there are n possibilities for the first choice, and $(n - 1)$ possibilities for the second choice, then there are $n(n - 1)$ possibilities for the first and second choices together. Extending this counting principle for all n areal units results in $n!$ possible spatial distributions; hence, one is looking at all possible permutations of the set $\{x_i\}$ over the map, a situation that is analogous to sampling without replacement, where order is important. This counting principle assumes that the set $\{x_i\}$ consists of n distinct values; appropriate adjustments can be made when this is not the case.

A third feature builds upon this permutation viewpoint, which is equivalent to an experimental approach using randomization.

This feature has to do with whether or not the allocation that is observed is representative of one resulting from a random allocation of the set $\{x_i\}$ to the n areal units. The statistic needed to measure spatial autocorrelation in this context, then, should have a sample mean that corresponds to zero autocorrelation. Furthermore, critical values for it need to be established in order to determine which spatial distributions of the x_i's are not representative of a random allocation. This classification of statistic values into the two groups of representative and non-representative (of zero spatial autocorrelation) is consistent with the establishment of 10%, 5% and 1% levels of significance in classical statistics.

A final salient feature refers to the nature of the map pattern that could be observed. On the one hand, if similar x_i values have a high propensity to cluster on the map, exceeding that due merely to chance, then the statistic in question should be indicative of positive spatial autocorrelation. On the other hand, if dissimilar x_i values have a high propensity to cluster, then negative spatial autocorrelation should be detected. Similarity of values in this sense is relative. It refers to ordering the x_i's in ascending order. If those x_i values that appear at the beginning of this ordering tend to occupy juxtaposed areal units, those that appear in the middle of this ordering tend to occupy juxtaposed areal units, and those that appear at the end of this ordering tend to occupy juxtaposed areal units, then as the value of x_i increases the values of the surrounding x_j's tend to increase. Hence, positive spatial autocorrelation exists in the corresponding map pattern. In contradistinction, if alternating areal units are allocated, in descending order, those x_i values that appear in the first half of the aforementioned ordering, while the remaining areal units are allocated, in ascending order, those x_i values that appear in the last half of this ordering, then as the value of x_i increases, the values of the surrounding x_j's tend to decrease. Hence, negative spatial autocorrelation exists in the corresponding map pattern.

The second connotation for spatial autocorrelation refers to the probability distribution that underlies a spatial distribution. As such, a random variable X is assumed to exist for each areal unit. The n variables are assumed to have the same mean, the same variance, and the same frequency distribution, say a normal distribution. Next, the observed map pattern is considered to be a random selection from a multivariate distribution. Consequently, if zero spatial autocorrelation is present amongst the random variables X_i, then a population correlation matrix **R** for these variables would be equal to the identity matrix.

Once again, if spatial autocorrelation is positive, then similar x_i values would tend to cluster on the map. If spatial autocorrelation is negative, then dissimilar x_i values would tend to cluster of the map. This perspective is overshadowed by the inability to observe more than one surface, or more than one realization of variable X, unless a simulation experiment is being performed or a space-time series is being studied. Accordingly, usually a sample of size one must be used to evaluate the nature and degree of spatial autocorrelation. Because of this constraint, the assumption of identically distributed X_i's is a fundamental requirement for obtaining analytical results. If a large number of two-dimensional surfaces can be observed, which are representative samples of a random relation between X_i's, then the average level of spatial autocorrelation measured for these surfaces should be approximately zero. As the number of observed surfaces increases to infinity, this average value should converge upon zero. Hopefully, then, when only one spatial distribution is observed, it is representative of the total number of possible spatial distributions.

This time the question that may be asked has to do with the joint distribution of the X_i variates, and may be phrased as follows:

> if the n realizations of the random variables X_i are mutually independent, is the single sample of joint realizations constituting the observed map pattern a representative one?

In order to answer this question, several features of this problem need to be recognized. Again it is necessary to define a statistic that can discriminate between different natures and degrees of spatial autocorrelation. In addition, the underlying spatial distribution needs to be tested for. If a finite number of two-dimensional realizations is available, such as in a simulation experiment, each set of areal unit X values can be tested. If only one map pattern is available, though, the assumption of identically distributed X_i's needs to be invoked. By doing so, the set of observed x_i's can be pooled into a single frequency distribution, and then this frequency distribution can be tested. Evidence exists to suggest that this latter test may be erroneous if the parent correlation matrix is not the identity matrix (see Griffith, 1980).

The spatial autocorrelation statistics being referred to here are of two types. First, a statistic can be employed that indexes the nature and degree of sampled spatial autocorrelation, much like a classical correlation coefficient. Second, a statistic can be utilized that allows the dependency structure to be modelled, much like a regression coefficient. Historically this first category of statistic was studied before modelling efforts were undertaken.

1.1. A BRIEF HISTORICAL BACKGROUND

Decades ago spatial analysts suspected that some sort of relationship existed between measurements taken on observations geographically distributed over a surface, and the underlying configuration of these observations (see, for instance, Student, 1914). This idea was counter to the fundamental statistical assumption of independent observations. Two major thrusts were pursued in an attempt to handle this assumption violation complication. On the one hand, point pattern analysis emerged as a topical area in spatial statistics. On the other hand, geostatistics emerged as a second topical area. A more recent refinement of these two approaches is the set of directional statistics. Because of the dissatisfaction with solutions offered by these approaches, considerable attention has been focused on the spatial autocorrelation approach in recent years.

Much has been written about point pattern analysis, especially during the past two decades (see Upton and Fingleton, 1985). Among its basic drawbacks are that (1) it is concerned with areal unit observations that are punctiform in nature, rather than areal units that are the outcome of a partitioning of some geographic surface into mutually exclusive and collectively exhaustive observations, and (2) it is concerned with the spacing or density of punctations, rather than quantitative measures that can be made at such discrete points. Haining, Griffith and Bennett (1984) illustrate one method for circumventing this first limiting feature.

Nevertheless, these two drawbacks have hampered the general usefulness of point pattern analysis as a solution to the dependent observations problem.

For the most part, geostatistics has been an attempt to employ the classical extension of univariate statistical results to the bivariate case as a model for spatial analysis. With its inception concepts such as the spatial mean, spatial median, standard distance, and two-dimensional normal curve were added to the common parlance of spatial statistics. Unfortunately, bivariate statistical models fail to characterize two-dimensional distributions in the real world. Furthermore, the classical bivariate model also assumes independent observations (see Griffith, 1984), meaning that geostatistics fail to provide a solution to the dependent observations problem, too. More recent extensions and developments, especially in terms of kriging and the theory of regionalized variables, offer far more promise.

Consequently, the pertinent statistical setting may be viewed in a contrasting fashion. Classical statistics often uses the urn model for demonstrative purposes. This particular model consists of an urn containing balls, and these balls both display some set of discrete traits and are randomly mixed. The drawing of a ball is considered an experiment, and sampling with replacement fully exemplifies the notion of independent observations. In contradistinction, this urn model would need to be modified so that its drawings more closely resembled an experiment in which clusters of grapes, rather than separate balls, are drawn from an urn (after Stephan, 1934). This latter description better characterizes the situation faced in spatial analysis.

Interest in directional data analysis in two dimensions is somewhat of an outgrowth of point pattern and geostatistical analyses. Now the emphasis is on the orientation of data, rather than distance, density or quantitative locational information. A useful review of this literature is provided by Gaile and Burt (1980), and by Schuenemeyer (1984). But directional statistics deal with a primitive spatial concept, rather than a general solution to the dependent observations assumption. And by now diminishing returns can be seen as having set in long ago on this strategic path for finding the desired solution; devising new procedures and exploring new techniques simply does not seem to be profitable.

Rather, it appears to spatial analysts as though the more fruitful approach of parameterizing spatial autocorrelation within the context of classical statistical models should be pursued.

Returning to the earlier definition and description of spatial autocorrelation, the need was established for an index that would discriminate between various map patterns. Three such indices can be found in the literature, namely the Moran Coefficient, the Geary Ratio, and the Cliff–Ord statistic. All three of these indices have the two previously mentioned properties of (1) an expected value that corresponds to the case of zero spatial autocorrelation, and (2) the ability to discriminate between different arrangements of some set of values $\{x_i\}$ on a two-dimensional surface.

The Moran Coefficient (MC) has an expected value of $-(n-1)^{-1}$ (see Appendix 1A), has an asymptotically normal sampling distribution given either of the two connotations for spatial autocorrelation, has an analytically derived standard error (see Cliff and Ord, 1981), and may be defined as

$$\mathrm{MC} = \left(n \Big/ \sum_{i=1}^{i=n} \sum_{j=1}^{j=n} c_{ij} \right) \sum_{i=1}^{i=n} \sum_{j=1}^{j=n} c_{ij}(x_i - \bar{x})(x_j - \bar{x}) \Big/ \sum_{i=1}^{i=n} (x_i - \bar{x})^2, \tag{1.1}$$

where c_{ij} are weights applied to neighbors of areal unit i. Specifying the neighboring structure is a problem that will be discussed in the next chapter. For now it suffices to point out that weights denoted by c_{ij} are binary, taking on the values of one if two areal units are juxtaposed, and zero otherwise. Clearly MC is a coefficient of covariation, similar to the classical product moment correlation coefficient. It has a direct relationship with the prevailing level of autocorrelation.

The Geary Ratio (GR) has an expected value of unity (see Appendix 1B), has an asymptotically normal sampling distribution given either of the two connotations for spatial autocorrelation, has an analytically derived standard error (see Cliff and Ord, 1981), and may be defined as

$$\mathrm{GR} = \left[(n-1) \Big/ \left(2 \sum_{i=1}^{i=n} \sum_{j=1}^{j=n} c_{ij} \right) \right] \sum_{i=1}^{i=n} \sum_{j=1}^{j=n} c_{ij}(x_i - x_j)^2 \Big/ \sum_{i=1}^{i=n} (x_i - \bar{x})^2. \tag{1.2}$$

Clearly GR is a paired comparisons coefficient. It has an indirect relationship with the prevailing level of autocorrelation, and relates to MC as follows:

$$\text{GR} = \left[(n-1)\Big/\left(2\sum_{i=1}^{i=n}\sum_{j=1}^{j=n} c_{ij}\right)\right]\sum_{i=1}^{i=n}\sum_{j=1}^{j=n} c_{ij}[(x_i - \bar{x}) - (x_j - \bar{x})]^2 \Big/ \sum_{i=1}^{i=n}(x_i - \bar{x})^2$$

$$= \left[(n-1)\Big/\left(2\sum_{i=1}^{i=n}\sum_{j=1}^{j=n} c_{ij}\right)\right]\sum_{i=1}^{i=n}\sum_{j=1}^{j=n} c_{ij}[(x_i - \bar{x})^2 + (x_j - \bar{x})^2]\Big/ \sum_{i=1}^{i=n}(x_i - \bar{x})^2 - [(n-1)/n]\,\text{MC}. \quad (1.3)$$

The Cliff–Ord statistic merely is equation (1.1) rewritten such that weights w_{ij} replace the c_{ij} values, where

$$\sum_{j=1}^{j=n} w_{ij} = 1 \quad \text{and} \quad w_{ij} \geqslant 0.$$

One major problem with this spatial autocorrelation index approach is that while it does address the problem of dependent observations, it only allows for hypothesis testing of the presence/absence of non-zero spatial autocorrelation. Although this is the correct direction to follow, one should recall that the aim here is to parameterize spatial autocorrelation within the context of classical statistical models, hence altering the assumption of independent observations. Cliff and Ord (1973, pp. 102–104) present one means of achieving this end when the errors of a multiple linear regression model are spatially autocorrelated. Their procedure will serve as a guide to many of the undertakings in this book.

1.2. THE PRINCIPAL PROBLEM OF SPATIAL STATISTICS

The principal problem, then, is one of deriving useful statistical measures for inferential purposes in the presence of dependent observations. More specifically, if the only question were one of sample specific description, then classical statistical measures would do. The actual problem is unveiled when inferences are to be drawn. In other words, violation of the independent observations assumption erodes an inferential basis, mainly by distorting the Type I and Type II error probabilities (this should be obvious from the specification of the probability density function, as will be seen subsequently), and some means of restoring this basis constitutes the problem at hand.

Prominent features of this problem may be described in terms of statistical properties of parameter estimates, such as sufficiency, unbiasedness, consistency, and efficiency (Lindgren, 1976). Each of these four elementary statistical properties will be discussed at some time in this monograph. For now, only the notion of sufficient statistic will be explored.

A statistic is sufficient if it summarizes all of the information contained in a sample. If a sample is geographic in nature, and non-zero spatial autocorrelation is present, then part of the information that it contains should have to do with the relative

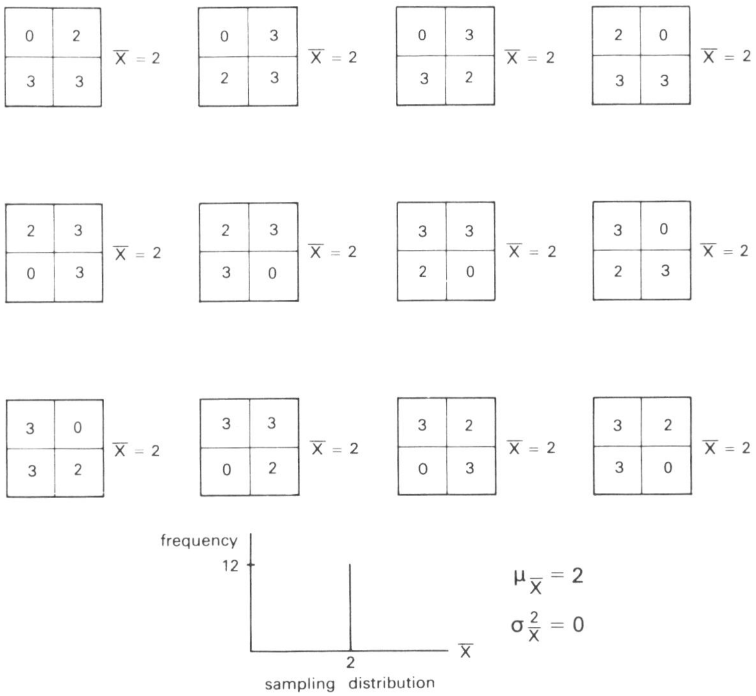

Fig. 1.1. All possible geographic distributions based upon the permutation of the set {0, 2, 3, 3} for a simple planar surface.

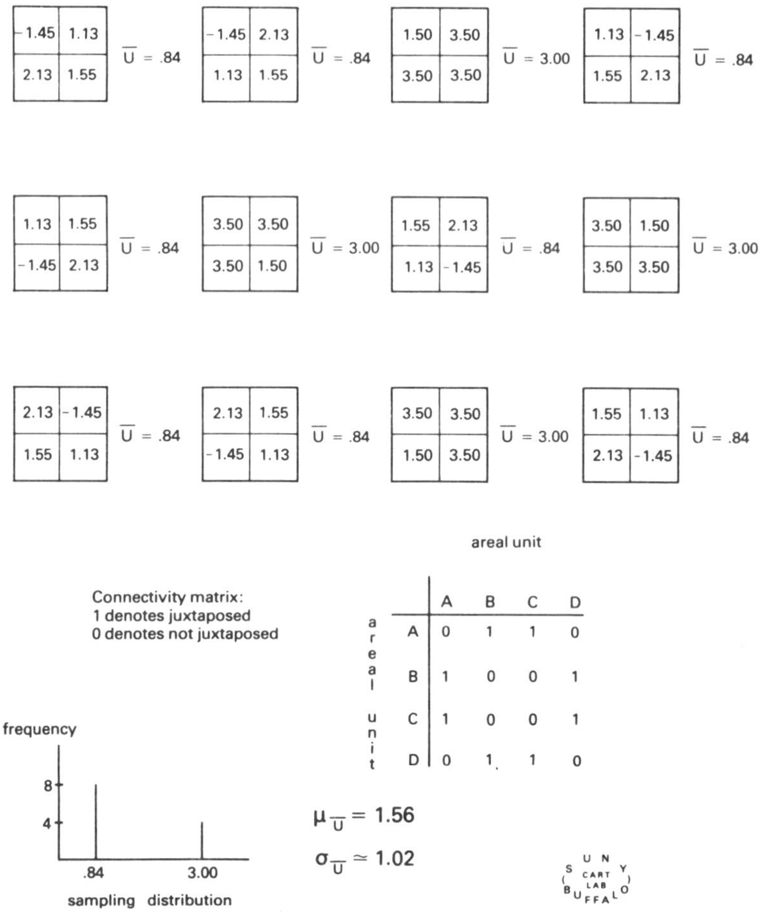

Fig. 1.2. The sampling distribution of random error constructed from Figure 1.1.

location of the areal units under study. Consider the simple geographic example appearing in Figure 1.1. Here all possible geographic distributions were constructed for the set of values {0, 2, 3, 3}. In actuality, two different spatial distributions result, namely one in which the two 3's are adjacent, and one in which the two 3's are not adjacent. There are eight geographic distributions in this first category, and four in this second category. The multiple appearances of these two basic map patterns are nothing more than rotations or reflections of any selected pair from the two groups. The accompanying sampling distribution of the sample mean is a spike (see Figure 1.1), indicating that the classical sample mean fails to embrace any locational information. Hence, it is not a sufficient statistic.

Once spatial autocorrelation is taken into account for those maps appearing in Figure 1.1, the corresponding set of geographic distributions appearing in Figure 1.2 results. Now spatial distributions in which two 3's are adjacent have an adjusted sample mean of 0.84, whereas ones where the two 3's are non-adjacent have an adjusted sample mean of 3.00. Further, the sampling distribution no longer is a spike (see Figure 1.2).

Consequently, the spatial autocorrelation approach to spatial statistics yields sufficient statistics, and holds promise as a fruitful avenue of research. This conjecture will be proven in a subsequent section.

1.3. SPATIAL SAMPLING PERSPECTIVES

The construction of sampling distributions in the preceding section alludes to another problematic issue in spatial statistics, namely the distinction between a population and a sample. This general problem has been placed in a proper context by Upton and Fingleton (1985, p. 325), who note that '[g]reat imagination has gone into turning what appears to be a population into a sample . . .'. The following six fundamental sample justifications have been put forth repeatedly in the literature: (1) a subset of areal units, (2) a permutation/randomization of values over a set of areal units, (3) subsets of items within areal units, (4) an underlying stochastic process (i.e., random noise in the geographic landscape), (5) a random surface partitioning, and (6) a time slice or space displacement.

A population is the total set of items for which some aggregate measure is to be obtained. Once a locational context is attached to these items, such that areal units become the observations, the total set constitutes a statistical geographical population. The first two sampling justifications refer to perspectives from a classical sampling framework. Suppose that there are N areal units, with n to be selected for a sample. Then sampling with replacement where order is important results in N^n possible samples. But the overriding reason for sampling in the first place is to simultaneously maximize information while minimizing costs. Because N in geography almost always is quite small (usually at most in the low hundreds), rarely can substantial cost savings be realized with this first sampling perspective. Sampling also can be done without replacement where order is unimportant, resulting in $C(N, n)$ possible samples. Again, though, substantial cost savings for geographic settings are rare.

Although with their traditional uses these two preceding sampling perspectives have not proven to be useful in geography, they do show some promise within the context of resampling theory. More specifically, sampling with replacement with order being important furnishes a foundation for the bootstrap procedure. The major problem with this resampling method with respect to spatial statistics is that the configuration of areal units is not preserved. Meanwhile, sampling without replacement with order being unimportant supplies a foundation for the jackknife procedure. Here samples are usually of size $(n - 1)$ or $(n - 2)$ and, as will be argued in a subsequent section on mising data, provides a powerful resampling method for spatial statistics. Actually these are the sorts of techniques that Costanzo (1983) had in mind.

Another classical sampling perspective is where there is no replacement, but order is important. In the limit, when $n = N$, the permutation/randomization justification occurs, resulting in $P(N, N) = N!$ possible samples. This perspective was used to construct the sampling distribution appearing in Figure 1.1. Summerfield (1983) argues that this perspective is a misapplication of the Fisherian statistical idea of 'randomization.' Rather than referring to the randomization of uncontrolled variables, all permutations of a set of N values over a geographic surface are considered to be equally likely. One appealing feature of this perspective is that the configuration of areal units is preserved. Another is that much of the current understanding about spatial autocorrelation rests upon it, as was seen earlier in the defining of spatial autocorrelation.

A third justification of sampling in spatial statistics is firmly grounded in classical sampling techniques, and refers to subsets of items within areal units. This is the classical case of stratified random sampling, where stratification is by areal unit. An illustration of this situation appears in Figure 1.3. Here the total number of possible samples is $\Pi_{i=1}^{i=k} N_i^{n_i}$. Two useful features of this perspective are that the configuration of areal units is preserved, and that even with small numbers of items and few areal units one often is able to assume an underlying normal distribution. Another is that much of the current understanding about spatial autocorrelation can be cast in terms of this perspective, as will be seen shortly.

The stochastic processes approach to justify sampling in spatial statistics is illuminating in two respects. First of all, it defines a random variable as the weighted sum of two components, namely a spatial structure component (i.e., locational information) and an independent random error component, with these errors being statistically distributed in accordance with some underlying probability model. Second, it allows Monte Carlo techniques to be employed in a geographic context, allowing the possibility of repeated measurements under the same or controlled conditions. If either case, '. . . causal processes generating the observed data can be explored by comparison with stochastically generated data or a theoretical sampling distribution' (Summerfield, 1983). The sampling distribution constructed in Figure 1.2 was done so utilizing the first of these two viewpoints. Furthermore, in part this stochastic processes perspective will form the basis of much of the spatial autocorrelation analysis undertaken in this book.

Another justification pertains to random surface partitioning, or the manner in which a planar surface is partitioned into areal units. A considerable amount of work has been published on this general topic (for an overview see Getis and Boots, 1978; Ahuja and Schachter, 1983). This approach has received substantial treatment in geography under the heading of the modifiable areal unit problem. 'The particular spatial categorization of data encountered in the [sample] is only one among an almost infinite number that are conceptually or actually possible' (Summerfield, 1983, p. 146). Summerfield (1983) incorrectly argues, however, that this sort of analogy holds only if the selection of an areal units network is random, then stating that this assumption of randomness is virtually never satisfied in practice. On the one hand, surely some systematic criterion was used to establish areal

A	B
C	D

General: $\{a_1,a_2\}$, $\{b_1,b_2\}$, $\{c_1,c_2\}$, $\{d_1,d_2\}$
Example: {0,2}, {2,4}, {0,4}, {4,6}

individual sample spaces (n = 2):

a_1,a_1	a_2,a_1	b_1,b_1	b_2,b_1	c_1,c_1	c_2,c_1	d_1,d_1	d_2,d_1
a_1,a_2	a_2,a_2	b_1,b_2	b_2,b_2	c_1,c_2	c_2,c_2	d_1,d_2	d_2,d_2

individual sampling distribution of $\bar{X}$

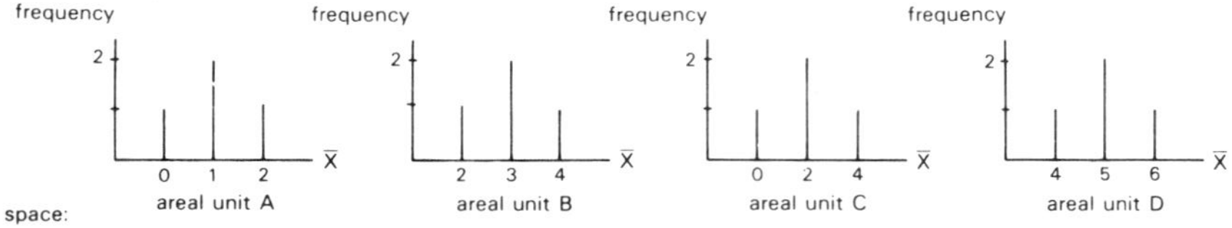

sample space:

0 2 / 0 4, $\bar{X}$ = 1.5	0 2 / 2 4, $\bar{X}$ = 2	0 2 / 4 4, $\bar{X}$ = 2.5	0 2 / 0 5, $\bar{X}$ = 1.75	0 2 / 2 5, $\bar{X}$ = 2.25	0 2 / 4 5, $\bar{X}$ = 2.75	0 2 / 0 6, $\bar{X}$ = 2	0 2 / 2 6, $\bar{X}$ = 2.5	0 2 / 4 6, $\bar{X}$ = 3
0 3 / 0 4, $\bar{X}$ = 1.75	0 3 / 2 4, $\bar{X}$ = 2.25	0 3 / 4 4, $\bar{X}$ = 2.75	0 3 / 0 5, $\bar{X}$ = 2	0 3 / 2 5, $\bar{X}$ = 2.5	0 3 / 4 5, $\bar{X}$ = 3	0 3 / 0 6, $\bar{X}$ = 2.25	0 3 / 2 6, $\bar{X}$ = 2.75	0 3 / 4 6, $\bar{X}$ = 3.25
0 4 / 0 4, $\bar{X}$ = 2	0 4 / 2 4, $\bar{X}$ = 2.5	0 4 / 4 4, $\bar{X}$ = 3	0 4 / 0 5, $\bar{X}$ = 2.25	0 4 / 2 5, $\bar{X}$ = 2.75	0 4 / 4 5, $\bar{X}$ = 3.25	0 4 / 0 6, $\bar{X}$ = 2.5	0 4 / 2 6, $\bar{X}$ = 3	0 4 / 4 6, $\bar{X}$ = 3.5
1 2 / 0 4, $\bar{X}$ = 1.75	1 2 / 2 4, $\bar{X}$ = 2.25	1 2 / 4 4, $\bar{X}$ = 2.75	1 2 / 0 5, $\bar{X}$ = 2	1 2 / 2 5, $\bar{X}$ = 2.5	1 2 / 4 5, $\bar{X}$ = 3	1 2 / 0 6, $\bar{X}$ = 2.25	1 2 / 2 6, $\bar{X}$ = 2.75	1 2 / 4 6, $\bar{X}$ = 3.25

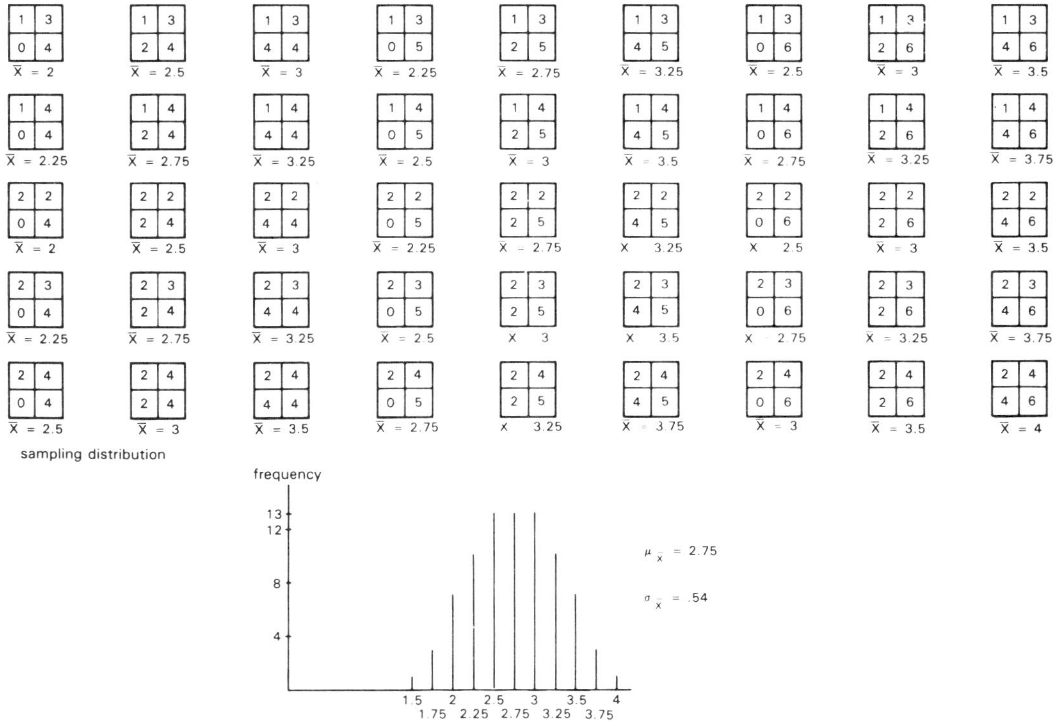

Fig. 1.3. A sampling distribution constructed with the stratified random sampling perspective.

TABLE 1.1

The Number of Ways Eight Items Can be Allocated to Four Areal Units

Composition	Number of Combinations	Frequencies Classical Model	Frequencies Spatial Distributions
8,0,0,0	1	4	4
7,1,0,0	8	12	12
6,2,0,0	28	12	12
6,1,1,0	56	12	24
5,3,0,0	56	12	12
5,2,1,0	168	24	24
5,1,1,1	336	4	24
4,4,0,0	70	6	12
4,3,1,0	280	24	24
4,2,2,0	420	12	24
4,2,1,1	840	12	24
3,3,2,0	560	12	24
3,3,1,1	1120	6	24
3,2,2,1	1680	12	24
2,2,2,2	2520	1	24
TOTAL	8143	65536	193468

NOTE: After Griffith (1984).

unit boundaries. But as long as those phenomena upon which the network is superimposed are not the ones initially used to establish the boundaries in question, then a random partitioning of the surface seems to be a reasonable argument. This justification is affiliated with the classical sampling perspective of replacement with order being unimportant. Consider the total population appearing in Figure 1.1, namely $0 + 2 + 3 + 3 = 8$. The relevant question asks how many ways eight items located on a two-dimensional surface can be grouped into four areal units. The answer is presented here in Table 1.1. One should note that the classical sampling solution has 8143 possibilities, which form a subset of the 4^8 total possibilities, while geographically there are 193,468 possible maps. Steinnes (1980) highlights many of the problematic attributes associated with this perspective.

The final legitimate kind of sampling justification to be reviewed here is the selection of a time slice or a sample region. Summerfield (1983) correctly argues against this perspective. Basically the limitation is that a multivariate sample of size one is being drawn.

As will be seen later in this book, this is not an uncommon situation for a spatial analyst to be in. But the assumption of a homogeneous process operating through time or over the planar surface is crucial here, and more than likely can be demonstrated not to hold for time slice or space displacement sampling.

Attempts have been made to relate spatial statistics to classical statistical analysis (see Smith, 1980). But, as has been emphasized in this section, the fundamental question still being tackled asks what the population is to which inferences are to be drawn.

1.4. MODELS OF SPATIAL AUTOCORRELATION

As was mentioned in the preceding section, a locationally specified realization x_i of the random variable X for areal unit i may be partitioned into a spatial structure component and an independent random error component. The two models used to achieve this decomposition are the conditional model and the simultaneous model. Both are based upon the ideas of a spatial weights matrix and a spatial linear operator.

Consider a set of n areal units. Then spatial autocorrelation means

$$P(X_i = x_i \mid X_1, X_2, \ldots, X_n) = P(X_i = x_i \mid \{X_j, j \in N_i\}), \tag{1.4}$$

where N_i denotes the set of areal units whose x_j values are correlated with x_i. Clearly equation (1.4) constitutes a Markov property. Specification of N_i in part is the problem of establishing what is known as a weights matrix. For the simultaneous model, then, and a second-order Markov structure, the decomposition becomes

$$\mathbf{X} = \varrho \mathbf{W}\mathbf{X} + \xi, \tag{1.5}$$

where $\mathbf{X}$ is the n-by-1 vector of observed values, ϱ is the autocorrelation parameter, $\mathbf{W}$ is the n-by-n weights matrix, $\mathbf{WX}$ is the n-by-1 spatial structure component, and ξ is the n-by-1 independent random error component. Equation (1.5) may be rewritten as

$$(\mathbf{I} - \varrho\mathbf{W})\mathbf{X} = \xi \quad \text{or} \quad \mathbf{X} = (\mathbf{I} - \varrho\mathbf{W})^{-1}\xi.$$

The matrix $(\mathbf{I} - \varrho\mathbf{W})$ is the spatial linear operator. Now if $\mathbf{W}$ is a stochastic matrix, such that

$$\sum_{j=1}^{j=n} w_{ij} = 1 \quad \text{and} \quad w_{ij} \geqslant 0,$$

then $|\varrho| < 1$ to ensure stationarity (this will be seen subsequently with regard to the accompanying spectral density function), and invertibility between this autoregressive model and its moving average counterpart.

Suppose $\xi \sim \text{MVN}(\mu\mathbf{1}, \sigma^2\mathbf{I})$, since for the univariate case here one would assume that a stationary process would be operating over the planar surface, rendering a constant μ and σ. Then the probability density function for the vector random variable $\mathbf{X}$ becomes

$$(2\pi)^{-n/2}\sigma^{-n}\left[\prod_{i=1}^{i=n}(1 - \varrho\lambda_i)\right]$$
$$\times \exp\{-(\mathbf{X} - \mu\mathbf{1})^T(\mathbf{I} - \varrho\mathbf{W})^T(\mathbf{I} - \varrho\mathbf{W})(\mathbf{X} - \mu\mathbf{1})/(2\sigma^2)\}. \tag{1.6}$$

Obviously matrix $\mathbf{W}$ need not be symmetric. By Neyman's factorization theorem (Lindgren, 1976), for the case of unknown μ, σ and ϱ, the sufficient statistic is the four-dimensional statistic

$$\left(\sum_{i=1}^{i=n} x_i, \sum_{i=1}^{i=n} x_i^2, \sum_{i=1}^{i=n}\sum_{j=1}^{j=n} w_{ij}x_j, \sum_{i=1}^{i=n} x_i \sum_{j=1}^{j=n} w_{ij}x_j\right). \tag{1.7}$$

By the theorem on completeness for the exponential family (Lindgren, 1976),

$$\mu/\sigma^2 \in R \quad \varrho/(2\sigma^2) \in R \quad -1/(2\sigma^2) \in R^- \quad -\mu\varrho/\sigma^2 \in R,$$

implying that $R \times R^- \times R \times R$ contains open sets in the four-dimensional space, and hence expression (1.7) is complete. Therefore, as was promised earlier, expression (1.7) proves that in the presence of locational information $\bar{x}$ and s are not sufficient statistics.

A first-order Markov structure is affiliated with the conditional spatial model for equation (1.4). Now the decomposition becomes

$$\mathbf{X} = \mathbf{E}\Omega^{-1/2}\xi, \tag{1.8}$$

where **E** is the n-by-n matrix of eigenvectors from the underlying weights matrix, and $\mathbf{\Omega}$ is the corresponding diagonal n-by-n matrix of eigenvalues. Denote the spatial linear operator here such that $\mathbf{E\Omega E}^T = (\mathbf{I} - \varrho\mathbf{C})$. Matrix **C** is used as the weights matrix because it must be a symmetric matrix, whereas matrix **W** was not so constrained. Stationarity and invertibility are ensured in equation (1.8) by restricting the autocorrelation parameter ϱ such that $|\varrho| < 1/\lambda_1$, where λ_1 is the principal eigenvalue of matrix **C**. Often matrix **C** is binary, taking on the c_{ij} value of 1 if areal units i and j are juxtaposed, and 0 otherwise, as well as matrix **C** often supplies the basis for calculating matrix **W**. If $\xi \sim \text{MVN}(\mu\mathbf{1}, \sigma^2\mathbf{I})$, then $\mathbf{X} \sim \text{MVN}(\mu\mathbf{1}, \sigma^2[\mathbf{I} - \varrho\mathbf{C}]^{-1})$.

Equation (1.8) is referred to as the conditional scheme because

$$E(X_i|\mathbf{x}) = \mu + \sum_{j=1}^{j=n} \omega_{ij}(x_j - \mu) \quad \text{and} \quad \text{VAR}(X_i|\mathbf{x}) = \sigma_i^2.$$

If the conditional distributions are normal, then the joint distribution is such that $\mathbf{X} \sim \text{MVN}(\mu\mathbf{1}, \sigma^2[\mathbf{I} - \varrho\mathbf{C}]^{-1})$. This scheme is fundamental in that all higher order schemes can be rewritten as conditional ones. Consider the simultaneous model given by equation (1.5). Then

$$\begin{aligned} &\sigma^2[(\mathbf{I} - \varrho\mathbf{W})^T(\mathbf{I} - \varrho\mathbf{W})]^{-1} \\ &= \sigma^2[(\mathbf{I} - \varrho\mathbf{W}^T - \varrho\mathbf{W} + \varrho^2\mathbf{W}^T\mathbf{W})]^{-1} \\ &= \sigma^2[\mathbf{I} - \varrho(\mathbf{W}^T + \mathbf{W} - \varrho\mathbf{W}^T\mathbf{W})]^{-1}, \end{aligned}$$

which is of the form of the conditional variance-covariance matrix.

1.5. TOWARDS A THEORY OF SPATIAL STATISTICS

The motivation for this book stems from the three basic problems discussed thus far that are encountered when a researcher analyzes spatial data. To briefly recapitulate, the classical independence of observations assumption is violated. Complex locational information is latent in these sorts of data. The complexity arises because interdependencies are multidirectional and two-dimensional in mature, and hence, unlike in time-series analysis, the likelihood equation includes a Jacobian term that does not simplify. And, classical statistics such as the sample mean do not capture this locational information, resulting in these statistics being insufficient.

But books such as the one by Cliff and Ord (1981), or Upton and Fingleton (1985), address these three issues in an illuminating, comprehensive and provocative manner. The point of departure for his book, then, lies in more advanced problems like those outlined in Griffith (1980, 1987). Moreover, a theory of spatial statistics necessitates a more thorough understanding of the structure of a two-dimensional surface partitioning. It requires a better articulation of relationships between both univariate and multivariate classical linear statistical models, on the one hand, and the notion of spatial autocorrelation, on the other hand. Further, it behooves a deeper understanding of spatial processes in terms of complications such as the edge effects problem, the proper construction of Monte Carlo simulation experiments for spatial data sets, and the missing data problem for a two-dimensional surface. Small sample theory relevant to spatial statistics needs to be more fully developed. Comparisons between the asymptotic behavior of parameter estimates and small lattice behavior need to be explored. Methods of model identification, estimation, and diagnosis need to be better formulated. The role of variable transformations of geographic data requires closer inspection. The finding of solutions to these more advanced problems will permit a theory of spatial statistics to be established. This book surveys progress to this end.

1.6. REFERENCES

Ahuja, N., and B. Schachter, 1983, *Pattern Models*. New York: Wiley.

Bhat, B., 1974, On the method of maximum-likelihood for dependent observations, *Journal of the Royal Statistical Society* B, Vol. 36: 48–53.

Cliff, A., and J. Ord, 1973, *Spatial Autocorrelation*. London: Pion.

Cliff, A., and J. Ord, 1981, *Spatial Processes*. London: Pion.

Costanzo, C., 1983, Statistical inference in geography: modern approaches spell better times ahead, *The Professional Geographer*, Vol. 35: 158–165.

Crowder, M., 1976, Maximum likelihood estimation for dependent observations, *Journal of the Royal Statistical Society* B, Vol. 38: 45–53.

Gaile, G., and J. Burt, 1980, *Directional Statistics*. Norwich, England: Geo Abstracts.

Getis, A., and B. Boots, 1978, *Models of Spatial Processes*. New York: Cambridge University Press.

Gould, P., 1970, Is *statistix inferens* the geographical name for a wild goose?, *Economic Geography*, Vol. 46 (supplement): 439–448.

Griffith, D., 1980, Towards a theory of spatial statistics, *Geographical Analysis*, Vol. 12: 325–339.

Griffith, D., 1984, Theory of spatial statistics, in *Spatial Statistics and Models*, edited by G. Gaile and C. Willmott. Boston: D. Reidel, pp. 3–15.

Griffith, D., 1987, Toward a theory of spatial statistics: another step forward, *Geographical Analysis*, Vol. 19: 69–82.

Haining, R., 1980, Spatial autocorrelation problems, in *Geography and the Urban Environment: Progress in Research and Application*, vol. 3, edited by D. Herbert and R. Johnston. New York: Wiley, pp. 1–44.

Haining, R., D. Griffith and R. Bennett, 1983, Simulating two-dimensional autocorrelated surfaces, *Geographical Analysis*, vol. 15: 247–255.

Haining, R., D. Griffith and R. Bennett, 1984, A statistical approach to the problem of missing spatial data using a first-order Markov model, *The Professional Geographer*, Vol. 36: 338–345.

Lindgren, B., 1976, *Statistical Theory*, 3rd ed. New York: MacMillan.

Schuenemeyer, J., 1984, Directional data analysis, in *Spatial Statistics and Models*, edited by G. Gaile and C. Willmott. Boston: D. Reidel, pp. 253–270.

Silvey, S., 1961, A note on maximum likelihood in the case of dependent random variables, *Journal of the Royal Statistical Society* B, Vol. 23: 444–452.

Smith, T., 1980, A central limit theorem for spatial samples, *Geographical Analysis*, Vol. 12: 299–324.

Steinnes, D., 1980, Aggregation, gerrymandering, and spatial econometrics, *Regional Science and Urban Economics*, Vol. 10: 561–569.

Stephan, F., 1934, Sampling errors and the interpretation of social data ordered in time and space, *Journal of the American Statistical Association*, Vol. 29: 165–166.

Student, 1914, The elimination of spurious correlation due to position in time or space, *Biometrika*, Vol. 10: 179–180.

Summerfield, M., 1983, Populations, samples and statistical inference in geography, *The Professional Geographer*, Vol. 35: 143–149.

Upton, G., and B. Fingleton, 1985, *Spatial Data Analysis by Example: Point Pattern and Quantitative Data*, Vol. 1. New York: Wiley.

APPENDIX 1A

Derivation of the Expected Value of MC

$$
\begin{aligned}
E(\mathrm{MC}) &= E\left[\left(n\Big/\sum_{i=1}^{i=n}\sum_{j=1}^{j=n} c_{ij}\right)\sum_{i=1}^{i=n}\sum_{j=1}^{j=n} c_{ij}(x_i - \bar{x})\right. \\
&\quad \left.\times (x_j - \bar{x})\Big/\sum_{i=1}^{i=n}(x_i - \bar{x})^2\right] \\
&= \left(n\Big/\sum_{i=1}^{i=n}\sum_{j=1}^{j=n} c_{ij}\right) E\left[\sum_{i=1}^{i=n}\sum_{j=1}^{j=n} c_{ij}(x_i - \bar{x})\right. \\
&\quad \left.\times (x_j - \bar{x})\right]\Big/ E\left[\sum_{i=1}^{i=n}(x_i - \bar{x})^2\right] \\
&= \left(n\Big/\sum_{i=1}^{i=n}\sum_{j=1}^{j=n} c_{ij}\right)\sum_{i=1}^{i=n}\sum_{j=1}^{j=n} c_{ij} E\{[(x_i - \mu) \\
&\quad - (\bar{x} - \mu)]\,[(x_j - \mu) - (\bar{x} - \mu)]\}/\{n[(n - 1)/n]\sigma^2\} \\
&= n(0 - \sigma^2/n - \sigma^2/n + \sigma_{\bar{x}}^2)/[(n - 1)\sigma^2] \\
&= -1/(n - 1).
\end{aligned}
$$

Step 2 of this derivation results from the Pitman–Koopmans theorem (Cliff and Ord, 1981).

APPENDIX 1B

Derivation of the Expected Value of GR

Using equation (1.3),

$$
\begin{aligned}
E(\mathrm{GR}) &= E\left\{\left[(n - 1)\Big/\left(2\sum_{i=1}^{i=n}\sum_{j=1}^{j=n} c_{ij}\right)\right]\sum_{i=1}^{i=n}\sum_{j=1}^{j=n} c_{ij}[(x_i - \bar{x})^2\right. \\
&\quad \left. + (x_j - \bar{x})^2]\Big/\sum_{i=1}^{i=n}(x_i - \bar{x})^2\right\} - [(n - 1)/n]E(\mathrm{MC}) \\
&= \left[(n - 1)\Big/\left(2\sum_{i=1}^{i=n}\sum_{j=1}^{j=n} c_{ij}\right)\right]\sum_{i=1}^{i=n}\sum_{j=1}^{j=n} c_{ij}\{E[(x_i - \bar{x})^2] \\
&\quad + E[(x_j - \bar{x})^2]\}/E\left[\sum_{i=1}^{i=n}(x_i - \bar{x})^2\right] + 1/n
\end{aligned}
$$

$$
\begin{aligned}
&= [(n - 1)/2]\,\{[(n - 1)/n]\sigma^2 \\
&\quad + [(n - 1)/n]\sigma^2\}/\{n[(n - 1)/n]\sigma^2\} + 1/n \\
&= (n - 1)/n + 1/n \\
&= 1
\end{aligned}
$$

CHAPTER 2

Developing a Theory of Spatial Statistics

The last chapter set the stage for a discussion of those ingredients necessary for the formulation of a theory of spatial statistics, namely the concept of spatial autocorrelation and the inferential basis built upon a multivariate normality assumption or randomization permutations. Section 1.5 furnishes a brief glimpse into the problematic issues plaguing the formulation of a theory of spatial statistics. A more comprehensive review of these issues can be found in Griffith (1980, 1981, 1987). The general purpose of this chapter is to illuminate avenues for this theory formulation to follow, and in doing so to supply an introduction to much of the content of this book. Moreover, the formulation of a theory of spatial statistics in part requires a number of obstructing problems to be solved.

2.1. THE SMALL SAMPLE SIZE PROBLEM

Probably the most bothersome problem for spatial statistical theory is that of sample size. A set of sampling perspectives was outlined in the previous chapter. But these perspectives simply articulate the relationship between samples and populations. They never actually address the issues of sample size, which is twofold. First, as can be seen from equation (1.6), when multivariate normality is assumed, then the map is the sampling unit. Hence a sample size of 1 is drawn. Even when sampling perspectives are used that consider areal units as the sampling units, n usually is quite small for geographical studies, often less than 30 and rarely greater than 100. Accordingly almost all spatial data series fall into the realm of small sample theory.

Two prominent problems are affiliated with this situation. First, the asymptotic variances of autoregressive model parameters specified by Upton and Fingleton (1985), and presented here in Chapter 4, are of extremely limited value. These asymptotic variances are for the limiting case of $n \to \infty$, and if $n < 30$ then they may well fail to provide any useful information about properties of the sampling distribution in question. The second problem refers to the Jacobian term appearing in equation (1.6)

$$\left[\prod_{i=1}^{n} (1 - \varrho\lambda_i)\right]^{-2/n}. \tag{2.1}$$

Taking the logarithmic version of equation (1.6) renders

$$-(2/n) \sum_{i=1}^{n} \ln (1 - \varrho\lambda_i). \tag{2.2}$$

Table 2.1 illustrates the distribution of eigenvalues λ_i of matrix **C** for regular square lattice situations of 2-by-2 through 10-by-10, together with a test statistic for normality. By construction $\Sigma_{i=1}^{i=n} \lambda_i = 0$, so that the mean $\mu_\lambda = 0$. In addition, as $n \to \infty$ the interval $[-4, 4]$ is being partitioned into increasingly finer intervals. In the limit, then, the summation term appearing in equation (2.2) can be approximated by

$$\int_{-4}^{4} \ln (1 - \varrho\lambda) f(\lambda) \, \mathrm{d}\lambda.$$

where $f(\lambda)$ is the asymptotic frequency distributions of λ, which appears to be 2 for matrix **C**, except for the case of 0, which is irrelevant since $\ln (1 + 0) = 0$. Therefore,

$$\begin{aligned} &\lim_{n\to\infty} - (2/n) \sum_{i=1}^{i=n} \ln (1 - \varrho\lambda_i) \\ &\quad = -(2/n) \int_{-4}^{4} \ln (1 - \varrho\lambda) f(\lambda) \, \mathrm{d}\lambda = 0. \end{aligned} \tag{2.3}$$

This result can be further illustrated by considering a regular pq-by-pq lattice, whose eigenvalues are given by (see Ord, 1975)

$$\lambda_{rs} = 2\{\cos [r\pi/(p + 1)] + \cos [s\pi/(q + 1)]\},$$

$$r = 1, 2, \ldots, p \text{ and } s = 1, 2, \ldots, q.$$

TABLE 2.1

Eigenvalues of matrix $\underline{\mathbf{C}}$ for the Conditional Model and an n-by-n Lattice

					n^2										n^2				
$\lvert\lambda\rvert$	4	9	16	25	36	49	64	81	100	$\lvert\lambda\rvert$	4	9	16	25	36	49	64	81	100
4.00										1.73				2					
3.84									1	1.69					2				
3.80								1		1.66									1
3.76							1			1.63									2
3.70						1				1.62								2	
3.60					1				2	1.59									2
3.52								2		1.53						1	2		
3.46				1						1.41		2				2			
3.41							2			1.40									2
3.37									1	1.36					2				
3.26						2				1.35							2		
3.24			1					1		1.28								2	
3.23									2	1.24			1					1	
3.08								2		1.18							2	2	
3.06							1			1.12									2
3.05					2					1.09									2
2.99									2	1.08						2			
2.88							2			1.03									2
2.83						1				1.00			2	2				2	
2.82		1								.89					1				
2.79								2		.88							2		
2.75									2	.85									2
2.73				2						.80					2				
2.62									1	.77						2			
2.61						2				.73				2				2	
2.53							2			.69							1		
2.52								2		.65						2	2		
2.51									2	.62								2	
2.49					1					.61									2
2.35								1		.57									1
2.24			2		2			2		.56								2	
2.23							2			.55					2				2
2.20									2	.53							2		
2.18						2				.48									2
2.14									2	.44								2	
2.00	1			1			1			.43						2			
1.97									2	.37									2
1.90								2		.35							2		
1.88							2			.28								2	
1.85						2				.24									2
1.79								2		0.00	2	3	4	5	6	7	8	9	10

Now the approximation for the summation term in equation (2.2) becomes

$$\int_0^{\pi}\int_0^{\pi} \ln\,[1 - 2\varrho(\cos\theta_1 + \cos\theta_2)]\,d\theta_1\,d\theta_2$$

$$= \pi \int_0^{\pi} \ln\,\{(1 - 2\varrho\cos\theta_1) + [(1 - 2\varrho\cos\theta_1)^2 - 4\varrho^2]^{1/2}\} - \ln\,(2)\,d\theta_1, \qquad (2.4)$$

where in the limit θ_1 and θ_2 are contained in the interval $[0, \pi]$, since

$$\lim_{x \to \infty} \pi/(x + 1) = 0, \quad \text{and} \quad \lim_{x \to \infty} \pi x/(x + 1) = \pi.$$

Considering the case of $\varrho = 0.249$, the numerical solution to equation (2.4) is -2.103. Consequently, equations (2.3) and (2.4) imply that for very large n the Jacobian term can be ignored. Accordingly OLS results will be reasonably accurate, providing dramatic simplifications of many results, as long as ϱ is not too close to the boundary of its parameter space $(-1, 1)$. But since $n < 30$ in most cases, this finding is of little value, for the Jacobian term always will play an important role in small sample situations.

Three advantages accompany the small sample situation referred to here. First, as Haining (1978) so aptly demonstrates, if the sample size is sufficiently small, then a sampling distribution for the autoregressive structure in question can be easily simulated using Monte Carlo techniques. Hence even though the asymptotic theory may not be very useful, good approximations to standard errors still can be attained inexpensively and efficiently. Second, the principal eigenvector of matrix **C**, which will be shown to be a measure of overall configurational structure in Chapter 3, will retain this interpretation. Apparently λ_p, the principal eigenvalue, is a quadratic function of sample size, but with a very shallow slope for the n^2 term. Accordingly, when n is relatively small the segment of this quadratic curve is approximated quite closely by a linear function of λ_p, meaning that λ_p can continue to be interpreted as a simple measure of areal unit structure. Third, resampling techniques can be employed to explore the population values and sampling distributions in a more comprehensive way. More specifically, the jackknife becomes a relevant tool.

The major drawback with resampling methods is that matrix **C** needs to be preserved. Clearly with the bootstrap method, where sampling occurs with replacement and order is important, retaining matrix **C** tends to become impossible. As such the autoregressive structure of the data is destroyed. In contradistinction, the jackknife method shows promise for spatial data analyses. This technique requires all possible subsets of a given size to be analyzed. If for example, all possible samples of size $(n - 1)$ were drawn, then the procedure would involve systematically dropping one observation, and then analyzing the remaining $(n - 1)$

observations. In this way a sampling distribution based upon n subsamples can be constructed. Again, though, the structure of matrix $\mathbf{C}$ is of concern. One way to retain this structure is to treat the deleted observation as a missing value, and then analyse each sample as though it consisted of $(n - 1)$ observed values and 1 missing value. The format and further details for this procedure will be presented here in Chapter 6.

2.2. FINITE VERSUS INFINITE SURFACES

Closely related to the problem of small sample size is the issue of finite landscapes. The benchmark models developed in this book and elsewhere assume that an infinite planar surface is being treated. If a spatial autocorrelation mechanism is operating over an infinite lattice, but only a finite subregion is analyzed, then the dependencies that normally cross the borders of this subregion are truncated and hence ignored in the analyses. Although the influences going out of the subregion that are being dismissed are of little concern, truncation of those influences stemming from the rest of the infinite landscape and coming into the subregion has profound and complicating consequences. This theme will be explored in greater depth in Chapter 7.

Two points concerning the boundedness of a geographic region merit attention now, namely homogeneity assumptions, and the use of a torus mapping to emulate an unbounded surface. The homogeneity assumption underlies much of the empirical research that has been conducted on spatial autocorrelation. Clearly if a drift exists in the spatial pattern under study, the constant mean assumption can be replaced with a trend surface specification. Moreover, $\mu\mathbf{1}$ is replaced by $\mathbf{X}\boldsymbol{\beta}$, where $\mathbf{X}$ is an n-by-p matrix of powers of the absolute co-ordinates for the areal units in question, except that $\mathbf{X}_1 \equiv \mathbf{1}$, an n-by-1 vector of unity values. Clearly then the constant mean is just a special case of the more general trend surface model. Another source of inhomogeneity has to do with whether or not the regional boundaries are arbitrary, or if in fact they represent some sort of physical barriers. As in time series analysis (see Wonnacott and Wonnacott, 1970), the homogeneity assumption assumes that a given spatial process has been operating over a considerable sized surface beyond the study region. If this spatial process begins at the boundaries of the region, this assumption is not true of course, and a heterogeneous process should be posited.

A second theme concerning infinite surfaces pertains to the common practice of emulating an infinite surface by mapping a finite regular lattice onto a torus. Basically this mapping involves the joining of those areal units located along the 'north' border of the lattice to those areal units located along the 'south' border, converting the finite lattice into a cylinder. Next, those areal units located along the 'east' border of the cylinder are joined to those areal units located along the 'west' border, converting this cylinder into a doughnut or torus. In Chapter 5 equations for calculating the theoretical autocorrelation values at each spatial lag will be presented. Table 2.2 presents selected values of these for various torus sizes, as well as their theoretical counterparts. Clearly two troublesome tendencies can be identified in Table 2.2. First, the spatial autocorrelation values calculated for a torus that emulates an infinite surface do not equal their theoretical infinite surface counterparts. For instance, $\varrho_{0,0}$ for a 3-by-3 torus is 0.7523, whereas $\varrho_{0,1}$ for an infinite surface is 0.4341. In fact, Griffith (1985) has found that across increasing lattice size two of the sets of estimates appearing in Table 2.2 can be approximated quite accurately by the negative exponential equations

$$\varrho_{10} = 0.4290 + 1.2577 \exp(-0.4460n),$$

and

$$\varrho_{11} = 0.2863 + 1.5209 \exp(-0.4328n).$$

The limits of these two equations as $n \to \infty$ are, respectively, 0.4290 and 0.2863. Therefore, the torus approximations do not even converge asymptotically upon their infinite surface counterparts.

A second troublesome tendency conspicuously displayed by the results presented in Table 2.2 is that the correlograms do not have tails that converge to zero. Rather, cycles occur. These cycles arise from the torus mapping itself. At first as lags increase, movement is in a direction away from an areal unit. But since the finite lattice has been wrapped around in order to join its edges, at midpoint

TABLE 2.2

Selected Approximations of Spatial Autocorrelation at Different Lags, From Torus Mappings, for $\rho = .24$

Spatial Autocorrelation	Lag i	3-by-3	4-by-4	5-by-5	6-by-6	7-by-7	8-by-8	9-by-9	10-by-10
ρ_{i0}	0	1.0000	1.0000	1.0000	1.0000	1.0000	1.0000	1.0000	1.0000
	1	.7523	.6439	.5643	.5135	.4824	.4636	.4523	.4454
	2	.7523	.5610	.4288	.3463	.2964	.2664	.2483	.2374
	3	1.0000	.6439	.4288	.3033	.2302	.1872	.1617	.1463
	4	.7523	1.0000	.5643	.3463	.2302	.1653	.1278	.1057
	5	.7523	.6439	1.0000	.5135	.2964	.1872	.1278	.0939
	6	1.0000	.5610	.5643	1.0000	.4824	.2664	.1617	.1057
	7	.7523	.6439	.4288	.5135	1.0000	.4636	.2483	.1463
	8	.7523	1.0000	.4288	.3463	.4824	1.0000	.4523	.2374
	9	1.0000	.6439	.5643	.3033	.2964	.4636	1.0000	.4454
	10	.7523	.5610	1.0000	.3463	.2302	.2664	.4523	1.0000
ρ_{ii}	0	1.0000	1.0000	1.0000	1.0000	1.0000	1.0000	1.0000	1.0000
	1	.6990	.5610	.4613	.3967	.3569	.3327	.3181	.3092
	2	.6990	.5039	.3662	.2755	.2183	.1831	.1615	.1482
	3	1.0000	.5610	.3662	.2504	.1788	.1342	.1065	.0891
	4	.6990	1.0000	.4613	.2755	.1788	.1227	.0884	.0670
	5	.6990	.5610	1.0000	.3967	.2183	.1342	.0884	.0614
	6	1.0000	.5039	.4613	1.0000	.3569	.1831	.1065	.0670
	7	.6990	.5610	.3662	.3967	1.0000	.3327	.1615	.0891
	8	.6990	1.0000	.3662	.2755	.3569	1.0000	.3181	.1482
	9	1.0000	.5610	.4613	.2504	.2183	.3327	1.0000	.3092
	10	.6990	.5039	1.0000	.2755	.1788	.1831	.3181	1.0000

across the lattice movement shifts to a direction that is toward rather than away from the areal unit in question. Therefore similarity of areal units declines at first, and then begins to increase.

2.3. DATA TRANSFORMATIONS

One common interpretation complication of spatial data analysis arises from the use of data transformations. Usually for inferential purposes assumptions such as variance stability, normality and linearity need to be satisfied. If these assumptions are violated then the underlying inferential basis is eroded. Various tradeoffs between satisfying these assumptions are well recognized. The question to be addressed here asks whether or not if some autoregressive variable **X** indicates the need for a data transformation, is this same transformation justifiable for ξ of equations (1.5) and (1.8). In other words, since studies dealing with geographical

TABLE 2.3

Results of Diagnostic Tests for the 1969 Puerto Rican Agricultural Data

Data		Autocorrelated Data		Unautocorrelated Data	
		F-ratio for Variance Homogeneity	Shapiro-Wilk Statistic for Multivariate Normality	F-ratio for Variance Homogeneity	Shapiro-Wilk Statistic for Multivariate Normality
Raw	San Juan	3353.65***	.5347***	3308.50***	.6542***
	Arecibo		.4066***		.3948***
	Mayaguez		.3419***		.4156***
	Ponce		.4425***		.4433***
	Caguas		.5460***		.4996***
Discriminant function scores	San Juan	11.82***	.9399	1.63	.9312
	Arecibo		.8312**		.9787
	Mayaguez		.9402		.9092
	Ponce		.9699		.9681
	Caguas		.8495**		.9573

** Denotes significance at the 5 percent level.
*** Denotes significance at the 1 percent level.

NOTE: After Griffith (1981).

distributions suggest that an overwhelming number of these distributions contain some degree of positive spatial autocorrelation, it is important to understand the relation between violation of the independence of observations assumption and assumptions such as variance stability, linearity and normality.

Considering the Puerto Rican example alluded to in Chapter 1, test statistics for multivariate variance homogeneity and normality for 1969 agricultural production across a set of 5 regions are presented in Table 2.3. The raw, autocorrelated data exhibit multivariate variance inhomogeneity across the 5 regions as well as non-normality within each of the five regions. In fact, the calculated F-ratio of 3354 is dramatically greater than its null hypothesis value of one. Similarly, the Shapiro–Wilk statistics, which range between 0.34 and 0.55, are substantially less than their null hypothesis value of one. Utilizing linear discriminant functions to transform these raw data considerably improves the multivariate normality test statistic, as would be expected given the central limit theorem. But the discriminant function scores still display marked variance inhomogeneity, as is indicated by the associated F-ratio value of 12.

Employing the simultaneous model given by equation (1.8), for the analysis of the raw unautocorrelated data (i.e., the multivariate matrix of ξ values) variance inhomogeneity is still substantial, but has dropped since the F-ratio decreased by 4.5. The multivariate normality situation has improved slightly, too, although all five regional data sets still deviate considerably from multivariate normality. However, discriminant scores based upon the multivariate matrix ξ display the desired and sought after properties. Consequently, by removing spatial autocorrelation from the data, and employing linear transformations, the multivariate variance homogeneity F-ratio has decreased by 3353, and on the average the Shapiro–Wilk statistic has more than doubled.

Now one could have concocted a complex transformation to attain these same results. But complex transformations produce measurement scales that are difficult to interpret (Gould, 1972). And, the trial and error procedure connected with such transformation selections does not guarantee success.

This empirical example links the issue of data transformations to that of homogeneity assumptions, and in part demonstrates that spatial statistical results tend to be highly sensitive to deviations from the homogeneity assumption, a feature uncharacteristic of many classical statistics. Suppose the process at work is normally distributed. Consider a set of independent and identically distributed (iid) white noise random variables $\xi \sim (\mu_\xi \mathbf{1}, \sigma_\xi^2 \mathbf{I})$, which are indistinguishable from n iid $\xi_i \sim N(\mu_\xi, \sigma_\xi^2)$ random variates. Then from classical statistics, $\hat{\mu}_\xi = \bar{\xi}$ and $E(\hat{\mu}_\xi) = \mu_\xi$. Next let a spatial autocorrelation mechanism operate on these variables, one described by the simultaneous model of equation (1.5). Hence

$$\mathbf{\Sigma}^{-1} = \sigma^{-2}(\mathbf{I} - \varrho\mathbf{W})^T(\mathbf{I} - \varrho\mathbf{W}) \quad \text{and} \quad \hat{\mu}_x = \bar{\xi}/[n(1 - \varrho)].$$

Now $E(\hat{\mu}_x) \neq E(\hat{\mu}_\xi)$ unless $\varrho = 0$. Meanwhile,

$$\begin{aligned} E(\mathbf{X}) &= E[(\mathbf{I} - \varrho\mathbf{W})^{-1}\xi] \\ &= (\mathbf{I} - \varrho\mathbf{W})^{-1}E(\xi) \\ &= \mu_\xi(\mathbf{I} - \varrho\mathbf{W})^{-1}\mathbf{1}. \end{aligned} \tag{2.5}$$

If matrix **W** is for an infinite lattice, then it is symmetric, and so

$$\mathbf{W} = \mathbf{E}\mathbf{\Omega}\mathbf{E}^T,$$

and $\mathbf{e}_p$, the principal eigenvector, equals $k\mathbf{1}$. Since $\mathbf{e}_p^T\mathbf{e}_j = 0, j \neq p$, then $\mathbf{1}^T\mathbf{e}_j = 0$, for all $j \neq p$. Returning to equation (2.5),

$$\begin{aligned} E(\mathbf{X}) &= \mu_\xi \mathbf{E}(\mathbf{I} - \varrho\mathbf{\Omega})^{-1}\mathbf{E}^T\mathbf{1} \\ &= \mu_\xi \mathbf{E}(\mathbf{I} - \varrho\mathbf{\Omega})^{-1}\begin{pmatrix} k \\ \hline \mathbf{0} \end{pmatrix} \\ &= \mu_\xi[k_2/(1 - \varrho\lambda_p)]\mathbf{1}. \end{aligned}$$

Consequently expectations are constant over the lattice.

But for a finite, regular lattice of areal units, matrix **W** no longer is symmetric. Accordingly $\mathbf{e}_p \neq k\mathbf{1}$, and so $\mathbf{e}_j^T\mathbf{1}$ may not equal zero. Hence $E(\mathbf{X})$ is no longer a constant vector. This result implies that an evaluation of the univariate frequency distribution for variable **X** might imply that these data need to be transformed in order to satisfy normality assumptions, even though the underlying ξ_i's do conform to a normal distribution. This complication is strictly due to the treatment of a finite lattice, though.

This issue of data transformations will be returned to in Chapter 7.

2.4. MULTIVARIATE ANALYSIS

As Table 2.3 suggests, spatial autocorrelation has a dramatic impact upon multivariate data structures. This contention is attested to by the MANOVA results reported in Griffith (1981), where significant regional differences consistently were found with autocorrelated data, while such differences were not found where the unautocorrelated $\boldsymbol{\xi}$ components were evaluated. This contention also is supported by the results appearing in Table 2.4, which show that about 33% of the ability to successfully classify areal units into regions may be attributed to locational information latent in the spatial data series under study. Table 2.4 summarizes the results of classifying areal units on the basis of discriminant functions that were extracted from a set of eight variables. The autocorrelated data contain locational information, and the

TABLE 2.4

Discriminant Function Classification Results
for the 1969 Puerto Rican Agricultural Data

Region the Areal Unit Actually Belongs to	Region the Areal Unit Was Classified Into									
	Autocorrelated Data					Unautocorrelated Data				
	SJ	A	M	P	C	SJ	A	M	P	C
San Juan (SJ)	17	0	0	0	2	13	2	1	1	2
Arecibo (A)	1	4	4	1	1	4	2	4	0	1
Mayaguez (M)	0	2	10	3	1	1	1	11	3	0
Ponce (P)	1	1	2	10	0	2	2	2	8	0
Caguas (C)	1	0	2	0	10	1	2	3	0	7

connectivity matrix for regions

	SJ	A	M	P	C
San Juan (SJ)	0	1	0	1	1
Arecibo (A)	1	0	1	1	0
Mayaguez (M)	0	1	0	1	0
Ponce (P)	1	1	1	0	1
Caguas (C)	1	0	0	1	0

NOTE: After Griffith (1984).

classificatory success rate is roughly 75 percent. Furthermore, many of the misclassifications actually are allocations to adjacent regions. In contrast, the unautocorrelated data are based upon the ξ's from equation (1.5), essentially calculated on data having all first- and second-order locational information removed. Here the classificatory success rate has dropped to 50 percent, and the misclassifications are more evenly spread across all regions. Bivand (1980) reached a similar sort of conclusion in evaluating product moment correlation coefficients with Monte Carlo simulation techniques.

Little research has been devoted to this theme. In part it has been overlooked because of initial results reported by Lebart (1969) for factor analysis. In his study he used equation (1.5), but set $\varrho = 1$ for all variables. Not surprisingly Lebart found no difference in the factor structure he uncovered for the autocorrelated and unautocorrelated data. This result is not surprising on two counts. First, he applied the same linear transformation to all variables. This restriction would be analogous to assuming that $\sigma_j^2 = \sigma^2$ for all variables. Clearly applied statisticians would find such an assumption quite unpalatable in most situations, even though such an assumption would help to simplify the multivariate equations involved, as well as considerably reduce the number of parameters that would need to be estimated. Second, the Jacobian term appearing in equation (1.6) becomes zero if $\varrho = 1$, since the first entry in the product is $(1 - 1 \cdot 1)$. Hence the multivariate probability density function in Lebart's analysis becomes zero at all points, suggesting that he could have obtained almost any solution.

Both Griffith (1980) and Bivand (1980) suggest that the product moment correlation coefficient takes on a generalized least squares form, such that

$$r = (\mathbf{X} - \mu_x \mathbf{1})^T (\mathbf{I} - \varrho_x \mathbf{W})^T (\mathbf{I} - \varrho_y \mathbf{W}) (\mathbf{Y} - \mu_y \mathbf{1}) / [(\mathbf{X} - \mu_x \mathbf{1})^T (\mathbf{I} - \varrho_x \mathbf{W})^T (\mathbf{I} - \varrho_x \mathbf{W}) (\mathbf{X} - \mu_x \mathbf{1}) \times (\mathbf{Y} - \mu_y \mathbf{1})^T (\mathbf{I} - \varrho_y \mathbf{W})^T (\mathbf{I} - \varrho_y \mathbf{W}) (\mathbf{Y} - \mu_y \mathbf{1})]^{1/2}. \quad (2.6)$$

Consider the principal components analysis of the data matrix **X** versus the data matrix ξ. Equation (2.6) suggests that the nature and degree of multicollinearity will tend to be different in these two cases. If all the spatial autocorrelations are positive, which usually will be the case, then the extent of multicollinearity will be greater for matrix **X** than for matrix ξ. From equation (2.6) it should be somewhat obvious that if all spatial autocorrelations are the same, as poisited by Lebart, then essentially the nature and degree of multicollinearity should be unaltered.

The example of principal components analysis was selected here because it is the technique examined by Wartenberg (1985). But the matrix that he subjected to a principal components analysis was not the usual correlation matrix. Rather it was based on a bivariate extension of the Moran Coefficient (MC), such that

$$\mathrm{MC}_{jk} = \mathbf{1}^T \mathbf{1} (\mathbf{X}_j - \mu_j \mathbf{1})^T \mathbf{C} (\mathbf{X}_k - \mu_k \mathbf{1}) / \{\mathbf{1}^T \mathbf{C} \mathbf{1} [(\mathbf{X}_j - \mu_j \mathbf{1})^T \times (\mathbf{X}_j - \mu_j \mathbf{1}) (\mathbf{X}_k - \mu_k \mathbf{1})^T (\mathbf{X}_k - \mu_k \mathbf{1})]^{1/2}\}. \quad (2.7)$$

The null hypothesis of independent variables and independent observations yields as the expected value the matrix $-(n - 1)^{-1}\mathbf{I}$. Further, as outlined in Chapter 3, the extreme values here depend only upon matrix **C**, and not upon data vectors $\mathbf{X}_j$ and $\mathbf{X}_k$. The two drawbacks of Wartenberg's approach are that negative eigenvalues can be obtained, since λ_i equals the trace of the MC matrix, and the entries in the MC matrix can be outside the interval $[-1, 1]$, as shown in Chapter 3. Encountering either or both of these two properties makes the interpretation of the resulting principal components impossible. Unfortunately Wartenberg avoids these problems by judiciously selecting his examples.

Chapter 8 will be devoted to extending models such as equation (2.6), especially in order to circumvent those problems affiliated with the work of Lebart (1969) and Wartenberg (1985). It follows the formats established by Upton and Fingleton (1985), and Streitberg (1979). Upton and Fingleton have discussed the extension of univariate spatial autocorrelation analysis to situations in which a fixed-effects regression model is appropriate. This extension is similar to that briefly outlined previously for trend surface models. Streitberg addresses the problem of multivariate data analysis for cases where the observations cannot be assumed to be independent. In a regression context, this would be analogous to extending univariate spatial autocorrelation analysis to situations in which a variable-effects, multivariate regression model is appropriate. Streitberg notes that there are three strategies that may be employed when exploring multivariate spatial data. First, the researcher could simply forget about any dependencies and apply standard, classical statistical procedures. As has been alluded to here in Chapters 1 and 2, this strategy results in tests losing power, incorrect levels of significance being applied, useful properties of estimators being lost, estimators behaving strangely, and the such. Second, the researcher could first use a standard procedure, and then test for the presence of non-zero spatial autocorrelation in regression residuals, principal component scores, discriminant function scores, canonical variates, and so forth. Unfortunately to date the only result for which a suitable test statistic has been developed is that of OLS regression residuals. In addition, this method works well only whcn the null hypothesis of independence is not rejected. Rejection of this null hypothesis means GLS estimates of regression parameters must be obtained. But the residuals generated by a GLS regression model cannot be tested using the theory that has been developed for OLS residuals. So this specific strategy becomes quite problematic. Third, as was done in the empirical example presented in the preceding section, a particular dependency structure can be posited from the beginning of an analysis, and standard techniques then can be redefined in accordance with the generalized variance type of framework needed to take account of this posited structuring of the data. This last strategy characterizes the one used throughout this book.

2.5. HIGHER ORDER AUTOREGRESSIVE MODELS

In Chapter 1 the first-order, conditional spatial autoregressive model, and the second-order, simultaneous spatial autoregressive model were discussed. Furthermore, the second-order model was rewritten as a first-order model with the purpose of illustrating some degree of generality for these sorts of autoregressive formulations. But few more complex error structures have been studied to date, in part because they intensify the accompanying numerical estimation problem.

Two exceptions regarding the use of higher order models can be found in the literature. One, alluded to by Martin (1974), closely parallels the Box–Jenkins modelling approach for time series data. It also relates to the trend surface specification mentioned earlier. If a drift exists in some spatial distribution of quantitative data, rather than fitting a trend surface one could let $\varrho_1 = 1$ and calculate spatial differences. Then the problem at hand would reduce to one of estimating a second spatial autocorrelation parameter, say ϱ_2, with the differenced date. The affiliated variance–covariance matrix for a linear spatial trend would be only for the differenced series, and would be written as

$$[(\mathbf{I} - \varrho_2\mathbf{W})^T(\mathbf{I} - \varrho_2\mathbf{W})]^{-1}\sigma^2,$$

which would be equivalent to a general variance–covariance matrix for the initial series of

$$[(\mathbf{I} - \mathbf{W})^T(\mathbf{I} - \varrho_2\mathbf{W})^T(\mathbf{I} - \varrho_2\mathbf{W})\,(\mathbf{I} - \mathbf{W})]^{-1}\sigma^2.$$

A quadratic trend would require a second differencing, introducing the term $(\mathbf{I} - \mathbf{W})^2$; a cubic trend would introduce $(\mathbf{I} - \mathbf{W})^3$. Moreover, spatial differencing would be used to induce a stationary mean. Unfortunately, because of the two-dimensional nature of geographic data, suitable statistical tests have yet to be developed for assessing these kinds of model identifications and specifications.

A second approach involves replacing the term $(\mathbf{I} - \varrho\mathbf{W})$ with the expanded term $(\mathbf{I} - \theta_1\mathbf{W}_1 - \theta_2\mathbf{W}_2)$. This approach has been investigated by Brandsma and Ketellapper (1979). But it is akin to the foregoing spatial differencing approach, with $\theta_2 = 1$. Consider the linear trend situation. Then

$$\begin{aligned}(\mathbf{I} - \theta_1\mathbf{W})(\mathbf{I} - \mathbf{W}) &= \mathbf{I} - \theta_1\mathbf{W} - \mathbf{W} + \theta_1\mathbf{W}^2 \\ &= \mathbf{I} - (\theta_1 + 1)\mathbf{W} + \theta_1\mathbf{W}^2.\end{aligned}$$

This last equation has exactly the same form as that explored by Brandsma and Ketellapper. In fact, in general for ϱ_1 and ϱ_2 one would get

$$\mathbf{I} - (\varrho_1 + \varrho_2)\mathbf{W} + \varrho_1\varrho_2\mathbf{W}^2,$$

where $\theta_1 = \varrho_1 + \varrho_2$ and $\theta_2 = -\varrho_1\varrho_2$. This more general formulation is affiliated with a situation in which the dependency structure is a function both of adjacencies depicted by matrix $\mathbf{W}$ and adjacencies depicted by matrix $\mathbf{W}^2$, while the geographic data display a stationary mean. The associated variance-covariance matrix would be

$$[(\mathbf{I} - \varrho_1\mathbf{W})^T(\mathbf{I} - \varrho_2\mathbf{W})^T(\mathbf{I} - \varrho_2\mathbf{W})(\mathbf{I} - \varrho_1\mathbf{W})]^{-1}\sigma^2,$$

with both ϱ_1 and ϱ_2 needing to be estimated. In contrast, with spatial differencing ϱ_1 is set to unity and only ϱ_2 is estimated.

Consequently the conditional and simultaneous spatial autoregressive formulations furnish a vehicle for analyzing simple dependency structures. Findings for them also reflect to some degree upon more complex dependency structures. Because of their simplicity, and the fact that few empirical studies have found a need to employ more complex formulations, attention in this book will be restricted to results for these two models. One should note, however, that the recent work by Arbia (1986) suggests that the selection of these two models is somewhat of an artifact of the commonly used identification procedures.

2.6. CONCLUDING COMMENTS

Griffith (1980, 1981, 1987) has attempted to chart a course for quantitative geography that would aid in the development of intermediate and advanced spatial statistical theory. Numerous authors (e.g., Ord, 1975; Haining, 1978; Upton and Fingleton, 1985) have tackled in a comprehensive way the affiliated estimation problems. But other questions remain unanswered, such as how to effectively employ the jackknife method in order to better cope with small sample theory constraints, how to deal with finite lattices in a meaningful way, and how to formulate multivariate spatial autoregressive models. This book is devoted to these sorts of issues.

2.7. REFERENCES

Arbia, G., 1986, Problems in the estimation of the spatial autocorrelation function arising from the form of the weights matrix, in *Transformations Through Space and Time*, edited by D. Griffith and R. Haining, Dordrecht: Martinus Nijhoff, pp. 295–308.

Bivand, R., 1980, A Monte Carlo study of correlation coefficient estimation with spatially autocorrelated observations, *Quaestiones Geographicae*, Vol. 6: 5–10.

Brandsma, A., and R. Ketellapper, 1979, A biparametric approach to spatial autocorrelation, *Environment and Planning* A, Vol. 11: 51–58.

Gould, P., 1970, Is *statistix inferens* the geographical name for a wild goose?, *Economic Geography*, Vol. 46 (supplement): 439–448.

Griffith, D., 1980, Towards a theory of spatial statistics, *Geographical Analysis*, Vol. 12: 325–339.

Griffith, D., 1981, Towards a theory of spatial statistics: a rejoinder, *Geographical Analysis*, Vol. 13: 91–93.

Griffith, D., 1984, Are locations unique?, *Progress in Human Geography*, Vol. 8: 82–94.

Griffith, D., 1985, *Correcting for Edge Effects in Spatial Statistical Analysis*, unpublished master's thesis, Department of Statistics, The Pennsylvania State University, University Park, Pennsylvania.

Griffith, D., 1987, Toward a theory of spatial statistics: another step forward, *Geographical Analysis*, Vol. 19: 69–82.

Haining, R., 1978, A spatial model for high plains agriculture, *Annals* of the Association of American Geographers, Vol. 68: 493–504.

Lebart, L., 1969, Analyse statistique de la contiguite, *Publications de l'Institute Statistique de l'Universite de Paris*, Vol. 18: 81–112.

Martin, R., 1974, On spatial dependence, bias and the use of first spatial differences in regression analysis, *Area*, Vol. 6: 185–194.

Ord, K., 1975, Estimation methods for models of spatial interaction, *Journal of the American Statistical Association*, Vol. 70: 120–126.

Streitberg, B., 1979, Multivariate models of dependent spatial data, in *Exploratory and Explanatory Statistical Analysis of Spatial Data*, edited by C. Bartels and R. Ketellapper. The Hague: Martinus Nijhoff, pp. 143–177.

Upton, G., and B. Fingleton, 1985, *Spatial Data Analysis by Example: Point Pattern and Quantitative Data*, Vol. 1. New York: Wiley.

Wartenberg, D., 1985, Multivariate spatial correlation: a method for exploratory geographic analysis, *Geographical Analysis*, Vol. 17: 263–283.

Wonnacott, R., and T. Wonnacott, 1970, *Econometrics*, Toronto: Wiley.

CHAPTER 3

Areal Unit Configuration and Locational Information

The importance of the arrangement of areal units was established in the preceding chapter. The configuration of areal units furnishes an infrastructure upon which autocorrelation mechanisms operate; moreover, it provides the underlying ordering that leads to a violation of the assumption of independent observations. As such it tells how the 'grapes' of the modified urn model mentioned earlier are clustered together. In addition, through the notion of random partitionings, this feature of areal units can be linked to one of the primary justifications for treating spatial data within a sampling context. It appears in both the indices of spatial autocorrelation and the variance-covariance terms of spatial autoregressive models.

Because areal unit configuration plays a prominent role in spatial data analysis, namely since at least some information contained in a geographic fact x_i is a function of the arrangement of areal units, a thorough understanding of its properties is needed. This understanding is beneficial from at least two viewpoints. First, in order to relate an areal unit configuration to other attributes of spatial data, it needs to be quantified and hence summarized with some convenient measure. Second, the limits of autocorrelation indices are strictly a function of areal unit configuration. Therefore a more comprehensive knowledge of the prevailing areal unit configuration properties should afford a deeper understanding of information latent in spatial data series.

3.1. PLANAR TESSELLATIONS

As was explained in Chapter 1, areal units are either zones that result from a mutually exclusive and collectively exhaustive surface partitioning, or punctations. Since for all practical purposes the geographic surface in question is two-dimensional, then attention here will be restricted to planar surfaces. If there are n areal units on this surface, then for a completely articulated system of areal units the maximum number of juxtaposition links is $3(n - 2)$, meaning that for $n > 3$ matrix **C** will become increasingly sparce as $n \to \infty$.

In fact, in the limit a negligible percentage of c_{ij} values will be unity. For a nondecomposable system of areal units, however, the minimum number of juxtaposition links is $(n - 1)$.

A tessellation is a partitioning of a planar surface into non-overlapping regions, such as the preceding system of areal units. Only three regular tessellations, or surface partitionings that consist of regular polygons of a single type (Ahuja and Schachter, 1983), exist. The corresponding polygons are the equilateral, equiangular (or regular) triangle, the square, and the hexagon constructed from six regular triangles. As an aside, all three of these polygons have been linked to central place theory. Of these three regular tessellations, only the square one, commonly referred to as a lattice, has been fully exploited in spatial analysis. It has the appealing and simplifying attribute that the Manhattan distance metric is preserved.

Ahuja and Schachter (1983) have surveyed models that generate the three regular tessellations. Besag (1972, 1974) outlines properties of the autoregressive model that are conditional upon a lattice configuration of areal units. But there are an infinite number of possible planar surface partitionings, with only three being regular tessellations. The remaining ones may be classified as irregular tessellations. Ahuja and Schachter (1983) present one generating model for this second class of tessellation, namely the Poisson line tessellation. Besag (1975) argues that the parameter estimation techniques for autoregressive models employed for lattice systems apply to irregular tessellations, too. Results for a lattice configuration of areal units are better understood, though, and probably should be used as a benchmark for comparative purposes when one interprets findings for irregular tessellations. After all, the use of lattices in simulation experiments serves as a variance reduction technique.

If a punctiform surface is under study, then an artificial partitioning of the surface can be superimposed upon the surface using Thiessen or Voronoi polygons. Two attractive features of this sort of partitioning are that (1) the resulting polygons are convex, and (2) any point contained within a polygon is closer to that polygon's punctation that to any other punctation on the surface. Delaunay triangulation is in some ways a dual to this polygon. Due to the geometric nature of this duality, theorems about and algorithms for constructing Thiessen polygons are more easily established. Again, Ahuja and Schachter (1983) provide a review of these materials. As an aside, this is the approach employed by Haining, Griffith and Bennett (1984) in their analysis of rainfall data by weather station (a punctation). Matula and Sokal (1980) have offered an affiliated alternative to the Thiessen polygon.

Consequently, a common set of geometric properties can be established for any set of areal units.

3.2. EIGENFUNCTION ANALYSIS OF AREAL UNIT CONFIGURATION TESSELLATIONS

The weights matrix **C** depicting a configuration of areal units, which was presented in Chapter 1, is generated by a comparative judgement model, whose cell entries are coded unity for an affirmative judgement, and zero for a negative judgement, and commonly is referred to as a connectivity matrix (see Taaffe and Gauthier, 1973). It is constructed with a nominal measurement scale. Further, both a summary statistic and an interval scale can be recovered from the latent structure burried in this matrix. This summary statistic is supplied by the principal eigenvalue, whereas an arrangement measure is supplied by the accompanying principal eigenvector. Useful properties of eigenvalues and eigenvectors are presented here in Table 3.1. One example of how these properties reflect upon matrix **C** is provided by the Jacobian term like the one that appears in equation (1.6). For a conditional

TABLE 3.1

Useful Properties of Eigenvalues and Eigenvectors

Number	Property
1	The eigenvalues of a matrix and its transpose are the same.
2	The sum of the eigenvalues of a matrix equals the sum of that matrix's principal diagonal elements.
3	If a matrix is multiplied by some scalar constant, then its eigenvalues are also multiplied by this constant, but its eigenvectors remain unchanged.
4	The product of a matrix's eigenvalues equals the value of that matrix's determinant.
5	For a symmetric matrix, two eigenvectors associated with two distinct eigenvalues are mutually orthogonal.
6	The eigenvalues of a matrix and its inverse are inverses of each other, while the eigenvectors are the same.
7	If a matrix is powered by some positive integer value, each of its eigenvalues is powered by this same positive integer value, but its eigenvectors remain unchanged.
8	All the eigenvalues and eigenvectors of a real, symmetric matrix are real.
9	For some real, symmetric matrix $\underline{\mathbf{A}}$, if some non-singular matrix $\underline{\mathbf{C}}$ exists such that $\underline{\mathbf{A}} = \underline{\mathbf{C}}^T\underline{\mathbf{C}}$, then all of the eigenvalues of matrix $\underline{\mathbf{A}}$ are positive.
10	The eigenvalues of a triangular or diagonal matrix are the elements in its principal diagonal.
11	The dominant eigenvalue of a matrix is contained in that interval defined by the largest and smallest row sums for this matrix.
12	The dominant eigenvalue and its corresponding eigenvector of a non-negative, irreducible matrix are positive, with no other eigenvalue being greater in absolute value.
13	The principal eigenvector of a non-negative, symmetric matrix has all non-negative values.
14	If b$\underline{\mathbf{I}}$ is added to a matrix , where b is a real scalar and $\underline{\mathbf{I}}$ is the identity matrix, then b is added to each of the eigenvalues of that matrix, but that matrix's eigenvectors remain unchanged.

model, this Jacobian would be written as $\det(\mathbf{I} - \varrho\mathbf{C})$. But Ord (1975) shows that with the application of Properties # 3, # 4, # 6, # 8 and # 14 this term can be rewritten as it appears in equation (1.6). The important question, however, asks whether or not those two indices furnished by the principal eigenvalue and its associated eigenvector can discriminate between different configurations of areal units.

Boots (1982, 1984) notes that the principal eigenvalue λ_p appears to measure primarily the differences in areal unit configuration composition, and is capable of capturing at least some of both the spatial and aspatial aspects of individual configurations. More specifically, the principal eigenvalue is a measure of overall connection strength of a connectivity matrix, whereas the values of its associated eigenvector $\mathbf{E}_p$ show how this strength is distributed over the configuration of areal units. Property #13 ensures that each entry of this vector will be non-negative. Since a planar surface is being dealt with, then the sparsest connectivity matrix is associated with a linear arrangement of areal units. Accordingly, $\lambda_p \in [1, 2]$ (Property #11). Meanwhile, the densest connectivity matrix is affiliated with a totally connected partitioning having $n(n - 1)$ internal borders. If the planar constraint is imposed, though, the maximum number of borders is $3(n - 2)$. However, one areal unit could be adjacent to all others. Accordingly, $\lambda_p \in [1, n - 1]$. The expected tendency, then, is for λ_p to increase as the configuration of areal units becomes better articulated. The principal eigenvector has a straightforward meaning. Any non-negative integral power of matrix $\mathbf{C}$, say $\mathbf{C}^k$, has cell entries that count the number of ways one could move from a row heading areal unit to a column heading areal unit crossing exactly k internal borders. Let matrix $\mathbf{A} = \Sigma_{k=0}^{k=\infty} \mathbf{C}^k$; then $e_{ip} \propto \Sigma_{j=1}^{j=n} a_{ij}$. Clearly, then, these two indices can discriminate, at least in part, between different areal unit configurations.

Another useful property of the principal eigenvalue refers to the notion of a random partitioning of a surface. Within this context, for $n = 3$ the sampling distribution of λ_p is approximately normal, having a modified Wilk–Shapiro correlation of 0.971, which is not significantly different from unity. For $n = 4$ this correlation increases to 0.992, and for $n = 5$ it increases to 0.997. All possible configurations of areal units for these three cases appear in Table 3.2. The conjecture suggested here is as follows:

> as $n \to \infty$ the sampling distribution for λ_p converges upon a normal distribution, with the rate of convergence being very rapid.

TABLE 3.2

The Principal Eigenvalues for Small Geographic Configurations

n	Number of Connections for Individual Areal Units	λ_p	Frequency
3	0 0 0	0	1
	1 1 0	1	3
	2 1 1	1.41421	3
	2 2 2	2	1
4	0 0 0 0	0	1
	1 1 0 0	1	6
	1 1 1 1	1	3
	2 1 1 0	1.41421	12
	2 2 1 1	1.61803	12
	2 2 2 0	2	4
	2 2 2 2	2	3
	3 2 2 1	2.17009	12
	3 3 2 2	2.56155	6
	3 3 3 3	3	1
5	0 0 0 0 0	0	1
	1 1 0 0 0	1	10
	1 1 1 1 0	1	15
	2 1 1 0 0	1.41421	30
	2 1 1 1 1	1.41421	30
	2 2 1 1 0	1.61803	60
	2 2 2 0 0	2	10
	2 2 2 1 1	1.73205	60
	2 2 2 1 1	2	10
	2 2 2 2 0	2	15
	2 2 2 2 2	2	12
	3 1 1 1 0	1.73205	20
	3 2 1 1 1	1.84776	60
	3 2 2 1 0	2.17009	60
	3 2 2 2 1	2.21432	120
	3 3 2 1 1	2.30278	60
	3 3 2 2 0	2.56155	30
	3 3 2 2 2	2.48119	70
	3 3 3 2 1	2.64119	60
	3 3 3 3 0	3	5
	3 3 3 3 2	2.85577	30
	4 1 1 1 1	2	5
	4 2 2 1 1	2.34292	30
	4 2 2 2 2	2.56155	15
	4 3 2 2 1	2.68554	60
	4 3 3 2 2	2.93543	60
	4 3 3 3 1	3.08613	20
	4 3 3 3 3	3.23607	15
	4 4 2 2 2	3	10
	4 4 3 3 2	3.32340	30
	4 4 4 3 3	3.64575	10

Amrhein, Guevara and Griffith (1983) furnish additional supporting evidence from a Monte Carlo simulation experiment for $n = 90$ that the asymptotic distribution is normal. If this is truly an attribute of the principal eigenvalue of matrix **C**, then the entire body of normal curve theory can be employed in assessing the partitioning of a planar surface, given a random partitioning type of sampling perspective. Hence, this characteristic of λ_p would facilitate assessment of a given arrangement of areal units, since the quantitative measures of geometric configurations have been found to be a function of n, the number of areal units, as will be shown shortly.

The denseness of matrix **C** implies that as all areal units become mutually juxtaposed, then $\lambda_p \rightarrow (n - 1)$. As an increasing number of these units become juxtaposed with just a few areal units, then $\lambda_p \rightarrow 1$. But in a planar setting the constraint of no more than $3(n - 2)$ borders very quickly becomes binding, helping to suppress the maximum value λ_p could attain. Hence, as $n \rightarrow \infty$, λ_p should increase at first, and then decrease as the percentage of zeroes in matrix **C** becomes very large. Since the maximum number of ones is $6(n - 2)$, whereas the number of possible non-zero cells is $n(n - 1)$, then the fraction $6(n - 2)/[n(n - 1)]$ peaks at $n = 2 + (2)^{0.5}$, and converges upon zero in the limit. Similarly, the equation $6(n - 2) = n(n - 1)$ holds for $n = 3$ and $n = 4$. Thus, beyond the case of $n = 4$, λ_p will never be able to equal $(n - 1)$. The presence of countervailing forces means that λ_p is not a monotonically increasing function of n. Rather, its relation to n may be defined as follows:

$$\lambda_p = \alpha + \beta n + \gamma n^2. \qquad (3.1)$$

Any value of λ_p for an n approaching or exceeding $\partial\lambda_p/\partial n = 0$ suggests that latent structure in the configuration of areal units is being somewhat masked by the number of areal units into which a surface has been partitioned. In other words, λ_p is measuring both of these features, not just the configuration of areal units involved. Possibly the expression $\alpha + \beta n$ represents a measure of arrangement, whereas the term γn^2 represents a measure of the impact of size.

Equation (3.1) has been calibrated for 28 selected cities of the Canadian urban system. The OLS results of this calibration

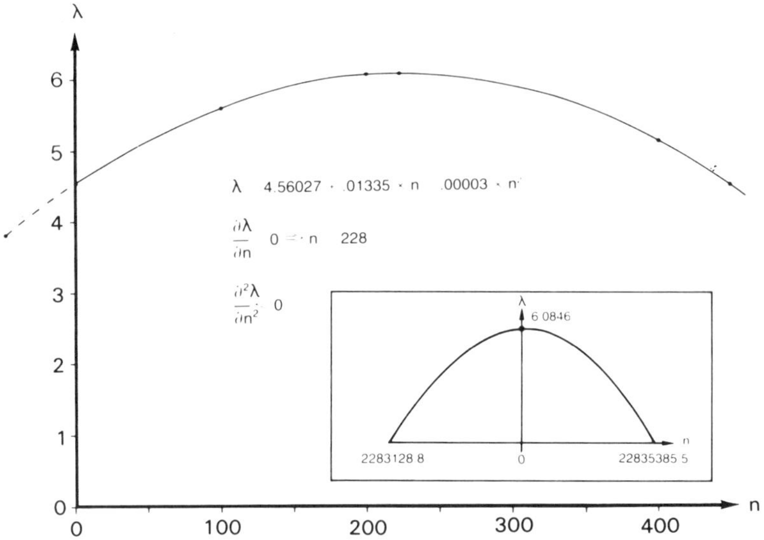

Fig. 3.1. Plot of the principal eigenvalue for 28 selected Canadian cities.

are as follows:

$$\lambda_p = \underset{(t\,=\,48.2)}{4.56027} + \underset{(t\,=\,5.8)}{0.01335n} - \underset{(t\,=\,4.3)}{0.00003n^2}, \tag{3.2}$$

where all of the parameter estimates are significantly different from zero at the 0.001 level, and the multiple correlation is 0.80371. A graph of this curve appears in Figure 3.1. According to these results, the term n accounts for 38.7% of the variation of λ_p over the urban areas, whereas the term n^2 accounts for 25.9% of this variation. Setting the first-order condition for equation (3.2) to zero, and then solving for n, implies that any value of λ_p for cities having $n \rightarrow 228$ and beyond are noticeably affected by the number of areal units involved.

Meanwhile, the geographic distribution of e_{ip} values over a partitioned surface tends to be such that similar values of e_{ip} have a high propensity to cluster together on the surface. In a second study of 24 Canadian cities, Griffith and Jones (1980) found that the spatial distributions of e_{ip} always were positively autocorrelated, ranging from weak to very strong, with an average for the cities indicating a moderate level of positive spatial autocorrelation.

This finding is intuitively appealing, since one would expect a concentration of larger e_{ip} values in finely partitioned regions on a surface, and lower e_{ip} values along the periphery and in more coarsely partitioned regions.

Unfortunately the non-principal eigenfunctions have not proven to be very illuminating measures (see Boots, 1985). In fact, the meaning of non-principal eigenvectors seems elusive. Their importance is demonstrated by the diagonalization involved; since $\mathbf{E}^T\mathbf{CE} = \mathbf{\Omega}$, where $\mathbf{\Omega}$ is the diagonal matrix of eigenvalues and $\mathbf{E}$ is a matrix whose columns are the individual eigenvectors of matrix $\mathbf{C}$, then $\mathbf{E}_p$ and λ_p do not necessarily summarize all of the information contained in the connectivity matrix. By Property # 5 one knows that each eigenvector is affiliated with an independent quantity of information, since the vectors are mutually orthogonal. And by Property # 13 $\mathbf{E}_p \geqslant \mathbf{0}$, meaning that there exists at least one e_{ik} such that e_{ik} and e_{jk} differ in sign ($i \neq j$ and $k \neq p$), since $\mathbf{E}_p^T\mathbf{E}_k = 0$. Boots and Tinkler (1983) have attempted to expand upon Tinkler's (1972) initial interpretation of non-principal eigenfunctions as designators of anti-rumor source locations. The argument put forth in these two papers rests upon an iterative arithmetic procedure that can be used to estimate these functions. In contrast with the explanation, non-principal eigenfunctions will be interpreted here in the spirit of principal components analysis. First of all, the equation $\mathbf{E}^T\mathbf{CE} = \mathbf{\Omega}$ represents a rigid, principal axes rotation of the matrix $\mathbf{C}$. The rows of this rotation matrix may be interpreted as the direction cosines for these principal axes, which in the context of principal components analysis alludes to the correlation between each original variable and the particular data dimension characterized by a given eigenvector. The matrix $\mathbf{E\Omega}^{0.5}$ produces these correlations. But matrix $\mathbf{C}$ is not a correlation matrix; it has negative eigenvalues associated with it, 0 diagonal entries, and intransitivities in that although $c_{ij} = 1$ and $c_{jk} = 1$, c_{ik} often may be zero, and vice versa.

The first two of these three problems can be remedied by determining $n_{\max} = \text{MAX}_i(\Sigma_{j=1}^{j=n} c_{ij})$, calculating $\mathbf{C}/n_{\max}$, and then constructing the matrix $(\mathbf{I} + \mathbf{C}/n_{\max})$. Since the maximum row sum for matrix $\mathbf{C}/n_{\max}$ is unity, then $\lambda_p \leqslant 1$, whereas $|\lambda_i| \leqslant \lambda_p$ for all $i \neq p$. Hence, adding the identity matrix to $\mathbf{C}/n_{\max}$ both codes the diagonal cells with ones, and ensures that all λ_i are non-negative. The eigenvalues for matrix $(\mathbf{I} + \mathbf{C}/n_{max})$ are related to those for matrix $\mathbf{C}$ as follows:

$$\mu_i = 1 + \lambda_i/n_{max} \quad \text{(Properties \#3, \#14).}$$

In both cases, though, the normalized eigenvectors are the same, namely $\mathbf{E}$, arranged by columns in matrix form. In addition, matrix $(\mathbf{I} + \mathbf{C}/n_{max})$ could be viewed as a variance–covariance matrix, since it is symmetric, non-negative, and at least semi-definite. Hence, matrix $(\mathbf{I} + \mathbf{C}/n_{max})$ can be decomposed in such a way that some matrix $\mathbf{L}$ can be found that satisfies the equation $\mathbf{L}^T\mathbf{L} = (\mathbf{I} + \mathbf{C}/n_{max})$ [Property #9]. This viewpoint permits one to interpret $\mathbf{E\Omega}^{0.5}$ as a principal components loading matrix, where $\mathbf{\Omega}$ now is the diagonal matrix whose ω_{ii} elements are μ_i. Similar to this mathematical manipulation, Boots and Tinkler (1983) note that the iterative process they discuss is numerically more stable when the main diagonal of matrix $\mathbf{C}$ is coded with ones.

Given this result, the entries of each non-principal eigenvector may be viewed, when post-multiplied by matrix $\mathbf{\Omega}^{0.5}$, as a set of pseudo-correlation coefficients. Accordingly, these $e_{ij}\mu_j^{0.5}$ values identify clusters of areal units, in a given surface partitioning sub-system, that display a similar pattern of local articulation within the given arrangement of areal units. These clusters will be equivalent to those identified by each vector $\mathbf{E}_j$ alone, since the matrix product $\mathbf{E\Omega}^{0.5}$ merely means that all entries e_{ij} of $\mathbf{E}_j$ are being multiplied by the same eigenvalue $\mu_j^{0.5}$. The only exception to this contention will occur when $\mu_j = 0$, an unlikely event. Because the corresponding eigenvectors are mutually orthogonal, independent sub-systems of areal units are being uncovered by each function convolution. In keeping with this perspective, then, negative eigenvector entries disclose areal units that are diametrically polar opposites to the configurational structure for a local sub-system of the partitioned surface that is demarcated by the marked positive $e_{ij}\lambda_j$ values.

The prominence of each eigenfunction can be evaluated utilizing a notion analogous to the percent of variance accounted for by each principal component. The summation term $\Sigma_{i=1}^{i=n}$ $(1 + \lambda_i/n_{max})$ equals n, since $\Sigma_{i=1}^{i=n} \lambda_i = 0$ (Property #2). Hence, the proportion of information from matrix $\mathbf{C}$ that is contained in eigenvector $\mathbf{E}_p$ is λ_p/n, which has a maximum value of $2/n$ for

$(\mathbf{I} + \mathbf{C}/n_{max})$. As $\lambda_i \rightarrow 0$ this fraction also approaches zero. Thus, a non-principal eigenvector whose associated fraction is close to $2/n$ would be considered prominent. This finding suggests that only a small part of the information contained in matrix **C** actually is captured by λ_p and $\mathbf{E}_p$. This idea is further supported by what is known about Roy's largest root criterion in multivariate analysis (see Tatsuoka, 1971). Furthermore, as Boots and Tinkler (1983) show, types of clusters that frequently are extracted map into spatial patterns such as a north–south or east–west bisection of the surface, or a northeast–southwest sectoring of the surface. Unfortunately, since the prominence index is sample specific, because n is in its denominator, it has the following undesirable attribute, considering even the best of all possibilities:

$$\lim_{n \to \infty} 2/n = 0.$$

Because $\lambda_p/n \leqslant 2/n$, presumably applying a simple structure rotation to the matrix $\mathbf{E}^*\mathbf{\Omega}^{0.5}$, where $\mathbf{E}^*$ is the matrix of selected eigenvectors whose eigenvalue counterparts are greater than or equal to some minimum value, will not be more enlightening. Such a rotation also might cause some entries in the principal eigenvector $\mathbf{E}_p$ to become negative, meaning that it would loose its combinatorial interpretation.

Consequently, since matrix **C** is non-negative, then the eigenfunction analysis of this matrix rests primarily on Properties #2, #5, #11 and #13. Although these properties are well known, the problem at hand focuses on how to use them in a creative and illuminating way so that meaningful and reflective interpretations will result from spatial data series.

3.3. SELECTED APPLICATIONS OF THE PRINCIPAL EIGENFUNCTIONS OF MATRIX C

Three examples will be presented here to illustrate applications of the principal eigenfunctions of matrix **C** or matrix **W**, as well as to address the two points raised at the outset of this chapter. First, the Jacobian term appearing in equation (1.6) will be scrutinized. Second, the principal eigenvector of matrix **C**, as a quantitative measure of areal unit configuration, will be related to selected attributes of a spatial data set. Third, the eigenvalues of matrix **C** will be used to establish limits for those values that can be taken on by MC and GR.

Equation (1.6) presents the Jacobian term as a function of matrix **W**, which can be the stochastic version of matrix **C**. One argument in favor of this form of the Jacobian is that it leads to a 'natural' interpretation of ϱ, since with it $|\varrho| < 1$, and so ϱ is like a correlation coefficient. But the principal eigenfunctions indicate that using matrix **W** has its drawbacks, too. First, since matrix **W** no longer is symmetric (except in the infinite lattice case), Property # 5 states that the eigenvector matrix **E** may consist of oblique eigenvectors. There no longer is an additive decomposition of information. Second, Property # 11 states that $\lambda_p = 1$, regardless of the underlying areal unit configuration or the size of n. Hence λ_p no longer offers an overall configuration structure summary statistic, and equation (3.1) no longer holds. Third, while Property # 12 ensures that $\mathbf{E}_p \geqslant \mathbf{0}$, the mathematics of the situation yields the solution $\mathbf{E}_p = n^{-1/2}\mathbf{1}$, again destroying the index meaning that was attached to the principal eigenvector of matrix **C**. Therefore an eigenfunction analysis of matrix **W** enlightens one about its advantages as well as its drawbacks.

In terms of a conventional data analysis, one interesting question revolves around whether or not results are attributable to the underlying surface partitioning. To this end, Griffith (1981) has presented an interesting empirical case study based upon data from Toronto. Historically, urban spatial structure has been divided into the four components of land use, social areas, spatial flows, and the configuration of areal units. Each of these components has been treated separately, with few attempts having been made to synthesize the results. Such a synthesis could be attained through the use of generalized canonical correlation analysis. This technique addresses the statistical problem of identifying dimensions that isolate and summarize linear relations between three or more sets of variables in a convenient and illuminating manner, such that the correlation between dimension pairs is maximized while the dimensions within each data set are mutually orthogonal.

These four data sets were compiled for Toronto for the year 1971. Since the investigation was exploratory in nature, a generalized variance solution was obtained. Results for this analysis showed that the surface partitioning dimension that ran across these four data sets accounted for 19.4% of the variation in the land use data, 7.7% of the variation in the social area data, and 3.3% of the variation in the journey-to-work data. The respective generalized canonical correlations for these data sets were 0.72885, 0.87014, 0.78985 and 0.31252. Consequently, this non-trivial dimension indicates that traditional spatial data analyses are at least weakly surface partitioning specific. These canonical correlations also suggest that a stronger sub-dimension probably crosses the three data sets of land use, social areas, and spatial flows. This example is discussed in greater length in Section 3.5.

Therefore, the measurement scale for areal unit configuration facilitates a determination of whether or not data analysis results may be thought of, at least in part, as being either robust or a function of the underlying surface partitioning. Some evidence does exist indicating that the answer to this question will be research problem specific.

In terms of MC and GR, de Jong, Sprenger and van Veen (1984) have derived limits based upon the connectivity matrix **C**. Heuristically speaking, since equation (1.1) is analogous to a product moment correlation coefficient, then one would expect its limits to be ± 1. Consider matrix **C** in the context of the following linear transformation:

$$(\mathbf{I} - \mathbf{1}\mathbf{1}^T/n)\mathbf{C}(\mathbf{I} - \mathbf{1}\mathbf{1}^T/n). \tag{3.3}$$

Then $MC_{\text{min}} = n\lambda_{\text{min}}/\mathbf{1}^T\mathbf{C}\mathbf{1}$ and $MC_{\text{max}} = n\lambda_{\text{max}}/\mathbf{1}^T\mathbf{C}\mathbf{1}$, where λ_{min} and λ_{max} are eigenvalues of expression (3.3). One should note that if either of these eigenvalues is the only *zero* eigenvalue, then the next smallest/largest one is selected. λ_{max} converges to λ_p as the configuration of areal units converges upon an infinite lattice.

In a similar heuristic fashion, equation (1.2) obviously has a minimum of zero. By analogy with the Durbin–Watson statistic in time-series analysis, it has a maximum value of about 2. Now consider matrix **C** in the context of the following linear transformation:

$$(\mathbf{I} - \mathbf{1}\mathbf{1}^T/n)\,[(\mathbf{C}\mathbf{1} + \mathbf{C}^T\mathbf{1})_{\text{diag}} - 2\mathbf{C}]\,(\mathbf{I} - \mathbf{1}\mathbf{1}^T/n), \tag{3.4}$$

where diag denotes a diagonal matrix. Expression (3.4) can be simplified if the weights matrix is symmetric, since the middle term will reduce to $2(\mathbf{C1}_{diag} - \mathbf{C})$, but has been written out here in its entirety in order to show the form when the weights matrix is not symmetric. Then $GR_{min} = (n - 1)\lambda_{min}/(2\mathbf{1}^T\mathbf{C1})$ and $GR_{max} = (n - 1)\lambda_{max}/(2\mathbf{1}^T\mathbf{C1})$, where λ_{min} and λ_{max} are eigenvalues of expression (3.4). Again if either of these eigenvalues is the only *zero* eigenvalue, then the next smallest/largest one is selected.

3.4. THE MODIFIABLE AREAL UNIT PROBLEM

When the sampling justification of a random surface partitioning is invoked, then an arbitrary aggregation of spatial information results. The dramatic impact this arbitrariness can have on the extraction of statistical information from spatial data is exemplified in the correlation study reported by Openshaw and Taylor (1979). Openshaw (n.d., p. 3) goes on to note that the definition of areal units

> . . . is arbitrary and (in theory) modifiable at choice; indeed, different researchers may well use different sets of units. This process of defining or creating areal units would be quite acceptable if it were performed using a fixed set of rules, or so that there was some explicit geographically meaningful basis for them. However, there are no rules for areal aggregation, no standards, and no international conventions to guide the spatial aggregation process.

This description is of the problem that has become known as the modifiable areal unit problem. Its existence lends support to Summerfield's (1983) argument against the random surface partitioning sampling perspective. Its existence also further undermines the validity of using non-lattice configurations, even though eigenfunctions for this set of configurations can be calculated and interpreted. Moreover, the modifiable areal unit problem implies that caution and critical appraisal should be exercised when drawing implications based upon non-lattice data.

3.5. THE IMPORTANCE OF CONFIGURATIONAL INFORMATION: A CASE STUDY OF TORONTO

As was argued in Chapter 1, one reason the configuration of areal units needs to be related to a quantitative index is because at least some information contained in each observation x_i is a function of the arrangement of the associated areal units. An instance of this contention will be demonstrated in this section using geographic data for Toronto.

Urban spatial structure has been divided into four components for the purpose of analysis. One component is land use. Geographic patterns of land use are studied in terms of transport costs and location rent curves. Basic types include government, office, commercial, industrial, and residential. Another component is socio-economic distributions that constitute social areas. The literature is replete with studies of geographic patterns of social areas in terms of factorial ecology, and relate to Burgess' concentric zone model, Hoyt's sector model, and Ullman and Harris' multiple nuclei model. A third component is spatial flows amongst urban locations. Geographic patterns of flows are studied in terms of push and pull factors associated with origins and destinations, and the attenuating nature of distance. This component has been modelled most successfully with the doubly-constrained entropy version of the gravity model. The final component is spatial infrastructure. Here the geographic configuration of locations is studied in terms of its geometric properties. This component also serves as the basis for spatial autocorrelation analyses.

Generally these four components are not treated simultaneously. For instance, Herbert and Stevens (1960) formulated a model that linked residential land use patterns to journeys-to-shop. Garin (1966) has generalized the Lowry model, which combines retail land use, residential land use, and industrial land use patterns with journey-to-work and journey-to-shop flows. Bourne and Murdie (1972) analyzed the relationship between spatial patterns of social areas and geographic land use patterns. A comprehensive study has yet to be completed. One objective of this section is to simultaneously analyze these four components of urban spatial structure. In so doing their articulation will be achieved. This synthesis will be attained by identifying common dimensions that are latent in these components. This goal will be attained through the use of generalized canonical correlation analysis.

3.5.1. *Generalized Canonical Correlation Analysis*

The classical canonical correlation technique addresses the statistical problem of identifying dimensions that isolate and summarize linear relations between two sets of variables in a convenient and illuminating manner, such that the correlation between dimension pairs is maximized while the correlations between dimensions of each data set are minimized. Hence this technique uncovers a set of dimensions common to the two data sets. When a research problem requires knowing whether or not three or more data sets have common dimensions across them, then the generalized canonical correlation technique is needed. Moreover, the statistical problem is to identify canonical variates that isolate and summarize linear relations among $m \geqslant 3$ sets of variables in a convenient and illuminating manner. Algebraic solutions to this generalization have been furnished by Kettenring (1971). Five specific solution methods are available. Four are based upon the factor analytic model, while the generalized variance method is related to principal components analysis. Since the present investigation is exploratory in nature, a generalized variance solution will be sought.

The initial step of generalized canonical correlation analysis is to reduce each set of variables to a set of principal components. This step permits analytical simplicity. In the case of urban spatial structure analysis, though, it takes on substantive meaning in that each of these types of variable sets already has undergone extensive study with respect to its latent principal components.

Let $\mathbf{\Omega}$ be the correlation matrix for the m-type canonical variates associated with a specific dimension across m sets of variables. By definition the generalized variance of this correlation matrix is det($\mathbf{\Omega}$). Since the objective is to maximize the correlations amongst the canonical variates, then in the limit every cell of matrix $\mathbf{\Omega}$ would contain unity. In contradistinction, if no linear combinations can be found then in the limit these m variates will be statistically independent, and hence $\mathbf{\Omega} = \mathbf{I}$. Let η_i be the i-th eigenvalue of matrix $\mathbf{\Omega}$. By definition $\det(\mathbf{\Omega}) = \Pi_{i=1}^{i=n} \eta_i$. For the case of perfect intercorrelations, $\eta_1 = 4$ and $\eta_2 = \eta_3 = \eta_4 = 0$. For the case of no intercorrelation, $\eta_1 = \eta_2 = \eta_3 = \eta_4 = 1$. Consequently, because those linear combinations that maximize the pairwise correlations amongst the canonical variates are sought, the generalized canonical correlation problem translates into the following optimization problem:

$$\text{MIN: } \det(\mathbf{\Omega}) = \prod_{i=1}^{i=n} \eta_i. \tag{3.5}$$

The solution to equation (3.5) is obtained by first constructing $m(m + 1)/2$ correlation matrices $\mathbf{R}_{jk}$ between all possible pairs of principal component sets ${}_j\mathbf{z}$. Let ${}_j\mathbf{b}$ be the normalized vector of canonical weights being sought. Then ${}_j\mathbf{b}^T{}_j\mathbf{b} = 1$, and ${}_j\mathbf{b}$ is the set of coefficients that is applied to principal component set j in order to calculate the desired linear combination. Further, let ${}_j\mathbf{N}$ be the partitioned matrix $\langle \mathbf{R}_{j1\,1}\mathbf{b}, \ldots, \mathbf{R}_{j,j-1\,j-1}\mathbf{b}, R_{j,j+1\,j+1}\mathbf{b}, \ldots, \mathbf{R}_{jm\,m}\mathbf{b}\rangle$, and let ${}_j\mathbf{\Omega}$ be a subset of matrix $\mathbf{\Omega}$ from which the j-th row and column have been deleted. Attaching constraint ${}_j\mathbf{b}^T{}_j\mathbf{b} = 1$ to equation (3.5) using Lagrangian multiplier ${}_j\theta$, and then taking partial derivatives with respect to ${}_j\mathbf{b}$, yields for the first canonical dimension

$$\{[\mathbf{I} - {}_j\mathbf{C}({}_j\mathbf{C}^T{}_j\mathbf{C}) - {}_j\mathbf{C}^T]\,[{}_j\mathbf{N}^T{}_j\mathbf{\Omega}^{-1}{}_j\mathbf{N} - ({}_j\theta - 1)\mathbf{I}]\}_j\mathbf{b} = \mathbf{0};$$
$$j = 1, 2, \ldots, m, \tag{3.6}$$

where ${}_j\mathbf{C}$ represents the constraint of pairwise uncorrelatedness of canonical variates for a specific data matrix. Equation (3.6) may be solved as an eigenfunction problem. Setting ${}_j\theta - 1 = \eta_{\max}$ and computing ${}_j\mathbf{b}$ as the corresponding eigenvector renders the desired solution to equation (3.6), since $\eta_{\max}$ dominates the behavior and structure of the matrix. Iteratively calculating ${}_j\mathbf{N}$, ${}_j\mathbf{\Omega}^{-1}$ and ${}_j\mathbf{b}$ renders a solution to equation (3.5) that asymptotically converges upon the minimum.

Kettenring (1971) notes that the term ${}_j\theta - 1 = \eta_{\max}$ equals ϱ_j^2, the squared generalized canonical correlation between variate ${}_jz$ and all other variates $({}_iz,\ i = 1, 2, \ldots, m_j,\ i \neq j\}$ for a given canonical dimension. This value also is the squared multiple correlation for variate ${}_jz$ regressed upon all other canonical variates.

Thus the significance of the dimension as well as the importance of each variate may be evaluated with a step-wise multiple regression technique. Clearly, though, the corresponding significance tests will need to be adjusted for the total number of variables under study.

For numerical purposes equation (3.6) must be evaluated iteratively, a seemingly straightforward task since the constrained version of objective function (3.5) is a monotonically decreasing function. In an attempt to minimize expression (3.5), then, each $({}_j\theta - 1)$ is moved as close to unity as possible. Kettenring (1971) suggests that a useful criterion function for the iteration procedure is

$$\sum_{j=1}^{j=m} |{}_j\theta_\tau - {}_j\theta_{\tau-1}| < \varepsilon,$$

where τ denotes the iteration number. He then ascribes a critical value of 0.0001 to ε. The problem with the generalized variance method arises from matrix $\mathbf{\Omega}^{-1}$. Numerical matrix inversion is plagued by rounding error problems. Consequently, $\varepsilon = 0.0001$ seems to be far too restrictive a critical value. Apparently the impact of rounding error can take over before this value is converged upon. A relaxation of this restriction is not too detrimental to the analysis, since one is interested in $\varrho = ({}_j\theta - 1)^{1/2}$. Assume that the true multiple correlation is $(1 - a)$, $0 \leqslant a \leqslant 1$. Then $\varrho = (1 - a)^{1/2}$. If an approximation is sought that is somewhat close to $(1 - a)^{1/2}$, then the value obtained is $[1 - (a + b)]^{1/2}$, $0 \leqslant a + b \leqslant 1$. This term can be expanded as

$$(1 - a)^{1/2} - [1 - (1 - b)^{1/2}] + (b/2)\,[(1 - a)^{-1/2} - 1] + (a/2)\,[(1 - b)^{-1/2} - (b \dotdiv 2)/2] + \delta, \qquad (3.7)$$

where δ represents higher order terms in the infinite series. As $a \rightarrow 0$, these first four terms provide a good approximation for $[1 - (a + b)]^{1/2}$, when b is small, since $\delta \simeq 0$. It permits an operationally useful value of b to be selected. For instance, if $\varrho = (0.999)^{1/2} \simeq 0.9995$, then values of b as large as 0.05 will change ϱ by only 0.0253. Because ε is spread over $m \geqslant 3\,\varrho$'s, however, such an impact will tend to be even less for a critical value of this magnitude. Hence, non-convergence due to rounding error exacerbated by inverting matrix $\mathbf{\Omega}$ can be avoided with relative ease. The usefulness of expression (3.7) diminishes as $a \to 1$, though. Simulation experiments to date indicate that a reasonable value of ε for the generalized variance problem is about 0.001 rather than 0.0001. Furthermore, this solution method appears to require more iterations than the four other methods discussed by Kettenring (1971).

Meanwhile, a canonical variate structure matrix may be constructed by correlating each variable set j with its corresponding canonical variate (see Ray and Lohnes, 1973). The sum of squares of these correlation coefficients divided by the number of variables in set j gives the proportion of generalized variance of the original variable set extracted by canonical variate j. This piece of information indicates the ability of the canonical dimension to summarize the original data. The usefulness of this information is twofold. First, since the minimization of expression (3.5) is achieved through a simultaneous maximization of the pairwise correlations of canonical variates for a given canonical dimension, every element of matrix $\mathbf{\Omega}$ could be unity or near unity, while little would be known about the variability accounted for in each data matrix by this canonical dimension. Second, canonical variates for a given canonical dimension will not necessarily extract the same proportions of variance from their respective domains.

Another useful piece of information is provided by the redundancy index (see Ray and Lohnes, 1973). The redundancy of a particular canonical variate in one data set when its counterpart for one of the other data sets is available is given by the product of the proportion of variance extracted by this canonical variate times the squared canonical correlation between these two canonical variates. Redundancy, just like the proportion of variance accounted, is an important index because, once again, a canonical dimension may display strong canonical correlations amongst its variates, but be a trivial explanatory construct for some or all of the data sets it is based on. The total redundancy of canonical variates from one data set given the availability of those calculated for one of the other data sets is simply the sum of the redundancies for this first set of canonical variates.

The final interpretative consideration revolves around matrix $\mathbf{\Omega}$. As was mentioned earlier, a significance test may be applied to the canonical correlations by viewing them within a multiple regression framework. A generalized canonical dimension would be significant, then, if all of its canonical correlations were significant. In keeping with this perspective, sub-dimensions could be extracted for which $1 \leqslant q < m(m - 1)/2$ canonical correlations are significant. A researcher may find these sub-dimensions of interest, whereas the generalized canonical variates would cease to be extracted as soon as at least one of the correlation coefficients was deemed insignificant, or when the total number extracted equals the minimum p_j, $j = 1, 2, \ldots, m$.

Interpretation can be further enhanced by studying attributes of matrix $\mathbf{\Omega}$. Kettenring (1971) achieves this goal by extracting the eigenvalues and eigenvectors of $\mathbf{\Omega}$. Actually he obtains an unrotated principal components solution, since $\mathbf{\Omega}$ is a correlation matrix. Component scores derived from the first principal component will permit a single synthetic variable to be analyzed that summarizes the canonical dimension in question that runs through the m data sets. More than likely, the power of this single summary will diminish as higher order canonical dimensions are extracted.

3.5.2. *Land Use Structure*

Turning now to the urban spatial structure problem, land use inventories primarily refer to the percentage of land comprising an areal unit devoted to some usage. Hence, the sum of these numerical values for a given areal unit equals 100%, resulting in a closed number system. Several problems plague analyses that utilize data measured with this type of number system. These problems are particular to properties of a closed number system itself, and not of the objects being measured.

First, if these percentages are being predicted from some set of variables, then the logit model calibrated by a weighted least-squares estimation procedure is required [Wrigley (1973)]. Second, correlation coefficients may be found to be spuriously high estimates of true relationships [Kuh and Meyer (1955)]. Third, Krumbein (1962) has shown that for a three-variable case involving values that sum to 100%, two of the correlations will always be negative, and one positive. Chayes (1971) has generalized this finding. Further, for a k-variable situation the k-th variable can always be expressed as a linear combination of the remaining

$(k - 1)$ variables. Consequently, great care is needed when interpreting correlations based upon the percentage of land devoted to a particular use.

In the case of Toronto, the Metropolitan Toronto Planning Board has defined 11 major land use categories, which were further subdivided into a total of 61 categories of land use. Bourne and Griffith (1975) have reported in detail on the basic statistical dimensions of these data, and so only a brief summary of results will be presented here.

This section is concerned with the 11 generic land use types. These include not only office, retail, industrial and residential, but also recreational, governmental, transportation, utilities/communication, and vacant categories. The problem of closure may not be as severe for the case of Toronto, though, since a miscellaneous category has been ignored. This deletion is a marginal loss, with the amount of acreage it embraces accounting for a very small proportion of the city's total land area.

The eleven land use variables were subjected to a principal components analysis coupled with a varimax rotation. The results of this statistical analysis appear in Table 3.3 Four underlying dimensions were uncovered, accounting for 64.4% of the variation exhibited by the 11 original variables. The first principal component, PCLU I, identifies the separation of residential and industrial geographic distributions. Not surprising, high percentages of industrial acreage concomitantly exist with transportation and storage. Meanwhile, high percentages of residential acreage are segregated from industry. This dimension signifies a separation of home and workplace, and accounts for 21% of the total land use variation. This dimension more than likely is dominant because the foundation of urban areas is constructed with basic employment activities and a labor market.

The second principal component, PCLU II, identifies a bundle of land uses that benefits from local agglomeration economies. In a monocentric city, retail, office, indoor recreational, and institutional activities concentrate together in the CBD, where 'face-to-face' contacts are maximized while average distance to the patron and retail consumer is minimized. In a polycentric city, these land use types disperse from the CBD, but still cluster together in subcenters in order to continue benefiting from

TABLE 3.3

Varimax Rotated Principal Components Structure for 1971 Toronto Land Uses

Land Use Variable	Varimax Rotated Principal Components Loadings			
	PCLU I	PCLU II	PCLU III	PCLU IV
% residential	*-.67284*	-.28659	-.45934	.47074
% retail	.22148	*.58398*	-.36658	.31174
% office	.26942	*.66437*	-.19374	-.18765
% indoor recreational	-.12356	*.77545*	-.07852	.12983
% outdoor recreational	-.11004	-.15845	.01956	*-.89367*
% institutional	-.19454	*.68773*	.25967	.06430
% industrial	*.78810*	-.15033	.13024	.19486
% transportation	*.66286*	.08939	-.09075	-.15803
% utilities	.05637	-.03274	*.68893*	.13360
% storage	*.75699*	-.05357	.16270	.21685
% vacant/agricultural	.14216	-.07924	*.68020*	-.19738
Eigenvalue	2.296	2.005	1.446	1.340
Percent of Variance Accounted	20.9	18.2	13.1	12.2

NOTE: Prominent correlation coefficients have been italicized.

agglomeration economies. Thus, this dimension signifies the presences of urban nodality.

The third principal component, PCLU III, identifies the land extensive attributes of utilities, vacant land, and agricultural uses. Because of land costs, land extensive activities have an incentive to locate in the suburbs. Along the urban fringe land is used for agricultural purposes, or often held vacant by speculators. Here, too, is where transformer power stations, and the like, tend to locate. Land in the CBD is too costly for any of these three uses. Hence, this dimension signifies the urban-periphery dichotomy.

The fourth principal component, PCLU IV, identifies outdoor recreational uscs. This category includes parks, stadia, golf courses, cemeteries, and the such. Presumably its geographic pattern is statistically independent of the other land use patterns because it embraces large acreages and sizeable capital investments whose locations take on meaning only in an historical context. For example, a cemetery may be established on the urban fringe in 1850. By 1970, though, if the city has enjoyed considerable growth and expansion, then this land use now may be relatively close to the CBD. Consequently, this dimension signifies investment inertia, or a temporal characteristic of Toronto.

Although these four dimensions seem interpretable, Bourne and Griffith (1975, p. 25) should be consulted for a more thorough treatment of this topic.

3.5.3. *Social Area Structure*

The urban literature is replete with comparative factorial ecologies. Generally speaking a myriad of census variables is subjected to a principal components analysis, and each prominent component is labelled with regards to those census variables that are strongly correlated with it. The first principal component tends to refer to a socio-economic dimension. Those variables highly correlated with this component include income, education, and occupation. The geographic pattern affiliated with this component is sectoral in nature. The second component tends to refer to a demographic dimension. Those variables highly correlated with this component include age structure, family size, and family composition. The geographic pattern associated with this component is that of concentric rings. The third component is usually minor, and tends to refer to a segregation dimension. Those variables highly correlated with this component include ethnic status, and housing attributes. The geographic pattern connected with this component is that of multiple nuclei. Palm and Caruso (1972) provided a timely review of much of the factorial ecology literature. This pursuit has been severely criticized because it is descriptive without being illuminating. Nevertheless, it does represent an in depth investigation of one aspect of urban spatial structure.

In the case of Toronto, Murdie (1980) has published an insightful analysis of 1971 social areas. His results are summarized here in Table 3.4. Although he uncovered six social area dimensions, he only interpreted five. Together the six dimensions accounted for 73.2% of the variation exhibited by 48 socio-economic variables. The first dimension, PCSA I, refers to Italian ethnicity, blue-collar

TABLE 3.4

Summary of Murdie's Social Area Analysis for 1971 Toronto Data

Factor	Prominent Loading	Canadian Census Variable	Percent of Variance Accounted
PCSA I: Ethnicity/economic status	|.8|	Neither English nor French speaking	18.2
		Roman Catholic religion	
		British ethnicity	
		Italian ethnicity	
		Anglican religion	
	|.7|	Education: elementary only	
		Two-family households	
		Manufacturing occupation: male	
	|.5|	Managerial/professional occupation: male	
		German ethnicity	
PCSA II: Family status I	|.9|	Population: age 5-9 years	16.3
		Population: age 10-14 years	
		Households with 4-5 persons	
	|.8|	Non-family households	
	|.7|	Apartments	
	|.6|	Single population	
		Female labor force participation	
		Rooms per dwelling	
	|.5|	Households with 6 or more persons	
		Automobile	
		Both English and French	
PCSA III: Economic status/ethnicity	|.7|	Unemployment: male	15.7
		Service Occupation: female	
		Households with private bathroom	
	|.6|	Lodgers	
		Service occupation: male	
		Clerical occupation: female	
		Automobile	
		Attached (row) houses	
		Sales occupation: male	
	|.5|	Households with 6 or more persons	
		Single population	
		Asian ethnicity	
		French ethnicity	
		Income: male median	

TABLE 3.4 (continued)

PCSA IV:			
Family status II	|.8|	Dwelling occupied 5 years or more by present occupant	11.3
		Population: age 45-54 years	
		Population: age 25-34 years	
	|.7|	Dwelling occupied 2 years or less by present occupant	
		Children: 15-18 years	
		Population change	
	|.4|	Children: under 6 years	
PCSA V:			
Economic status	|.6|	Managerial/professional occupation: male	7.2
		Clerical occupation: male	
		Rooms per dwelling	
	|.5|	Manufacturing occupation: male	
		French ethnicity	
PCSA VI:			
unnamed	|.6|	Jewish religion	4.5
		Lutheran religion	
	|.5|	German ethnicity	
		Ukrainian ethnicity	

Source: Murdie (1980, pp. 302-303) and personal correspondence.

occupations and two-family households versus British ethnicity coupled with managerial/professional occupations. The second dimension, PCSA II, refers to small households and apartments together with high female labor force participation versus large households with single detached housing and children. The third dimension, PCSA III, refers to unemployment and service occupations combined with lodgers, row housing, lack of household amenities and French-Asian ethnicity versus female clerical and male sales occupations. Dimension four, PCSA IV, refers to older families and low residential mobility versus younger families and high residential mobility. The fifth dimension refers to blue-collar occupations and smaller houses versus higher status occupations coupled with larger houses. Although Murdie failed to interpret Dimension Six, PCSA VI, he found that employing it for over-rotation produced the necessary simple structure for interpreting dimension PCSA I through PCSA V. It may well be a segregation dimension, since it refers to German-Ukranian ethnicity with Lutheran religion versus Jewish religion. Once again, the reader should consult Murdie (1980) for greater detail.

3.5.4. *Spatial Interaction Structure*

Analysis of the structure of spatial interaction patterns has been undertaken in one of four ways. The most descriptive is a cartographic representation of individual flows with arrows. A second fits some variant of the gravity model to the flows data. A third describes the underlying network with graph theory indices. The fourth identifies common flow patterns within a system determined on the basis of similarities among individual flows out of origins, or into destinations. This last approach yields a parsimonious description of flow patterns, and may be referred to as a factorial ecology of flows [see Black (1973); Davies and Thompson (1980)].

A factorial ecology of flows can be used to study two types of structure. First, the bulk of the flow data can be reduced to a somewhat smaller number of components, while the statistical information in these data is preserved. A high correlation between two sets of flows implies similar flow patterns for their corresponding areal units. Second, the complex flow system can be reduced to a set of regionalized components. Here a dimension from *R*-mode factor analysis identifies a group of destinations, in the traditional origin–destination (O–D) matrix, that are characterized by a common pattern of flows from the origins. In the *Q*-mode analysis groups of origins are identified that are characterized by a common pattern of flows into the destinations [Britton (1971)]. Simmons (1970) has found that *Q*-mode results tend to be far more complex than *R*-mode results, and hence far more difficult to interpret. One example of the factorial ecology of intraurban flows has been provided by Wiseman (1976).

Although a methodology exists for evaluating the statistical structure of flows, the problem of defining an appropriate type of flow remains. Intraurban spatial interaction is of four primary kinds, namely journey-to-work, journey-to-shop, journey-to-recreate, and household migration. Intuitively speaking, urban spatial structure is most influenced by journey-to-work flows, followed by journey-to-shop flows. This intuition seems to be corroborated by the Herbert-Stevens and Lowry models. Journey-to-work flows will be employed here.

In the case of Toronto, an origin-destination table for 341 census tracts was constructed. Nine census tracts had no information

TABLE 3.5

Dominant Varimax Rotated Principal Components
and Prominent Loadings for 1971 Toronto Journey-to-work Flows

Factor	Prominent Loading	Destination Census Tracts	Percent of Variance Accounted
PCSI I:			
East Toronto	|.8|	CT347, CT368, CT370	4.91
	|.7|	CT342, CT344, CT348, CT350, CT353, CT354, CT355, CT356	
	|.6|	CT261, CT343, CT363, CT365	
	|.5|	CT357, CT360, CT364	
	|.4|	CT192, CT262, CT331, CT341, CT346, CT349, CT352, CT361, CT362, CT366, CT367, CT376	
PCSI II:			
Downtown Toronto	|.8|	CT35, CT62	2.83
	|.7|	CT61, CT63, CT64, CT88, CT89	
	|.6|	CT34, CT37, CT91	
	|.4|	CT32, CT36, CT66	
PCSI III:			
West Toronto-Lakeshore	|.9|	CT210	2.50
	|.8|	CT207, CT208, CT211	
	|.7|	CT205, CT206, CT213, CT214	
	|.6|	CT212, CT215	
	|.5|	CT204	
	|.4|	CT200, CT219	
PCSI IV:			
Northwest Toronto	|.8|	CT312, CT315	2.84
	|.7|	CT292, CT313, CT316	
	|.6|	CT290, CT311	
	|.5|	CT293, CT314	
	|.4}	CT247, CT250, CT280, CT281, CT282, CT283, CT286, CT294, CT295	
PCSI V:			
Central Toronto-Yonge/ Eglinton	|.7|	CT128, CT136 2.15	
	|.6|	CT122, CT129	
	|.5|	CT121, CT124, CT135	
	|.4|	CT127, CT196, CT260, CT263, CT273	
PCSI VI:			
North-Central Toronto-Yonge/Finch/Sheppard	|.8|	CT299, CT307, CT308, CT318	2.19
	|.6|	CT300, CT321, CT322	
	|.5|	CT319, CT320, CT323	
	|.4|	CT274, CT317	
PCSI VII:			
Central Toronto-Southeast	|.7|	CT182 2.45	
	|.6|	CT1, CT184, CT182	
	|.4|	CT28, CT80, CT183, CT184, CT189, CT194, CT261	

TABLE 3.5 (continued)

PCSI VIII:			
Keele/St. Clair	|.6|	CT106, CT107, CT112	2.12
	|.5|	CT97, CT153, CT158	
	|.4|	CT98	
PCSI IX:			
Central Toronto-Southwest	|.8|	CT48, CT49	1.55
	|.6|	CT10, CT52	
	|.5|	CT7	
	|.4|	CT6, CT9, CT47, CT53	
PCSI X:			
West Toronto	|.6|	CT221, CT248	1.67
	|.5|	CT247	
	|.4|	CT213, CT222, CT233, CT234, CT236, CT244, CT245, CT246, CT250	
PCSI XI:			
unnamed	|.6|	CT11	1.40
	|.5|	CT277	
	|.4|	CT16, CT276	
		Cumulative percent of variance accounted = 26.61	

available because of the confidentiality of census information, while one census tract had no flows into or out of it. Eleven principal components accounting for 25.5% of the variation in the O–D table were extracted. Summary results for these dimensions are presented in Table 3.5, and a basic map of the results appears in Figure 3.2.

The first principal component, PCSI I, refers to common flow patterns for much of eastern Toronto. This primarily is a residential area, and a prominent source of labor for the CBD area. The second principal component, PCSI II, refers to the CBD, which is the source for an overwhelming number of jobs in Toronto. Principal component three, PCSI III, refers to southwest Toronto, a mixed residential and industrial area. The fourth principal component, PCSI IV, refers to northwest Toronto, a residential and land extensive employment activities area. Principal component five, PCSI V, refers to central Toronto, a residential area that is emerging as a subcenter and increasingly a source of employment. The sixth principal component, PCSI VI, refers to north-central Toronto, a newer residential area and designated location for the cultivation of an urban subcenter. Principal component seven, PCSI VII, refers to east-central Toronto, a residential area. The

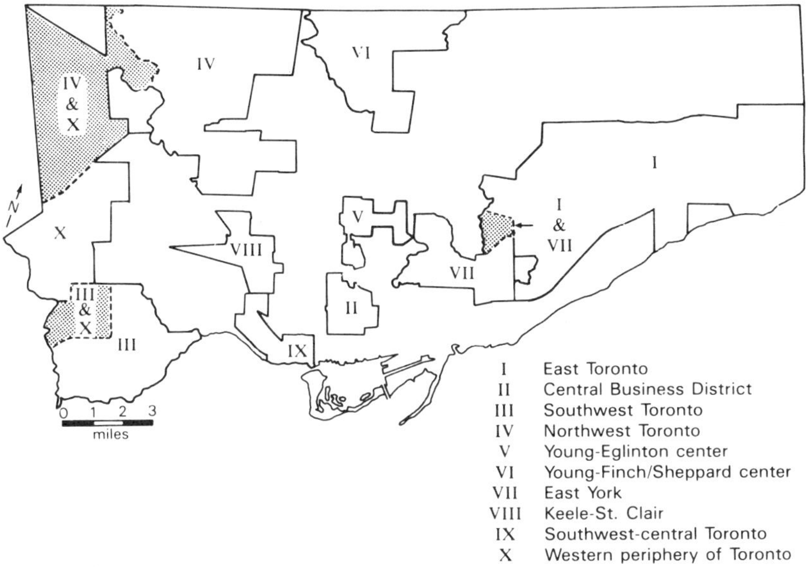

Fig. 3.2. Principal components regionalization based upon the journey-to-work spatial interaction matrix.

eighth principal component, PCSI VIII, refers to west-central Toronto, a residential and retail area. Principal component nine, PCSI IX, refers to southwest-central Toronto, a residential and manufacturing area. The tenth principal component, PCSI X, refers to west Toronto, primarily a residential area that is in juxtaposition with the dormitory town of Mississauga. Principal component eleven had no meaningful spatial expression, but permitted over-rotation that rendered more interpretable components PCSI I through PCSI X.

Although these 11 components account for only about one-fourth of the variation in the flows data, one should keep in mind that they are based upon a 341-by-341 matrix. Accordingly, the reduction is very satisfying.

3.5.5. *Spatial Infrastructure*

Consider an urban region that has been partitioned into n mutually exclusive and collectively exhaustive areal units. Spatial infrastructure refers to the configuration of these areal units, and may

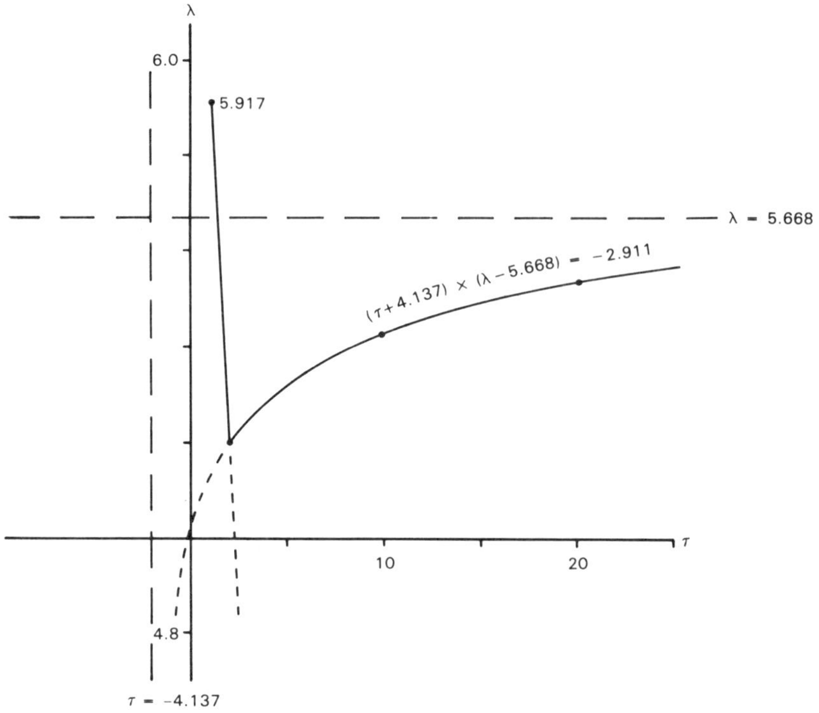

Fig. 3.3. Convergence trajectory for equation (3.8).

be defined more specifically as the geometric arrangement of areal units that place phenomena in juxtaposition with one another, bringing about spatial interrelationships. This infrastructure may be depicted by the binary connectivity matrix **C** already discussed in this chapter that represents areal unit juxtapositions.

In the case of Toronto the principal eigenvalue measuring areal unit configuration was extracted from a 351-by-351 matrix **C**. Since this matrix is quite sparce, the iteration technique described by Tinkler (1972) was used to calculate λ_{max} and $\mathbf{E}_{max}$, where the diagonal was coded with zeros. At iteration 790 the resulting estimate of λ_{max} was significant to 15 digits. A graph of the asymptotic convergence of $\hat{\lambda}$ on λ is presented in Figure 3.3. The resulting curve is an hyperbola. Thus, the relationship between $\hat{\lambda}$ and the number of iterations τ is given by

$$(\lambda - \alpha)(\tau - \beta) = \gamma, \quad \text{or} \quad \hat{\lambda} = \alpha + \gamma/(\tau - \beta). \tag{3.8}$$

TABLE 3.6

Regression for Convergence of the Principal Eigenvalue for Toronto's Connectivity Matrix

Variable	Regression Coefficient	t-value	Multiple Correlation	Percent of Variance Accounted
$1/(\tau - 4.13711)$	-2.91118	227.7***	.99249	98.50
constant	5.66834	29.7***		

*** Denotes a significant difference from zero at the .01 level, with 788 degrees of freedom.

NOTE: The original model

$$(\tau - \alpha)(\lambda - \beta) = \delta$$

was estimated with the non-linear IMSL subroutine ZXSSQ. Once a was obtained, then the model was rewritten so that

$$\lambda = \beta + \delta/(\tau - a)$$

This second model was estimated with OLS.

The rate of convergence upon γ is given by

$$\partial\hat{\lambda}/\partial\tau = \gamma/(\tau - \beta)^2.$$

Consequently, one may easily evaluate the number of iterations necessary to achieve a particular degree of accuracy.

Equation (3.8) was calibrated to see how well it described the Toronto situation. First non-linear least squares was employed to estimate β of equation (3.8). This yielded a value fo 4.13711. Next, $\hat{\lambda}$ was regressed on $1/(\tau - 4.13711)$ using OLS. The results of this regression are presented in Table 3.6. Not only does the OLS solution yield a value of 5.66834 for $\hat{\lambda}$, but it also indicates a very close correspondence between equation (3.8) and the interaction data, accounting for 98.5 percent of the variance.

As was mentioned earlier, the $e_{i,\max}$ values of eigenvector $\mathbf{E}_{\max}$ are directly related to geometric centrality with respect to the configuration of areal units. Accordingly, values close to zero refer to physically peripheral areal units, within the context of the given arrangement of units, while values approaching the maximum refer to physical centrality. The range of values observed for Toronto is between 0.0003 and 44.01. Mapping these eigenvector values (see Figure 3.4) reveals that the east side of Toronto constitutes the physical periphery, whereas the physical core is located

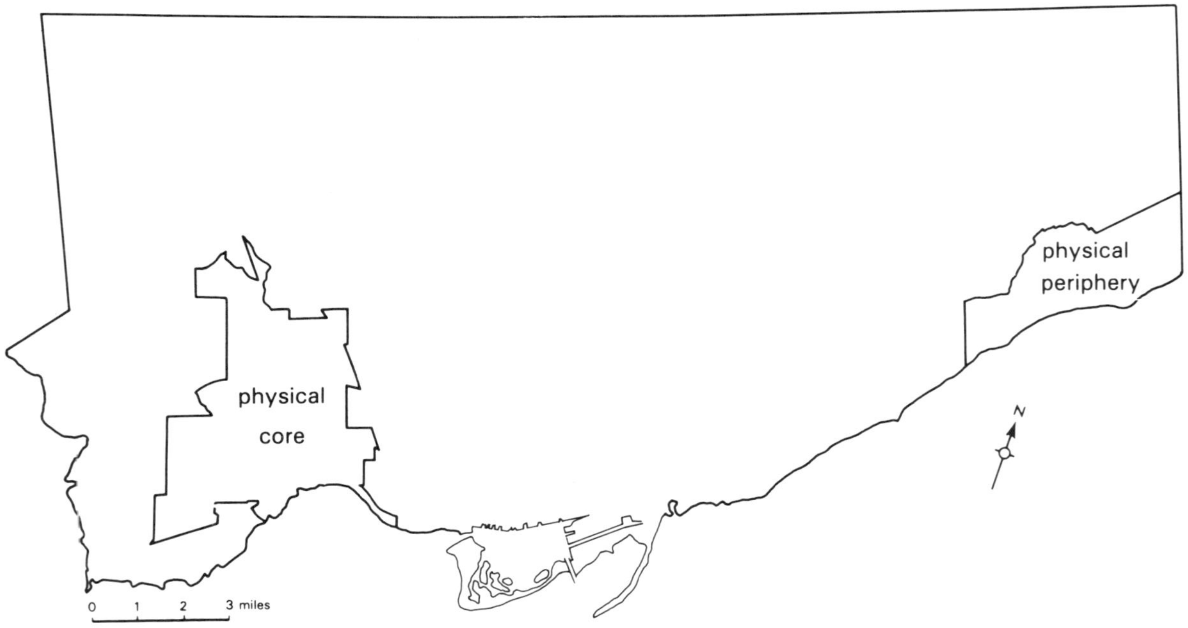

Fig. 3.4. Clusters of census tracts based upon the principal eigenvector extreme values of the Toronto census tract connectivity matrix.

in southwest Toronto. One should note that $\lambda_{\max}$ implies that Toronto has a low degree of connectivity amongst its areal units primarily because of size complexities.

3.5.6. *A Generalized Canonical Correlation Solution for the Toronto Data*

Characteristics of the four data sets have been described here using factor analytic types of techniques. Bringing these data together for analysis required that observations aggregated in any one data set had to be aggregated in all data sets, while those missing from any one set had to be deleted from all sets. This modification reduced the number of observations to 334, for 7 pairs of areal units were consolidated and 10 areal units were deleted. These modifications should not influence the results drastically, since they involve less than 5% of the total number of areal units.

Principal components solutions reduced the land used data to a 334-by-4 matrix, the social area data to a 334-by-6 matrix, the journey-to-work flows to a 334-by-11 matrix, and the juxtaposition matrix to a 334-by-1 matrix. Consequently, only one generalized canonical variate can be extracted from these four matrices, since the number extracted cannot exceed the minimum p_j for the n-by-p_j matrix dimensions.

In the first iteration all ${}_j\mathbf{b}$ vectors were initialized to unity vectors. Once again asymptotic convergence was observed. The trajectory of this convergence is summarized in Table 3.7. By iteration 501 the squared canonical correlations had a combined absolute difference from iteration 500 values of 0.00228, which basically is a negligible amount. The significance of this common dimension is indicated by the stepwise multiple regression results presented in Table 3.8. The generalized land use canonical variate, GNCVLU, is related significantly to the remaining three variates, and has a joint correlation with them of 0.729. The generalized spatial interaction canonical variate, GNCVSI, is related significantly to the remaining three variates, and has a joint correlation of 0.790 with them. The generalized social area canonical variate is related significantly only to GNCVLU and GNCVSI, and has a joint correlation of 0.870 with the remaining three variates.

TABLE 3.7

Summary of Iteration Statistics for the Generalized Canonical Dimensions

Sample Iteration	Dominant Characteristic Root for Derivative PHILU	PHISA	PHISI	PHICN	Value of Criterion Function
1	.08063	.28315	.38336	1.00000	2.25286
51	.52046	.74726	.62389	1.00000	.02731
101	.52568	.75204	.62474	1.00000	.01629
151	.52770	.75360	.62478	1.00000	.01156
201	.52873	.75447	.62475	1.00000	.00874
251	.52936	.75503	.62469	1.00000	.00682
301	.52978	.75543	.62464	1.00000	.00540
351	.53008	.75572	.62458	1.00000	.00433
401	.53029	.75595	.62454	1.00000	.00349
451	.53045	.75612	.62450	1.00000	.00282
501	.53057	.75626	.62446	1.00000	.00228

NOTE 1: The criterion value for the first iteration is quite large because all canonical variates are initialized with an arbitrary correlation of $1/p^{1/2}$, where p denotes the number of principal components for a data set.

NOTE 2: Criterion function $\sum_{j=1}^{j=4} = |\lambda_{j\tau} - \lambda_{j,\tau-1}|$

critical value = 0.001 , and

maximum number of iterations = 500.

The generalized configuration canonical variate, GNCVCN, which is of paramount importance to the theme of this section, is related significantly only to GNCVLU and GNCVSI, and has a joint correlation of 0.313 with the remaining three variates. Basically, then, the canonical dimension seems to be significant, although more prominent sub-dimensions may well exist for the land use, social area, and spatial interaction matrices.

Subjecting the matrix of correlations amongst the four canonical variates, $\mathbf{\Omega}$, to a principal components analysis shows that the basic dimension across these data accounts for 56.2% of the variation displayed by the individual canonical variates. The second component reinforces the previous finding that GNCVCN is noticcably independent of the other three variates. The dominant canonical dimension is roughly proportional to $5 \times$ GNCVLU $- 6 \times$ GNCVSA $+ 5 \times$ GNCVSI + GNCVCN.

Table 3.9 indicates that the generalized land use canonical variate, GNCVLU, is strongly and directly related to the urban

TABLE 3.8

Generalized Canonical Correlation Structure for 1971 Toronto Data

Individual Canonical Variate	Remaining Variates	Stepwise Multiple Correlation	Generalized Canonical Correlation
GNCVLU: generalized land use canonical variate	GNCVSA***	.69544	.72885
	GNCVSI***	.72438	
	GNCVCN**	.72885	
GNCVSA: generalized social areas canonical variate	GNCVSI***	.75166	.87014
	GNCVLU***	.86970	
	GNCVCN	.87014	
GNCVSI: generalized spatial interaction canonical variate	GNCVSA***	.75166	.78985
	GNCVCN***	.77479	
	GNCVLU***	.78985	
GNCVCN: generalized configuration canonical variate	GNCVSI***	.23698	.31252
	GNCVLU**	.30795	
	GNCVSA	.31252	

Canonical Variate	Unrotated Principal Components Loadings			
	PC I	PC II	PC III	PC IV
GNCVLU	*.78071*	-.37289	.47833	.14999
GNCVSA	*-.95048*	.07194	.06083	.29614
GNCVSI	*.84416*	.25361	.42893	.19773
GNCVCN	.15456	*.94104*	.30049	-.01649
Eigenvalue	2.24957	1.09409	.50677	.14957
Percent of Variance Accounted	56.2	27.4	12.7	3.7

NOTE: Prominent principal components loadings are italicized.

** Denotes a significant difference from zero at the .05 level, for a two-tail situation, with 330 degrees of freedom.

*** Denotes a significant difference from zero at the .01 level, for a two-tail situation, with 330 degrees of freedom.

TABLE 3.9

Generalized Canonical Variates

Data Set	Principal Component [1]	Correlation With Canonical Variate
Land Use	PCLU I	- .00823
	PCLU II	.81012***
	PCLU III	- .50800***
	PCLU IV	.29252***
Social Area	PCSA I	- .27319***
	PCSA II	- .03874
	PCSA III	- .19148***
	PCSA IV	- .50401***
	PCSA V	.00497
	PCSA VI	.79571***
Spatial Interaction	PCSI I	- .56593***
	PCSI II	.49865***
	PCSI III	- .08069
	PCSI IV	- .26851***
	PCSI V	- .05191
	PCSI VI	- .27396***
	PCSI VII	- .13343**
	PCSI VIII	.07563
	PCSI IX	.38150***
	PCSI X	- .22199***
	PCSI XI	.23817***
Configuration	PCCN I	1.00000

1 The principal component acronyms are formulated such that the first two alphanumeric characters of PC indicate principal component, the second two characters indicate an abbreviation for the given data set, and the last characters are Roman numerals numbering the principal components of each data set.

** Denotes a significant difference from zero at the 0.05 level, for a two-tail situation, with 332 degrees of freedom.

*** Denotes a significant difference from zero at the 0.01 level, for a two-tail situation, with 332 degrees of freedom.

nodality land use component, PCLU II, and moderately and indirectly related to the urban-periphery dichotomy land use component, PCLU IV. The generalized social areal canonical variate, GNCVSA, is strongly and directly related to the spatial segregation social area component, PCSA VI, and moderately and indirectly related to the second family status social area component, PCSA IV. The generalized spatial interaction canonical variate, GNCVSI, is moderately and directly related to the Toronto CBD spatial interaction component, PCSI II, and moderately and indirectly related to the east Toronto spatial interaction component, PCSI I. Finally, GNCVCN $\equiv$ PCCN I.

TABLE 3.10

Generalized Canonical Correlation Structure Matrices

Data Set	Original Variable	Correlation With Variate	Percent of Variance Accounted
Land Use	% residential	.1445***	19.44
	% retail	.7487***	
	% office	.5794***	
	% indoor recreational	.7071***	
	% outdoor recreational	-.3993***	
	% institutional	.4456***	
	% industrial	-.1376**	
	% transportation	.0664	
	% utilities	-.3381***	
	% storage	-.0689	
	% vacant/agricultural	-.4694***	
Social Areas[1]	data not available		7.68
Configuration	PCCN I	1.0000	100.00
Spatial Interaction[2]	CT11	.4330***	3.29
	CT34	.4099***	
	CT35	.4650***	
	CT52	.4280***	
	CT61	.4468***	
	CT347	-.4696***	
	CT350	-.4740***	
	CT353	-.4030***	
	CT356	-.4121***	
	CT368	-.4456***	
	CT369	-.4480***	
	CT370	-.5076***	
	CT371	-.4761***	

1 Factor scores were obtained from Robert Murdie, York University, Toronto, Canada.

2 For economy of space only prominent loadings have been included here.

** Denotes a significant difference from zero at the 0.05 level, for a two-tail situation, with 332 degrees of freedom.

*** Denotes a significant difference from zero at the 0.01 level, for a two-tail situation, with 332 degrees of freedom.

Because the configuration index is a single measure, the statistical problem reduced here to one of finding a set of weights ${}_j\mathbf{b}$ for the land use, social area, and spatial interaction principal component sets that maximized correlations between the resulting canonical variates and PCCN I. Moreover, the question being addressed asks whether or not surface partitioning and areal unit configuration play an important role in the spatial organization of land use, socio-economic, and spatial flows patterns. Recalling that PCCN I refers to surface partitioning centrality, one would expect it to be an important correlate of agglomeration economy attributes, core-periphery dichotomization, spatial segregation, family status attributes representative of the urban core, and spatial flows that concentrate on the CBD and away from residential areas.

Table 3.10 further clarifies this picture by relating the generalized canonical correlation results to the original variables. GNCVLU accounts for 19.4% of the variation exhibited by the 11 land use categories. It relates strongly and directly to retail and indoor recreational land use patterns, moderately and directly to office and institutional land use patterns, and moderately and indirectly to vacant and agricultural land use patterns. Clearly urban nodality and a core-periphery discrimination are being detected. In addition, this variate is as powerful an identifier of conspicuous land use covariations as is the dominant land use principal component, PCLU I. GNCVSI accounts for 3.3% of the variation exhibited by the 341 journey-to-work flows variables, which is nearly as good as PCSI I. It relates directly to the departure flow patterns from the city center, and indirectly to the departure flow patterns for the residential east side of Toronto. Although the original socio-economic data were not correlated with GNCVSA, all indications observed here suggest that the uncovered canonical dimension is not trivial.

3.6. IMPLICATIONS

The purpose of this chapter has been to define and demonstrate the importance of a quantitative measure of the geometric arrangement of areal units on a planar surface. To briefly recapitulate the findings reported on here, such an index is furnished by extracting the principal eigenvalue and its respective eigenvector from the connectivity matrix **C**. Three salient properties of this measure are that (a) it is based upon a set of nominal measurements, (b) it is a function of the arrangement as well as the number of areal units on a surface, and (c) as n increases the frequency distribution of

λ_p for all surface partitionings of a given size apparently approaches a normal distribution. Presumably additional structural information may be gleaned from the non-principal eigenvalues and eigenvectors, if matrix **C** is treated as a variance–covariance type of matrix. This particular interpretation approach is the one proposed in Griffith (1984).

The importance of having a quantitative, interval scale measure of structure stems from its ability to help determine whether or not results of geographical data analyses are due, at least in part, to the underlying arrangement of areal units. Two examples presented here show both that there is no evidence to indicate the presence of this relationship in some cases, and that there is evidence to support the existence of this relation in other cases.

Finally, the modifiable areal unit problem supplies a cogent argument against the random surface partitioning sampling perspective. All in all, then, spatial researchers need to pay more attention to the geography of their data landscapes if they are to fully understand areal unit configuration features and locational information.

3.7. REFERENCES

Amrhein, C., J. Guevara and D. Griffith, 1983, The effect of random Thiessen structure and random processes on the measurement of spatial autocorrelation, *Proceedings*, 14th Annual Pittsburgh Modelling and Simulation Conference, University of Pittsburgh, pp. 585–589.

Besag, J., 1972, On the statistical analysis of nearest-neighbour systems, *Proceedings*, 9th European meeting of statisticians, Budapest, pp. 101–105.

Besag, J., 1974, Spatial interaction and the statistical analysis of lattice systems, *Journal of the Royal Statistical Society* B, Vol. 36: 192–236.

Besag, J., 1975, Statistical analysis of non-lattice data, *The Statistician*, Vol. 24: 179–195.

Black, W., 1973, Toward a factorial ecology of flows, *Economic Geography*, 49: 59–67.

Boots, B., 1982. Comments on the use of eigenfunctions to measure structural properties of geographic networks, *Environment and Planning* A, Vol. 14: 1063–1072.

Boots, B., 1983, Evaluating principal eigenvalues as measures of network structure, *Geographical Analysis*, Vol. 16: 270–275.

Boots, B., 1985, Size effects in the spatial patterning of nonprincipal eigenvectors of planar networks, *Geographical Analysis*, Vol. 17: 74–81.

Boots, B., and K. Tinkler, 1983, The interpretation of non-principal eigenfunctions of urban structure, *Proceedings*, 14th Annual Pittsburgh Modelling and Simulation Conference, University of Pittsburgh, pp. 715–719.

Bourne, L., and D. Griffith, 1975, The spatial organization of urban land use: a statistical evaluation of a classification, *Research Paper* No. 69, University of Toronto, Centre for Urban and Community Studies.

Bourne, L., and R. Murdie, 1972, Interrelationships of social and physical space in the city: a multivariate analysis of Metropolitan Toronto, *Canadian Geographer*, 16: 211–229.

Britton, J., 1971, Methodology in flow analysis, *East Lakes Geographer*, 7: 22–36.

Chayes, F., 1971, *Ratio Correlation: A Manual for Students of Petrology and Geochemistry*. Chicago: University of Chicago Press.

Cliff, A., P. Haggett, and J. Ord, 1979, Graph theory and geography, in *Applications of Graph Theory*, edited by R. Wilson and L. Beineke. New York: Academic Press, pp. 293–326.

Davies, W., and R. Thompson, 1980, The structure of interurban connectivity: a dyadic factor analysis of prairie commodity flows, *Regional Studies*, 14: 297–311.

de Jong, P., C. Sprenger and F. van Veen, 1984, On extreme values of Moran's I and Geary's c, *Geographical Analysis*, Vol. 16: 17–24.

Garin, R., 1966, A matrix formulation of the Lowry model for intrametropolitan activity allocation, *Journal of the American Institute of Planners*, 32: 361–364.

Griffith, D., 1981, Synthesizing urban spatial structure: a case study of Toronto,' paper presented at the 77th Annual Association of American Geographers meetings, Los Angeles.

Griffith, D., 1984, Measuring the arrangement property of a system of areal units generated by partitioning a planar surface, in *Recent Developments in Spatial Analysis: Methodology, Measurement and Models*, edited by G. Bahrenberg, M. Fischer and P. Nijkamp. Aldershot, U.K.: Grower, pp. 191–200.

Griffith, D., and K. Jones, 1980, Explorations into the relationship between spatial structure and spatial interaction, *Environment and Planning* A, Vol. 12: 187–201.

Hebert, J., and B. Stevens, 1960, A model for the distribution of residential activity in urban areas, *Journal of Regional Science*, 2: 21–36.

Kettenring, J., 1971, Canonical analysis of several sets of variables, *Biometrika*, 58: 433–451.

Krumbein, W., 1972, Open and closed number systems is stratigraphic mapping, *Bulletin of the American Association of Petroleum Geologists*, 46: 2229–2245.

Kuh, E., and J. Meyer, 1955, Correlation and regression estimates when the data are ratios, *Econometrica*, 23: 400–416.

Matula, D., and R. Sokal, 1980, Properties of Gabriel graphs relevant to geographic variation research and the clustering of points in the plane, *Geographical Analysis*, Vol. 12: 205–222.

Murdie, R., 1980, Factor scores: a neglected element of factorial ecology studies, *Urban Geography*, 1: 295–316.

Openshaw, S., n.d., *The Modifiable Areal Unit Problem*. Norwich, England: Geo Abstracts.

Openshaw, S., and P. Taylor, 1979, A million or so correlation coefficients: three experiments on the modifiable areal unit problem, in *Statistical Applications in the Spatial Sciences*, edited by N. Wrigley. London: Pion, pp. 127–144.

Ord, K., 1975, Estimation methods for models of spatial interaction, *Journal of the American Statistical Association*, Vol. 70: 120–126.

Palm, R., and D. Caruso, 1972, Factor labelling in factorial ecology, *Annals*, Association of American Geographers, 62: 122–133.

Ray, D., and P. Lohnes, 1971, Canonical correlation in geographical analysis, *Geographia Polonica*, 25: 49–65.

Simmons, J., 1970, Interprovincial interaction patterns in Canada, *Research Paper* No. 24, University of Toronto, Centre for Urban and Community Studies.

Taaffe, E., and H. Gauthier, 1973, *Geography of Transportation*. Englewood Cliffs, N. J.: Prentice-Hall.

Tatsuoka, M., 1971, *Multivariate Analysis: Techniques for Educational and Psychological Research*. New York: Wiley.

Tinkler, K., 1972. The physical interpretation of eigenfunctions of dichotomous matrices, *Transactions*, Institute of British Geographers. Vol. 55: 17–46.

Tinkler, K., 1974, On summing power series expansions of accessibility matrices by the inverse method, *Geographical Analysis*, 6: 175–178.

Tinkler, K., 1976, On functional regions and indirect flows, *Geographical Analysis*, 8: 205–213.

Wiseman, R., 1976, Interpretation of factored flow data: a comparison of two traffic systems, *Professional Geographer*, 28: 172–177.

Wrigley, N., 1973, The use of percentages in geographical research, *Area*, 5: 183–186.

CHAPTER 4

Reformulating Classical Linear Statistical Models

The problematic points in Chapter 1 can be reduced to a single issue, namely that inferential problems arise when statistical analyses are carried out on spatial data series having dependent observations. As Chapter 2 has suggested, this is a more troublesome issue than its counterpart found in time series, because spatial interdependencies are both multidirectional and two-dimensional. The resulting complications are ones of accuracy, bias and nuisance. For instance, in the usual situation of positive spatial autocorrelation, the information contained in some observed value x_i is less than its independent counterpart, since spillover effects produce communalities amongst juxtaposed areal unit measures. An analogy can be made here between spatial data and a contrived classical situation. Suppose a random sample of size n is drawn, and the set of measures $\{x_i; i = 1, 2, \ldots, n\}$ is made. This set of measures contains a certain amount of information. Now if the sample size were to be increased merely by repeating each x_i measure twice, then it would become $2n$. But the amount of information contained in the data would not change from that found in the original n observations. To some degree this is what happens with spatial data, for often the sample size increases much faster than does the information. For instance, in the more usual situation of positive spatial autocorrelation, a single observation carries less information than an independent observation, since its measure is partly predictable from neighboring measures. In addition, sampling variances for parameter estimates become distorted, causing a modification of the shape of the sampling distribution, and hence changing the distribution of probabilities and thus the probabilities of Types I and II errors. In an analogous fashion estimates of σ_{ξ}^{2} for the classical linear regression model can be shown to be biased in the presence of spatial autocorrelation.

The notion of nuisance is one that has received little attention in the spatial autocorrelation literature. Clearly within the context of classical statistics the presence of complications that hamper the drawing of valid inferences constitutes a nuisance. One reaction would be to remove the source of such complications, and then continue to operate within the realm of classical statistics. Griffith (1979) illustrates this approach in demonstrating the inferential problem affiliated with MANOVA; he finds significant regional differences based upon the raw autocorrelated data, and insignificant regional differences based upon the unautocorrelated data.

A major drawback with treating spatial autocorrelation as a nuisance is that removing it from data is equivalent to discarding the geographic content, and thus restricting attention to non-geographic features and relations latent in a data set. Moreover, the ultimate aim of spatial autocorrelation analysis should be to posit properly specified models for spatial data. These models should contain explicit autoregressive terms, and exploit the spatial structure that geographers believe is of importance and interest. This is exactly the approach taken in time series analysis, where the temporal structure is removed in order to obtain good parameter estimates, and then this structure is returned to the model in order to generate sound forecasts. To this end the present chapter will survey prominent themes in the literature about spatial autoregressive models.

4.1. AUTOCORRELATED ERRORS MODELS

Most of the attention to date has been focused on linear statistical models in which autocorrelation appears only in the error term. These include the pure autoregressive model and the autoregressive trend surface model. This first model is of the form, using a simultaneous specification,

$$\mathbf{Y} = \mu\mathbf{1} + \varrho\mathbf{W}(\mathbf{Y} - \mu\mathbf{1}) + \xi, \tag{4.1}$$

or

$$\mathbf{Y} = \mu\mathbf{1} + (\mathbf{I} - \varrho\mathbf{W})^{-1}\xi, \tag{4.2}$$

where $\xi \sim \text{MVN}(\mathbf{0}, \sigma^2\mathbf{I})$; equations (4.1) and (4.2) assume that a homogeneous process is operating over the geographical landscape. Accordingly the mean μ is constant over this landscape. Since $\xi \sim$ MVN its likelihood function is written as equation (1.6). The MLEs of equation (1.6) are

$$\hat{\mu} = \mathbf{1}^T(\mathbf{I} - \varrho\mathbf{W})^T(\mathbf{I} - \varrho\mathbf{W})\mathbf{Y}/\mathbf{1}^T(\mathbf{I} - \varrho\mathbf{W})^T(\mathbf{I} - \varrho\mathbf{W})\mathbf{1}, \tag{4.3}$$

$$\hat{\sigma}^2 = (\mathbf{Y} - \mu\mathbf{1})^T(\mathbf{I} - \varrho\mathbf{W})^T(\mathbf{I} - \varrho\mathbf{W})\,(\mathbf{Y} - \mu\mathbf{1})/n, \tag{4.4}$$

and

$$\underset{|\varrho|<1}{\text{MIN:}}\left[\prod_{i=1}^{n}(1 - \varrho\lambda_i)\right]^{-2/n}(\mathbf{Y} - \mu\mathbf{1})\,(\mathbf{I} - \varrho\mathbf{W})^T(\mathbf{I} - \varrho\mathbf{W})\,(\mathbf{Y} - \mu\mathbf{1}). \tag{4.5}$$

Expression (4.5) must be solved using nonlinear optimization techniques. Rather than iteratively estimating equation (4.3) and optimizing expression (4.5), though, equation (4.3) can be substituted directly into (4.5), yielding a single expression in one unknown to be optimized. Then, once $\hat{\varrho}$ is found both $\hat{\mu}$ and $\hat{\sigma}^2$ can be calculated.

One should note that equations (4.3) and (4.4) are equivalent to generalized least squares results, but with ϱ replaced by $\hat{\varrho}$. Further, as Upton and Fingleton (1985) illustrate, expression (4.5) should have a unique minimum.

The assumption of a constant mean can be relaxed by introducing a set of fixed effects X variables that are powers of the associated absolute coordinates of areal units. Accordingly, in equation (4.2) $\mu\mathbf{1}$ is replaced by $\mathbf{X}\boldsymbol{\beta}$, yielding the specification

$$\mathbf{Y} = \mathbf{X}\boldsymbol{\beta} + (\mathbf{I} - \varrho\mathbf{W})^{-1}\xi, \tag{4.6}$$

and a likelihood function of the form

$$\begin{aligned} L = {} & (2\pi)^{-n/2}\sigma^{-n}\det[(\mathbf{I} - \varrho\mathbf{W})^T(\mathbf{I} - \varrho\mathbf{W})]^{1/2} \\ & \times \exp\,[-(\mathbf{Y} - \mathbf{X}\boldsymbol{\beta})^T(\mathbf{I} - \varrho\mathbf{W})^T(\mathbf{I} - \varrho\mathbf{W}) \\ & \times (\mathbf{Y} - \mathbf{X}\boldsymbol{\beta})/(2\sigma^2)]. \end{aligned} \tag{4.7}$$

The MLEs of equation (4.7) are

$$\hat{\boldsymbol{\beta}} = [\mathbf{X}^T(\mathbf{I} - \varrho\mathbf{W})^T(\mathbf{I} - \varrho\mathbf{W})\mathbf{X}]^{-1}\mathbf{X}^T(\mathbf{I} - \varrho\mathbf{W})^T(\mathbf{I} - \varrho\mathbf{W})\mathbf{Y}, \tag{4.8}$$

$$\hat{\sigma}^2 = (\mathbf{Y} - \mathbf{X}\boldsymbol{\beta})^T(\mathbf{I} - \varrho\mathbf{W})^T(\mathbf{I} - \varrho\mathbf{W})(\mathbf{Y} - \mathbf{X}\boldsymbol{\beta})/n, \tag{4.9}$$

and

$$\underset{|\varrho|<1}{\text{MIN:}} \left[\prod_{i=1}^{n}(1 - \varrho\lambda_i)\right]^{-2/n}(\mathbf{Y} - \mathbf{X}\boldsymbol{\beta})^T(\mathbf{I} - \varrho\mathbf{W})^T \times (\mathbf{I} - \varrho\mathbf{W})(\mathbf{Y} - \mathbf{X}\boldsymbol{\beta}). \tag{4.10}$$

Again, expression (4.10) must be solved using nonlinear optimizing techniques, and equation (4.8) can be substituted directly into (4.10) to avoid excessive estimation iterations and numerous local optima due to rounding error.

Once more, one should note that equations (4.8) and (4.9) are equivalent to generalized least squares results, but with ϱ replaced by $\hat{\varrho}$. The constraint $|\varrho| < 1$ for expressions (4.5) and (4.10) ensures that $\hat{\varrho} \in (-1, 1)$, whereas the Jacobian term $\{\det[(\mathbf{I} - \varrho\mathbf{W})^T(\mathbf{I} - \varrho\mathbf{W})]\}^{1/2}$ furnishes a weighting that restricts meaningful estimates of $\hat{\varrho}$ to the interval $(-1, 1)$. If in the case of expression (4.5) $\hat{\varrho}$ approaches very close to the boundary of the parameter space, or its graph fails to display a parabolic curve in the interval $(-1, 1)$, then the trend surface specification (4.6) should be considered. An alternative approach is to employ spatial differencing, as Martin (1974) has suggested. In so doing, specification (4.2) is replaced with a specification like

$$(\mathbf{I} - \mathbf{W})\mathbf{Y} = \mu(\mathbf{I} - \mathbf{W})\mathbf{1} + (\mathbf{I} - \varrho\mathbf{W})^{-1}\xi = (\mathbf{I} - \varrho\mathbf{W})^{-1}\xi, \tag{4.11}$$

which as was mentioned earlier is the differencing form for a linear trend component. Because of the lack of understanding of spatial differencing, for the time being at least, the wealth of knowledge concerning trend surface models makes this approach a more appealing route to follow.

4.2. AUTOCORRELATED BIVARIATE MODELS

Very little discussion has been devoted to spatial autoregressive models other than the one for autocorrelated errors. Griffith (1980) and Bivand (1980) suggest that mixed and variable effects regression models merit treatment, too, by having argued that the sampling distribution of the correlation coefficient based upon spatially autocorrelated observations will be different. Upton and Fingleton (1985) present the autoregressive response model, which is a fixed effects regression model, and is a first step in this direction. This response model is of the form

$$\mathbf{Y} = \varrho\mathbf{W}\mathbf{Y} + \mathbf{X}\boldsymbol{\beta} + \xi, \tag{4.12}$$

and has a likelihood function of the form

$$\begin{aligned} L = {} & (2\pi)^{-n/2}\sigma^{-n} \det[(\mathbf{I} - \varrho\mathbf{W})^T(\mathbf{I} - \varrho\mathbf{W})]^{1/2} \\ & \times \exp\{-[(\mathbf{I} - \varrho\mathbf{W})\mathbf{Y} - \mathbf{X}\boldsymbol{\beta}]^T \\ & \times [(\mathbf{I} - \varrho\mathbf{W})\mathbf{Y} - \mathbf{X}\boldsymbol{\beta}]/(2\sigma^2)\}. \end{aligned} \tag{4.13}$$

The MLEs of equation (4.13) are

$$\hat{\boldsymbol{\beta}} = (\mathbf{X}^T\mathbf{X})^{-1}\mathbf{X}^T(\mathbf{I} - \varrho\mathbf{W})\mathbf{Y}, \tag{4.14}$$

$$\hat{\sigma}^2 = \mathbf{Y}^T(\mathbf{I} - \varrho\mathbf{W})^T[\mathbf{I} - \mathbf{X}(\mathbf{X}^T\mathbf{X})^{-1}\mathbf{X}^T]\,(\mathbf{I} - \varrho\mathbf{W})\mathbf{Y}/n, \tag{4.15}$$

and

$$\begin{aligned} \underset{|\varrho|<1}{\text{MIN:}} & \left[\prod_{i=1}^{n}(1 - \varrho\lambda_i)\right]^{-2/n} \mathbf{Y}^T(\mathbf{I} - \varrho\mathbf{W})^T[\mathbf{I} - \mathbf{X}(\mathbf{X}^T\mathbf{X})^{-1}\mathbf{X}^T] \\ & \times (\mathbf{I} - \varrho\mathbf{W})\mathbf{Y}. \end{aligned} \tag{4.16}$$

Yet again, expression (4.16) must be solved using nonlinear optimizing techniques.

Equation (4.6) represents a fixed effects model in that the **X** matrix is comprised of powers of areal unit coordinates that do not change with repeated sampling, whereas the elements of vector **Y** may well change. Equation (4.12) might be viewed as a modification of (4.6), but in (4.12) the X variables still are fixed. A generalization of equation (4.12) for a variable effects X variate, for the bivariate regression situation, will serve to exemplify mixed and variable effects spatially autoregressive models.

A pure spatially autoregressive model is of the form presented in equation (4.2). Modifying this equation for random variables Y and X yields

$$\mathbf{Y} = \mu_y \mathbf{1} + (\mathbf{I} - \varrho_y \mathbf{W})^{-1} \xi_y, \tag{4.17}$$

and

$$\mathbf{X} = \mu_x \mathbf{1} + (\mathbf{I} - \varrho_x \mathbf{W})^{-1} \xi_x, \tag{4.18}$$

where matrix $\mathbf{W}$ is the same for both variates since X and Y are measured on the same set of areal units. The joint distribution of these two variates has as its likelihood function

$$\begin{aligned} L = {} & (2\pi)^{-n} \left[\prod_{i=1}^{n} (1 - \varrho_x \lambda_i)(1 - \varrho_y \lambda_i) \right] \sigma_x^{-n} \sigma_y^{-n} (1 - \varrho^2)^{-n/2} \\ & \times \exp [\{-1/[2(1 - \varrho^2)]\} [(\mathbf{Y} - \mu_y \mathbf{1})^T (\mathbf{I} - \varrho_y \mathbf{W})^T \\ & \times (\mathbf{I} - \varrho_y \mathbf{W})(\mathbf{Y} - \mu_y \mathbf{1})/\sigma_y^2 - 2\varrho(\mathbf{Y} - \mu_y \mathbf{1})^T \\ & \times (\mathbf{I} - \varrho_y \mathbf{W})^T (\mathbf{I} - \varrho_x \mathbf{W})(\mathbf{X} - \mu_x \mathbf{1})/(\sigma_x \sigma_y) \\ & + (\mathbf{X} - \mu_x \mathbf{1})^T (\mathbf{I} - \varrho_x \mathbf{W})^T (\mathbf{I} - \varrho_x \mathbf{W})(\mathbf{X} - \mu_x \mathbf{1})/\sigma_x^2]], \end{aligned} \tag{4.19}$$

which can be rewritten for conditional regression purposes as

$$\begin{aligned} L = {} & (2\pi)^{-n} \left[\prod_{i=1}^{n} (1 - \varrho_x \lambda_i)(1 - \varrho_y \lambda_i) \right] \\ & \times \sigma_x^{-n} \sigma_y^{-n} (1 - \varrho^2)^{-n/2} \exp \{[-1/(1 - \varrho^2)] [(\mathbf{Y} - \mu_y \mathbf{1}) \\ & - \beta(\mathbf{I} - \varrho_y \mathbf{W})^{-1} (\mathbf{I} - \varrho_x \mathbf{W})(\mathbf{X} - \mu_x \mathbf{1})]^T (\mathbf{I} - \varrho_y \mathbf{W})^T \\ & \times (\mathbf{I} - \varrho_y \mathbf{W}) [(\mathbf{Y} - \mu_y \mathbf{1}) - \beta(\mathbf{I} - \varrho_y \mathbf{W})^{-1} (\mathbf{I} - \varrho_x \mathbf{W}) \\ & \times (\mathbf{X} - \mu_x \mathbf{1})]/(2\sigma_y^2)\} \exp [-(\mathbf{X} - \mu_x \mathbf{1})^T (\mathbf{I} - \varrho_x \mathbf{W})^T \\ & \times (\mathbf{I} - \varrho_x \mathbf{W})(\mathbf{X} - \mu_x \mathbf{1})/(2\sigma_x^2)]. \end{aligned} \tag{4.20}$$

Equation (4.20) can be compared with equations (4.6) and (4.12), the two previous regression forms studied here.

If the traditional correlation ϱ between variables X and Y is zero, then as would be expected equation (4.19) reduces to the product of two likelihood functions, each having the form presented in equation (1.6). The corresponding MLEs for μ_x, μ_y, σ_x, σ_y, ϱ_x and ϱ_y are given by (4.3)–(4.5). If $\varrho_x = 0$, then equation (4.20) reduces in form to an equation that is very similar to equation (4.13), except that X is a random variable so that the correlation ϱ between X and Y must be taken into account and μ_x and σ_x also become parameters to be estimated. In this case the MLEs of equation (4.19) are

$$\hat{\mu}_x = \bar{x} - (\varrho\sigma_x/\sigma_y)\mathbf{1}^T(\mathbf{I} - \varrho_y\mathbf{W})(\mathbf{Y} - \mu_y\mathbf{1})/n, \quad (4.21)$$

$$\begin{aligned}\hat{\mu}_y = {} & \mathbf{1}^T(\mathbf{I} - \varrho_y\mathbf{W})^T(\mathbf{I} - \varrho_y\mathbf{W})\mathbf{Y}/\mathbf{1}^T(\mathbf{I} - \varrho_y\mathbf{W})^T(\mathbf{I} - \varrho_y\mathbf{W})\mathbf{1} \\ & - (\varrho\sigma_y/\sigma_x)\mathbf{1}^T(\mathbf{I} - \varrho_y\mathbf{W})^T(\mathbf{X} - \mu_x\mathbf{1})/\mathbf{1}^T \\ & \times (\mathbf{I} - \varrho_y\mathbf{W})^T(\mathbf{I} - \varrho_y\mathbf{W})\mathbf{1}, \end{aligned} \quad (4.22)$$

$$\begin{aligned}\hat{\varrho} = {} & (\mathbf{Y} - \mu_y\mathbf{1})^T(\mathbf{I} - \varrho_y\mathbf{W})^T(\mathbf{X} - \mu_x\mathbf{1})/[(\mathbf{Y} - \mu_y\mathbf{1})^T \\ & \times (\mathbf{I} - \varrho_y\mathbf{W})^T(\mathbf{I} - \varrho_y\mathbf{W})(\mathbf{Y} - \mu_y\mathbf{1})(\mathbf{X} - \mu_x\mathbf{1})^T \\ & \times (\mathbf{X} - \mu_x\mathbf{1})]^{1/2}, \end{aligned} \quad (4.23)$$

the positive roots of

$$\begin{aligned}\sigma_x^2 & + \{\varrho/[n\sigma_y(1 - \varrho^2)]\}(\mathbf{Y} - \mu_y\mathbf{1})^T(\mathbf{I} - \varrho_y\mathbf{W})^T \\ & \times (\mathbf{X} - \mu_x\mathbf{1})\sigma_x - \{1/[n(1 - \varrho^2)]\}(\mathbf{X} - \mu_x\mathbf{1})^T(\mathbf{X} - \mu_x\mathbf{1}) = 0, \end{aligned} \quad (4.24)$$

and

$$\begin{aligned}\sigma_y^2 & + \{\varrho/[n\sigma_x(1 - \varrho^2)]\}(\mathbf{Y} - \mu_y\mathbf{1})^T(\mathbf{I} - \varrho_y\mathbf{W})^T(\mathbf{X} - \mu_x\mathbf{1})\sigma_y \\ & - \{1/[n(1 - \varrho^2)]\}(\mathbf{Y} - \mu_y\mathbf{1})^T(\mathbf{I} - \varrho_y\mathbf{W})^T(\mathbf{I} - \varrho_y\mathbf{W}) \\ & \times (\mathbf{Y} - \mu_y\mathbf{1}) = 0, \end{aligned} \quad (4.25)$$

and the value of ϱ_y is such that

$$\begin{aligned}\underset{|\varrho_y|<1}{\text{MAX:}} & \left[\prod_{i=1}^{n}(1 - \varrho_y\lambda_i)\right][\sigma_x^2\sigma_y^2(1 - \varrho^2)]^{-n/2} \\ & \times \exp\left[\{-1/[2\sigma_x^2\sigma_y^2(1 - \varrho^2)]\}[(\mathbf{X} - \mu_x\mathbf{1})^T(\mathbf{X} - \mu_x\mathbf{1})\sigma_y^2\right. \\ & - 2\varrho\sigma_x\sigma_y(\mathbf{X} - \mu_x\mathbf{1})^T(\mathbf{I} - \varrho_y\mathbf{W})(\mathbf{Y} - \mu_y\mathbf{1}) \\ & \left.+ (\mathbf{Y} - \mu_y\mathbf{1})^T(\mathbf{I} - \varrho_y\mathbf{W})^T(\mathbf{I} - \varrho_y\mathbf{W})(\mathbf{Y} - \mu_y\mathbf{1})\sigma_x^2]\right]. \end{aligned} \quad (4.26)$$

Clearly equations (4.21)–(4.25) and expression (4.26) will need to be estimated in an iterative fashion, and expression (4.26) must be maximized using nonlinear optimization techniques.

As equations (4.19)–(4.26) indicate, letting X be a random variable introduces a considerably greater level of complexity into the autoregressive analysis. Suppose, however, that $\varrho_x = \varrho_y \neq 0$. Then the joint conditional function becomes

$$\begin{aligned} L \;=\; & (2\pi)^{-n}\left[\prod_{i=1}^{n}(1 - \varrho_x\lambda_i)\,(1 - \varrho_x\lambda_i)\right]\sigma_x^{-n}\sigma_y^{-n}(1 - \varrho^2)^{-n/2} \\ & \times \exp\,[\{-1/[(1 - \varrho^2)]\}\,[(\mathbf{Y} - \mu_y\mathbf{1}) - \beta(\mathbf{X} - \mu_x\mathbf{1})]^T \\ & \times (\mathbf{I} - \varrho_x\mathbf{W})^T(\mathbf{I} - \varrho_x\mathbf{W})\,[(\mathbf{Y} - \mu_y\mathbf{1}) - \beta(\mathbf{X} - \mu_x\mathbf{1})]/(2\sigma_y^2)] \\ & \times \exp\,[-(\mathbf{X} - \mu_x\mathbf{1})^T(\mathbf{I} - \varrho_x\mathbf{W})^T(\mathbf{I} - \varrho_x\mathbf{W}) \\ & \times (\mathbf{X} - \mu_x\mathbf{1})/(2\sigma_x^2)], \end{aligned} \tag{4.27}$$

which may look very similar to the autocorrelated errors model equation (1.6). Furthermore, the MLEs of μ_x and μ_y are given by equation (4.3), the MLE of ϱ is given by

$$\begin{aligned} \hat{\varrho} \;=\; & (\mathbf{X} - \mu_x\mathbf{1})^T(\mathbf{I} - \varrho_x\mathbf{W})^T(\mathbf{I} - \varrho_y\mathbf{W})\,(\mathbf{Y} - \mu_y\mathbf{1})/ \\ & [(\mathbf{Y} - \mu_y\mathbf{1})^T(\mathbf{I} - \varrho_y\mathbf{W})^T(\mathbf{I} - \varrho_y\mathbf{W})\,(\mathbf{Y} - \mu_y\mathbf{1}) \\ & \times (\mathbf{X} - \mu_x\mathbf{1})^T(\mathbf{I} - \varrho_x\mathbf{W})^T(\mathbf{I} - \varrho_x\mathbf{W})\,(\mathbf{X} - \mu_x\mathbf{1})]^{1/2}, \end{aligned} \tag{4.28}$$

The form of equation (4.28) can be seen in the GLS estimates for variance terms and regression coefficients.

4.3. A SPATIALLY ADJUSTED ANOVA MODEL

In Section 4.1 the fixed effects regression model was reviewed in terms of X variates that denoted trend surface components, and an autoregressive mechanism that operated on the error terms. A second class of fixed effects regression model will be discussed now.

Consider a set of n areal units that has been divided into m regions, each consisting of a subset of coterminous areal units ($j = 1, 2, \ldots, m$). For some random variable Y repeated sampling may produce different sample vectors **Y** but the region to which each areal unit belongs will not change from sample to sample. One question that could be asked about the observed Y_{ij} values is whether or not their respective parent population regional means are equal. One-way ANOVA furnishes a standard method for assessing differences in the set of observed regional means $\bar{y}_j$, given their respective regional variances s_j^2 and some probability of observed differences accounted for strictly because of sampling error.

Let X_j^* be an indicator variable denoting whether or not an areal unit belongs to region j. Let $X_0 \equiv 1$, and let $X_j \equiv X_j^* - X_m^*$. Neter, Wasserman and Kutner (1985) have shown that the one-way ANOVA model may be written in regression form as the standard linear model $\mathbf{Y} = \mathbf{X}\boldsymbol{\beta} + \boldsymbol{\xi}$, where $\mathbf{X}$ is an n-by-m matrix of zeroes, ones and minus ones. The hypotheses here are

$$H_0\colon \beta_1 = \cdots = \beta_{m-1} = 0$$

$$H_a\colon \text{at least one } \beta_j \neq 0.$$

This null hypothesis is equivalent to the standard ANOVA one of

$$H_0\colon \mu_1 = \mu_2 = \cdots = \mu_m.$$

Since the null hypothesis for the regression formulation states that all coefficients in the regression model are zero, the test statistic becomes the usual one for testing whether or not there is a regression relation, namely the F-ratio statistic. As was seen for the previously discussed fixed-effects model, if variable Y is autocorrelated, then

$$\mathbf{Y} = \varrho\mathbf{WY} + \mathbf{X}\boldsymbol{\beta} + \boldsymbol{\xi}$$

$$\mathbf{Y} - \varrho\mathbf{WY} = \mathbf{X}\boldsymbol{\beta} + \boldsymbol{\xi}$$

$$\mathbf{Y} = (\mathbf{I} - \varrho\mathbf{W})^{-1}\mathbf{X}\boldsymbol{\beta} + (\mathbf{I} - \varrho\mathbf{W})^{-1}\boldsymbol{\xi} = \mathbf{X}'\boldsymbol{\beta} + \boldsymbol{\xi}'.$$

Moreover, since matrix $\mathbf{X}$ is fixed, $(\mathbf{I} - \varrho\mathbf{W})^{-1}\mathbf{X}$ is a simple linear transformation from $\mathbf{X}$ to $\mathbf{X}'$, and hence latent spatial autocorrelation is transferred to the errors. In other words one still has and autocorrelated errors model.

Because the basic model has not changed, equation (4.8) still furnishes the regression coefficient estimates $\mathbf{b}$. Furthermore, the analysis of variance sum-of-squares terms become (see Griffith, 1978)

$$\text{between region: } \mathbf{b}^T\mathbf{X}^T(\mathbf{I} - \varrho\mathbf{W})^T(\mathbf{I} - \varrho\mathbf{W})\mathbf{Y} \tag{4.29}$$

$$\text{within regions: } \mathbf{Y}^T(\mathbf{I} - \varrho\mathbf{W})^T(\mathbf{I} - \varrho\mathbf{W})\mathbf{Y} - \mathbf{b}^T\mathbf{X}^T(\mathbf{I} - \varrho\mathbf{W})^T(\mathbf{I} - \varrho\mathbf{W})\mathbf{Y}. \tag{4.30}$$

As $\varrho \to -1$, expressions (4.29) and (4.30) behave in such a way that although significant differences exist in the regional error means $\bar{\xi}_m$, a tendency exists for insignificant differences to be detected in the $\bar{y}_m$'s. In contrast, as $\varrho \to 1$, no differences in $\bar{\xi}_m$'s are converted into significant differences in $\bar{y}_m$'s. In other words if $\varrho > 0$, then the spatial linear operator $(\mathbf{I} - \varrho_y\mathbf{W})$ means that each areal unit value is being reduced by a weighted average of its neighboring values. Hence the variance terms will tend to have the following relation

$$\mathbf{Y}^T(\mathbf{I} - \varrho_y\mathbf{W})^T(\mathbf{I} - \varrho_y\mathbf{W})\mathbf{Y} < \mathbf{Y}^T\mathbf{Y}.$$

Intuitively this relationship makes sense because positive spatial autocorrelation dampens local variation. As $\varrho \to 1$, then, the regional differences between groups of y_i's tend to increase, even though the underlying unautocorrelated errors do not display significant regional differences. Moreover, because the total variation is being reduced, smaller deviations in the regional variances become more pronounced, relatively speaking. Thus, the probability of rejecting the null hypothesis when it is true tends to increase. On the other hand, as $\varrho \to -1$, the regional differences between groups of y_i's tend to decrease, even though the underlying unautocorrelated errors do display significant regional differences. Local map variation is being accentuated, for the spatial linear operator has become $(\mathbf{I} + |\varrho_y|\mathbf{W})$. Accordingly, each areal unit value is being incremented by a weighted average of its neighbors. Now the tendency for the relation between variance terms is

$$\mathbf{Y}^T\mathbf{Y} < \mathbf{Y}^T(\mathbf{I} - \varrho_y\mathbf{W})^T(\mathbf{I} - \varrho_y\mathbf{W})\mathbf{Y}.$$

Moreover, because the total variation is being increased, larger deviations in the regional variances become less pronounced, relatively speaking. Hence the probability of failing to reject the null hypothesis when it is false tends to increase.

Cliff and Ord (1975) have explored this problem in some depth for the special ANOVA case of two regions. They illustrate the disastrous consequences of spatial dependencies upon applications of tests for difference of means based upon Student's t distribution, and suggest procedures that enable such tests to be applied more appropriately in the presence of spatially autocorrelated samples. Their approach is slightly different in that it assumes that the two regions in question are not adjacent.

Suppose

$$\mathbf{Y}_j \sim \mathrm{MVN}\{\mu_j \mathbf{1}, \sigma^2[(\mathbf{I} - \varrho_j \mathbf{W}_j)^T(\mathbf{I} - \varrho_j \mathbf{W}_j)]^{-1}\} \quad (j = 1, 2),$$

so that each variable is normally distributed, has a common variance, and is described by a simultaneous spatial autoregressive process like equation (1.5). The point estimate of each μ_j is given by equation (4.3). The common estimate of σ^2 is given by

$$\begin{aligned}&[(\mathbf{Y}_1 - \mu_1\mathbf{1})^T(\mathbf{I} - \varrho_1\mathbf{W}_1)^T(\mathbf{I} - \varrho_1\mathbf{W}_1)(\mathbf{Y}_1 - \mu_1\mathbf{1})\\ &+ (\mathbf{Y}_2 - \mu_2\mathbf{1})^T(\mathbf{I} - \varrho_2\mathbf{W}_2)^T(\mathbf{I} - \varrho_2\mathbf{W}_2)\\ &\times (\mathbf{Y}_2 - \mu_1\mathbf{1})]/(n_1 + n_2 - 2).\end{aligned} \tag{4.31}$$

Next consider a test of the hypothesis

$$H_0\colon \mu_1 = \mu_2,$$

for ϱ_1, ϱ_2 and σ^2 unknown. The proposed test statistic is

$$\begin{aligned}t &= (\bar{y}_1 - \bar{y}_2)/\hat{\sigma}\{[\mathbf{1}^T(\mathbf{I} - \hat{\varrho}_1\mathbf{W}_1)^T(\mathbf{I} - \hat{\varrho}_1\mathbf{W}_1)\mathbf{1}]^{-1}\\ &\quad + [\mathbf{1}^T(\mathbf{I} - \hat{\varrho}_2\mathbf{W}_2)^T(\mathbf{I} - \hat{\varrho}_2\mathbf{W}_2)\mathbf{1}]^{-1}\}^{1/2}\\ &= (\bar{y}_1 - \bar{y}_2)/\hat{\sigma}\{[n_1(1 - \hat{\varrho}_1)^2]^{-1} + [n_2(1 - \hat{\varrho}_2)^2]^{-1}\}^{1/2}.\end{aligned} \tag{4.32}$$

Cliff and Ord found, through simulation experimentation, that if $n_1 > 25$, $n_2 > 25$, $\varrho_1 > 0$ or $\varrho_2 > 0$, then t follows a Student's t distribution with $(n_1 + n_2 - 2)$ degrees of freedom, if σ is unknown.

As an aside, Cliff and Ord's approach would be consistent with Griffith's approach if the matrix $(\mathbf{I} - \varrho\mathbf{W})$ used by Griffith was constructed from the two matrices $(\mathbf{I} - \varrho_1\mathbf{W}_1)$ and $(\mathbf{I} - \varrho_2\mathbf{W}_2)$

used by Cliff and Ord, setting $\varrho_1 = \varrho_2$, so that the following block diagonal matrix was obtained:

$$(\mathbf{I} - \varrho\mathbf{W}) = \begin{pmatrix} \mathbf{I} - \varrho_1\mathbf{W}_1 & \vdots & \mathbf{0} \\ \cdots & \cdots & \cdots \\ \mathbf{0} & \vdots & \mathbf{I} - \varrho_2\mathbf{W}_2 \end{pmatrix}.$$

Clearly the difference here is that the spatial autocorrelation parameter ϱ is allowed to be region specific.

4.4. THE TWO-GROUPS DISCRIMINANT FUNCTION MODEL

One popular multivariate analysis problem addresses the question of how groups differ, given a set of variables. This problem of studying the direction of group differences is equivalent to the problem of finding a linear combination of the original variables that exhibits maximum differences in the linear transformed group means. One method for determining such linear combinations is that of discriminant function analysis. When there are only two groups under study, the single discriminant function can be calculated in a greatly simplified way, namely multiple regression, since the eigenfunction problem reduces to the very mathematical problem addressed by OLS.

Let Y be a dichotomous binary 0–1 variable, denoting membership of areal units in one of the two groups. Then the model

$$\mathbf{Y} = \mathbf{X}\boldsymbol{\beta} + \boldsymbol{\xi}$$

yields a set of regression coefficients $\{\beta_1, \beta_2, \ldots, \beta_p\}$ that is proportional to the discriminant function weights obtained by solving an eigenvalue problem (Tatsuoka, 1971). One should note that β_0 need not be proportional, since it is an intercept.

In this setting matrix **X** still would be considered fixed, and variate Y would be random. Hence the X's would be free of spatial autocorrelation, and any autocorrelation that is observed would be affiliated with the manner in which zeroes and ones were distributed over the geographic landscape. This autocorrelated errors model would behave exactly like the aforementioned trend surface model.

When there are more than two groups under study, discriminant function analysis reduces to canonical correlation analysis. Further, treating matrix **X** as fixed effects is somewhat bothersome. These two situations require the treatment of multivariate spatial autocorrelation processes, a topic that will be reflected upon in Chapter 8.

4.5. HYPOTHESIS TESTING AND SPATIAL DEPENDENCE

Thus far in this chapter selected linear statistical models have been modified in order to accommodate non-zero spatial autocorrelation. In the case of the two-groups ANOVA model Cliff and Ord (1975) have demonstrated that traditional hypothesis testing statistics are not suitable for drawing proper inferences about parameters of these modified models. In many cases parameter estimates remain unbiased in the presence of non-zero autocorrelation. For example, from equation (4.3),

$$\begin{aligned} E(\hat{\mu}) &= E[\mathbf{1}^T(\mathbf{I} - \varrho\mathbf{W})^T(\mathbf{I} - \varrho\mathbf{W})\mathbf{Y}/\mathbf{1}^T(\mathbf{I} - \varrho\mathbf{W})^T(\mathbf{I} - \varrho\mathbf{W})\mathbf{1}] \\ &= \mathbf{1}^T(\mathbf{I} - \varrho\mathbf{W})^T(\mathbf{I} - \varrho\mathbf{W})E(\mathbf{Y})/\mathbf{1}^T(\mathbf{I} - \varrho\mathbf{W})^T(\mathbf{I} - \varrho\mathbf{W})\mathbf{1} \\ &= \mu\mathbf{1}^T(\mathbf{I} - \varrho\mathbf{W})^T(\mathbf{I} - \varrho\mathbf{W})\mathbf{1}/\mathbf{1}^T(\mathbf{I} - \varrho\mathbf{W})^T \\ &\quad \times (\mathbf{I} - \varrho\mathbf{W})\mathbf{1} = \mu. \end{aligned}$$

Similarly, from equation (4.8), the fixed-effects regression model result, one obtains

$$\begin{aligned} E(\hat{\boldsymbol{\beta}}) &= E\{[\mathbf{X}^T(\mathbf{I} - \varrho\mathbf{W})^T(\mathbf{I} - \varrho\mathbf{W})\mathbf{X}]^{-1}\mathbf{X}^T(\mathbf{I} - \varrho\mathbf{W})^T \\ &\quad \times (\mathbf{I} - \varrho\mathbf{W})\mathbf{Y}\} \\ &= E\{[\mathbf{X}^T(\mathbf{I} - \varrho\mathbf{W})^T(\mathbf{I} - \varrho\mathbf{W})\mathbf{X}]^{-1}\mathbf{X}^T(\mathbf{I} - \varrho\mathbf{W})^T \\ &\quad \times (\mathbf{I} - \varrho\mathbf{W})\, [\mathbf{X}\boldsymbol{\beta} + (\mathbf{I} - \varrho\mathbf{W})^{-1}\boldsymbol{\xi}]\} \\ &= [\mathbf{X}^T(\mathbf{I} - \varrho\mathbf{W})^T(\mathbf{I} - \varrho\mathbf{W})\mathbf{X}]^{-1}\mathbf{X}^T(\mathbf{I} - \varrho\mathbf{W})^T(\mathbf{I} - \varrho\mathbf{W}) \\ &\quad \times [\mathbf{X}E(\boldsymbol{\beta}) + (\mathbf{I} - \varrho\mathbf{W})^{-1}E(\boldsymbol{\xi})] \\ &= [\mathbf{X}^T(\mathbf{I} - \varrho\mathbf{W})^T(\mathbf{I} - \varrho\mathbf{W})\mathbf{X}]^{-1}\mathbf{X}^T(\mathbf{I} - \varrho\mathbf{W})^T \\ &\quad \times (\mathbf{I} - \varrho\mathbf{W})\mathbf{X}\boldsymbol{\beta} = \boldsymbol{\beta}. \end{aligned}$$

But the standard errors of these generalized least squares types of parameter estimates do change. In part this finding is what Cliff and Ord (1975) showed for the difference of means test, where $\mu_{\bar{x}_1 - \bar{x}_2}$ remained unchanged while $\sigma^2_{\bar{x}_1 - \bar{x}_2}$ dramatically changed.

The objective of this section is to discuss the standard errors needed to conduct proper hypothesis testing in the presence of spatial dependence. Two approaches can be taken to this problem. First, asymptotic variances can be calculated, with these results pertaining to large sample theory. For the simultaneous model, these results are outlined by Ord (1975), Dorien (1980, 1981, 1982), and Upton and Fingleton (1985). The standard errors are contained within the covariance matrix provided by the second partial derivative of a log-likelihood function, or

$$\mathbf{V} = \{-E[\partial^2 \ln (L)]/\partial\theta_r\partial\theta_s\}^{-1},$$

where θ_r and θ_s are unknown parameters in L. Second, as both the article by Haining (1978) and the one by Cliff and Ord (1975) indicate, because most geographical problems fall into the realm of small sample theory, asymptotic variances may not be the best estimates to use. Rather, small sample estimates can be obtained by conducting Monte Carlo simulation experiments. Such experimentation will be the topic of Chapter 9.

Returning to equations (1.6) and (4.2), the asymptotic variances for $\hat{\mu}$, $\hat{\sigma}$ and $\hat{\varrho}$ are needed. The first partial derivatives are

$$\begin{aligned}
\partial \ln (L)/\partial\mu &= -[-\mathbf{1}^T(\mathbf{I} - \varrho\mathbf{W})^T(\mathbf{I} - \varrho\mathbf{W})\mathbf{Y} + \mu\mathbf{1}^T \\
&\quad \times [(\mathbf{I} - \varrho\mathbf{W})^T(\mathbf{I} - \varrho\mathbf{W})\mathbf{1}]/\sigma^2 \\
\partial \ln (L)/\partial\sigma^2 &= -n/(2\sigma^2) + (\mathbf{Y} - \mu\mathbf{1})^T \\
&\quad \times (\mathbf{I} - \varrho\mathbf{W})^T(\mathbf{I} - \varrho\mathbf{W})\,(\mathbf{Y} - \mu\mathbf{1})/(2\sigma^4) \\
\partial \ln (L)/\partial\varrho &= \sum_{i=1}^{n} - \lambda_i/(1 - \varrho\lambda_i) - (\mathbf{Y} - \mu\mathbf{1})^T \\
&\quad \times (-\mathbf{W}^T - \mathbf{W} + 2\varrho\mathbf{W}^T\mathbf{W})\,(\mathbf{Y} - \mu\mathbf{1})/(2\sigma^2).
\end{aligned}$$

The second partial derivatives are

$$\begin{aligned}
\partial^2 \ln (L)/\partial\mu^2 &= -\mathbf{1}^T(\mathbf{I} - \varrho\mathbf{W})^T(\mathbf{I} - \varrho\mathbf{W})\mathbf{1}/\sigma^2 \\
\partial^2 \ln (L)/\partial(\sigma^2)^2 &= n/(2\sigma^4) - (\mathbf{Y} - \mu\mathbf{1})^T(\mathbf{I} - \varrho\mathbf{W})^T \\
&\quad \times (\mathbf{I} - \varrho\mathbf{W})\,(\mathbf{Y} - \mu\mathbf{1})/\sigma^6
\end{aligned}$$

$$\partial^2 \ln (L)/\partial \varrho^2 = \sum_{i=1}^{n} - \lambda_i^2/(1 - \varrho\lambda_i)^2$$
$$- (\mathbf{Y} - \mu\mathbf{1})^T\mathbf{W}^T\mathbf{W}(\mathbf{Y} - \mu\mathbf{1})/\sigma^2$$
$$\partial^2 \ln (L)/\partial\mu\partial\sigma^2 = -\mathbf{1}^T(\mathbf{I} - \varrho\mathbf{W})^T(\mathbf{I} - \varrho\mathbf{W})\,(\mathbf{Y} - \mu\mathbf{1})/\sigma^4$$
$$\partial^2 \ln (L)/\partial\mu\partial\varrho = [\mathbf{1}^T(-\mathbf{W}^T - \mathbf{W} + 2\varrho\mathbf{W}^T\mathbf{W})\mathbf{Y}$$
$$- \mu\mathbf{1}^T(-\mathbf{W}^T - \mathbf{W} + 2\varrho\mathbf{W}^T\mathbf{W})\mathbf{1}]/\sigma^2$$
$$= \mathbf{1}^T(-\mathbf{W}^T - \mathbf{W} + 2\varrho\mathbf{W}^T\mathbf{W})$$
$$\times (\mathbf{Y} - \mu\mathbf{1})/\sigma^2$$
$$\partial^2 \ln (L)/\partial\varrho\partial\sigma^2 = (\mathbf{Y} - \mu\mathbf{1})^T(-\mathbf{W}^T - \mathbf{W} + 2\varrho\mathbf{W}^T\mathbf{W})$$
$$\times (\mathbf{Y} - \mu\mathbf{1})/(2\sigma^4).$$

The respective expectations of these six equations are

$$E[\partial^2 \ln (L)/\partial\mu^2] = -\mathbf{1}^T(\mathbf{I} - \varrho\mathbf{W})^T(\mathbf{I} - \varrho\mathbf{W})\mathbf{1}/\sigma^2$$
$$E[\partial^2 \ln (L)/\partial(\sigma^2)^2] = n/(2\sigma^4) - n\sigma^2/\sigma^6 = -n/(2\sigma^4)$$
$$E[\partial^2 \ln (L)/\partial\varrho^2] = \sum_{i=1}^{n} - \lambda_i^2/(1 - \varrho\lambda_i)^2$$
$$- E[(\mathbf{Y} - \mu\mathbf{1})^T\mathbf{W}^T\mathbf{W}(\mathbf{Y} - \mu\mathbf{1})/\sigma^2]$$
$$= \sum_{i=1}^{n} - \lambda_i^2/(1 - \varrho\lambda_i)^2 - E[\boldsymbol{\xi}^T(\mathbf{I} - \varrho\mathbf{W}^T)^{-1}$$
$$\times \mathbf{W}^T\mathbf{W}(\mathbf{I} - \varrho\mathbf{W})^{-1}\boldsymbol{\xi}]/\sigma^2$$
$$= \sum_{i=1}^{n} - \lambda_i^2/(1 - \varrho\lambda_i)^2 - \mathrm{tr}\,[(\mathbf{I} - \varrho\mathbf{W}^T)^{-1}$$
$$\times \mathbf{W}^T\mathbf{W}(\mathbf{I} - \varrho\mathbf{W})^{-1}E(\boldsymbol{\xi}\boldsymbol{\xi}^T)]/\sigma^2$$
$$= \sum_{i=1}^{n} - \lambda_i^2/(1 - \varrho\lambda_i)^2 - \mathrm{tr}\,[(\mathbf{I} - \varrho\mathbf{W}^T)^{-1}$$
$$\times \mathbf{W}^T\mathbf{W}(\mathbf{I} - \varrho\mathbf{W})^{-1}],$$

where tr denotes the trace of the matrix,

$$E[\partial^2 \ln (L)/\partial\mu\partial\sigma^2] = 0, \quad \text{since } E(\mathbf{Y} - \mu\mathbf{1}) = 0,$$
$$E[\partial^2 \ln (L)/\partial\mu\partial\varrho] = 0, \quad \text{since } E(\mathbf{Y} - \mu\mathbf{1}) = 0,$$

and

$$\begin{aligned}
E[\partial^2 \ln (L)/\partial\varrho\partial\sigma^2] &= E[\boldsymbol{\xi}^T(\mathbf{I} - \varrho\mathbf{W}^T)^{-1}(\mathbf{W}^T - \mathbf{W} + 2\varrho\mathbf{W}^T\mathbf{W}) \\
&\quad \times (\mathbf{I} - \varrho\mathbf{W})^{-1}\boldsymbol{\xi}/(2\sigma^4)] \\
&= \{E[-\boldsymbol{\xi}^T(\mathbf{I} - \varrho\mathbf{W}^T)^{-1}\mathbf{W}^T(\mathbf{I} - \varrho\mathbf{W})^{-1}\boldsymbol{\xi}] \\
&\quad + E[-\boldsymbol{\xi}^T(\mathbf{I} - \varrho\mathbf{W}^T)^{-1}\mathbf{W}(\mathbf{I} - \varrho\mathbf{W})^{-1}\boldsymbol{\xi}] \\
&\quad + E[2\varrho\boldsymbol{\xi}^T(\mathbf{I} - \varrho\mathbf{W}^T)^{-1}\mathbf{W}^T\mathbf{W} \\
&\quad \times (\mathbf{I} - \varrho\mathbf{W})^{-1}\boldsymbol{\xi}]\}/(2\sigma^4) \\
&= -E[\boldsymbol{\xi}^T(\mathbf{I} - \varrho\mathbf{W}^T)^{-1}(\mathbf{I} - \varrho\mathbf{W})^T \\
&\quad \times \mathbf{W}(\mathbf{I} - \varrho\mathbf{W})^{-1}\boldsymbol{\xi}]/\sigma^4 \\
&= -E[\boldsymbol{\xi}^T\mathbf{W}(\mathbf{I} - \varrho\mathbf{W})^{-1}\boldsymbol{\xi}]/\sigma^4 \\
&= -\mathrm{tr}\,[\mathbf{W}(\mathbf{I} - \varrho\mathbf{W})^{-1}E(\boldsymbol{\xi}\boldsymbol{\xi}^T)]/\sigma^4 \\
&= -\mathrm{tr}\,[\mathbf{W}(\mathbf{I} - \varrho\mathbf{W})^{-1}]/\sigma^2.
\end{aligned}$$

Therefore, the asymptotic variances are given by the matrix

$$\mathbf{V} = \begin{pmatrix}
n/(2\sigma^4) & \mathrm{tr}\,[\mathbf{W}(\mathbf{I} - \varrho\mathbf{W})^{-1}]/\sigma^2 & 0 \\
\mathrm{tr}\,[\mathbf{W}(\mathbf{I} - \varrho\mathbf{W})^{-1}]/\sigma^2 & \begin{array}{l}\sum_{i=1}^{n} \lambda_i^2/(1 - \varrho\lambda_i)^2 \\ + \mathrm{tr}\,[(\mathbf{I} - \varrho\mathbf{W})^{-T} \\ \times \mathbf{W}^T\mathbf{W}(\mathbf{I} - \varrho\mathbf{W})^{-1}]\end{array} & 0 \\
0 & 0 & \mathbf{1}^T(\mathbf{I} - \varrho\mathbf{W})^T(\mathbf{I} - \varrho\mathbf{W})\mathbf{1}/\sigma^2
\end{pmatrix}^{-1}$$

Clearly the simplest hypothesis to test is one about $\hat{\mu}$. The necessary standard error for the particular test is

$$\begin{aligned}
\sigma_{\bar{y}} &= \{(\mathbf{Y} - \mu\mathbf{1})^T(\mathbf{I} - \varrho\mathbf{W})^T(\mathbf{I} - \varrho\mathbf{W})\,(\mathbf{Y} - \mu\mathbf{1})/ \\
&\quad [n\mathbf{1}^T(\mathbf{I} - \varrho\mathbf{W})^T(\mathbf{I} - \varrho\mathbf{W})\mathbf{1}]\}^{1/2}.
\end{aligned}$$

For a stochastic matrix **W** (i.e., all rows sum to unity), this is exactly the estimate obtained by Cliff and Ord (1975). Tests about $\hat{\sigma}^2$ and $\hat{\varrho}$ would require calculation of the inverse matrix

$$\begin{pmatrix}
n/(2\sigma^4) & \mathrm{tr}\,[\mathbf{W}(\mathbf{I} - \varrho\mathbf{W})^{-1}]/\sigma^2 \\
\mathrm{tr}\,[\mathbf{W}(\mathbf{I} - \varrho\mathbf{W})^{-1}]/\sigma^2 & \sum_{i=1}^{n} \lambda_i^2/(1 - \varrho\lambda_i)^2 + \mathrm{tr}\,[(\mathbf{I} - \varrho\mathbf{W})^{-T}\mathbf{W}^T\mathbf{W}(\mathbf{I} - \varrho\mathbf{W})^{-1}]
\end{pmatrix}^{-1}$$

If equations (4.6) and (4.7) are of interest, then the only entry into matrix **V** that changes is for $-E[\partial^2 \ln(L)/\partial^2 \boldsymbol{\beta}^2]$. This should be somewhat intuitively obvious, since $\mu\mathbf{1}$ is a special case of $\mathbf{X}\boldsymbol{\beta}$ for the trend surface model. Here

$$\begin{aligned}\partial \ln(L)/\partial\boldsymbol{\beta} &= -[-\mathbf{X}^T(\mathbf{I} - \varrho\mathbf{W})^T(\mathbf{I} - \varrho\mathbf{W})\mathbf{Y} \\ &\quad + \mathbf{X}^T(\mathbf{I} - \varrho\mathbf{W})^T(\mathbf{I} - \varrho\mathbf{W})\mathbf{X}\boldsymbol{\beta}]/\sigma^2 \\ \partial^2 \ln(L)/\partial\boldsymbol{\beta}^2 &= -\mathbf{X}^T(\mathbf{I} - \varrho\mathbf{W})^T(\mathbf{I} - \varrho\mathbf{W})\mathbf{X}/\sigma^2 \\ -E[\partial^2 \ln(L)/\partial\boldsymbol{\beta}^2] &= \mathbf{X}^T(\mathbf{I} - \varrho\mathbf{W})^T(\mathbf{I} - \varrho\mathbf{W})\mathbf{X}/\sigma^2.\end{aligned}$$

The corresponding matrix **V** for this fixed effects regression situation is presented in Upton and Fingleton (1985).

The conditional model also could be of interest, and so will be treated here since its asymptotic variances differ from the foregoing ones. Consider the likelihood function

$$\begin{aligned}L &= (2\pi)^{-n/2}\sigma^{-n} \det[(\mathbf{I} - \varrho\mathbf{C})]^{1/2} \\ &\quad \times \exp[-(\mathbf{Y} - \mu\mathbf{1})(\mathbf{I} - \varrho\mathbf{C})(\mathbf{Y} - \mu\mathbf{1})/(2\sigma^2)]. \qquad (4.33)\end{aligned}$$

The model for this likelihood function is given by equation (1.8). Here the first partial derivatives are

$$\begin{aligned}\partial \ln(L)/\partial\mu &= -[-\mathbf{1}^T(\mathbf{I} - \varrho\mathbf{C})\mathbf{Y} + \mu\mathbf{1}^T(\mathbf{I} - \varrho\mathbf{C})\mathbf{1}]/\sigma^2 \\ \partial \ln(L)/\partial\varrho &= \sum_{i=1}^{n} -\lambda_i/[2(1 - \varrho\lambda_i)] \\ &\quad - (\mathbf{Y} - \mu\mathbf{1})^T(-\mathbf{C})(\mathbf{Y} - \mu\mathbf{1})/(2\sigma^2) \\ \partial \ln(L)/\partial\sigma^2 &= -n/(2\sigma^2) + (\mathbf{Y} - \mu\mathbf{1})^T(\mathbf{I} - \varrho\mathbf{C})(\mathbf{Y} - \mu\mathbf{1})/(2\sigma^4).\end{aligned}$$

The second partial derivatives are

$$\begin{aligned}\partial^2 \ln(L)/\partial\mu^2 &= -\mathbf{1}^T(\mathbf{I} - \varrho\mathbf{C})\mathbf{1}/\sigma^2 \\ \partial^2 \ln(L)/\partial(\sigma^2)^2 &= n/(2\sigma^4) - (\mathbf{Y} - \mu\mathbf{1})^T(\mathbf{I} - \varrho\mathbf{C})(\mathbf{Y} - \mu\mathbf{1})/\sigma^6 \\ \partial^2 \ln(L)/\partial\varrho^2 &= \sum_{i=1}^{n} - \lambda_i^2/[2(1 - \varrho\lambda_i)^2] \\ \partial^2 \ln(L)/\partial\mu\partial\sigma^2 &= -\mathbf{1}^T(\mathbf{I} - \varrho\mathbf{C})(\mathbf{Y} - \mu\mathbf{1})/\sigma^4 \\ \partial^2 \ln(L)/\partial\mu\partial\varrho &= \mathbf{1}^T(-\mathbf{C})(\mathbf{Y} - \mu\mathbf{1})/\sigma^2 \\ \partial^2 \ln(L)/\partial\varrho\partial\sigma^2 &= -(\mathbf{Y} - \mu\mathbf{1})^T\mathbf{C}(\mathbf{Y} - \mu\mathbf{1})/(2\sigma^4).\end{aligned}$$

The respective expectations of these six equations are

$$E[\partial^2 \ln(L)/\partial\mu^2] = -\mathbf{1}^T(\mathbf{I} - \varrho\mathbf{C})\mathbf{1}/\sigma^2$$

$$\begin{aligned} E[\partial^2 \ln(L)/\partial\sigma^4] &= n/(2\sigma^4) - E[\xi^T\mathbf{D}^{-T}(\mathbf{I} - \varrho\mathbf{C})\mathbf{D}^{-1}\xi]/\sigma^6 \\ &= n/(2\sigma^4) - E(\xi^T\xi)/\sigma^6 = n/(2\sigma^4) - n\sigma^2/\sigma^6 \\ &= -n/(2\sigma^4), \end{aligned}$$

where $\mathbf{D}^T\mathbf{D} = \mathbf{I} - \varrho\mathbf{C}$, and $\mathbf{D}$ is defined as in equation (1.8),

$$E[\partial^2 \ln(L)/\partial\varrho^2] = \sum_{i=1}^{n} -\lambda_i^2/[2(1 - \varrho\lambda_i)^2]$$

$$E[\partial^2 \ln(L)/\partial\mu\partial\sigma^2] = 0$$

$$E[\partial^2 \ln(L)/\partial\mu\partial\varrho] = 0$$

$$\begin{aligned} E[\partial^2 \ln(L)/\partial\varrho\partial\sigma^2] &= -E[\xi^T\mathbf{D}^{-T}\mathbf{C}\mathbf{D}^{-1}\xi]/(2\sigma^4) \\ &= -\mathrm{tr}\,[(\mathbf{D}^{-T}\mathbf{C}\mathbf{D}^{-1})E(\xi\xi^T)]/(2\sigma^4) \\ &= -\mathrm{tr}\,(\mathbf{D}^{-T}\mathbf{C}\mathbf{D}^{-1})/(2\sigma^2). \end{aligned}$$

Now the asymptotic variances are given by

$$\mathbf{V} = \begin{pmatrix} n/(2\sigma^4) & \mathrm{tr}\,(\mathbf{D}^{-T}\mathbf{C}\mathbf{D}^{-1})/(2\sigma^2) & 0 \\ \mathrm{tr}\,(\mathbf{D}^{-T}\mathbf{C}\mathbf{D}^{-1})/(2\sigma^2) & \sum_{i=1}^{n} \lambda_i^2/[2(1 - \varrho\lambda_i)^2] & 0 \\ 0 & 0 & \mathbf{1}^T(\mathbf{I} - \varrho\mathbf{C})\mathbf{1}/\sigma^2 \end{pmatrix}^{-1}$$

Again if the trend surface version of this conditional scheme is considered, then only the entry of matrix $\mathbf{V}$ defined by $-E[\partial^2 \ln(L)/\partial\boldsymbol{\beta}^2]$ will change. The new entry will be

$$\mathbf{X}^T(\mathbf{I} - \varrho\mathbf{C})\mathbf{X}/\sigma^2.$$

Asymptotic variances for more complicated error structures can be derived in exactly the same fashion. Upton and Fingleton, for example, present matrix $\mathbf{V}$ for an autoregressive response model of the form

$$\mathbf{Y} = \varrho\mathbf{W}\mathbf{Y} + \mathbf{X}\boldsymbol{\beta} + \xi,$$

which already has been discussed in this text in terms of the spatial lag model.

4.6. EFFICIENCY OF SPATIAL STATISTICS ESTIMATORS

Establishing the asymptotic variances of spatial statistics estimators allows the property of efficiency to be explored. As was alluded to in Chapter 1, efficiency refers to the property of minimum mean square error. If a statistic, say T, has a mean square error that is smaller than the mean square error of an alternative statistic, say T^*, in estimating some parameter θ from a given sample $\{x_1, x_2, \ldots, x_n\}$, then the statistic T is considered to be more efficient than its competitor T^* because it is making more efficient use of information contained in the n sample observations. Hence the most efficient statistic is the one with the smallest variance (Lindgren, 1976).

A number of different efficiency measures have been proposed by statisticians. First, one could consider the relative efficiency of statistic T with respect to statistic T^*, which is measured as

$$E[(T - \theta)^2]/E[(T^* - \theta)^2].$$

If the statistic T has the smaller variance, then this ratio will be less than unity. To illustrate this measure for statistics calculated on spatial data series, let T denote that statistic recognizing the presence of spatial autocorrelation, let T^* denote that statistic ignoring the presence of spatial autocorrelation, and consider the constant spatial mean conditional autoregressive model. Now for T,

$$\begin{aligned} &E\{[\mathbf{1}^T(\mathbf{I} - \varrho\mathbf{C})\mathbf{Y}/\mathbf{1}^T(\mathbf{I} - \varrho\mathbf{C})\mathbf{1} - \mu]^2\} \\ &= \mathbf{1}^T(\mathbf{I} - \varrho\mathbf{C})E(\mathbf{Y}\mathbf{Y}^T)(\mathbf{I} - \varrho\mathbf{C})^T\mathbf{1}/ \\ &\quad \mathbf{1}^T(\mathbf{I} - \varrho\mathbf{C})\mathbf{1}\mathbf{1}^T(\mathbf{I} - \varrho\mathbf{C})\mathbf{1} - \mu^2 \\ &= \sigma^2/\mathbf{1}^T(\mathbf{I} - \varrho\mathbf{C})\mathbf{1}. \end{aligned} \qquad (4.34)$$

Meanwhile for statistic T^*,

$$\begin{aligned} E[(\mathbf{1}^T\mathbf{Y}/n - \mu)^2] &= \mathbf{1}^T E(\mathbf{Y}\mathbf{Y}^T)\mathbf{1}/n^2 - \mu^2 \\ &= \sigma^2\mathbf{1}^T(\mathbf{I} - \varrho\mathbf{C})^{-1}\mathbf{1}/n^2. \end{aligned} \qquad (4.35)$$

The ratio of expressions (4.34) and (4.35) equals unity if $\varrho = 0$, implying that these two statistics are equally efficient if the case of zero spatial autocorrelation. Otherwise the ratio is $n^2/[\mathbf{1}^T(\mathbf{I} - \varrho\mathbf{C})\mathbf{1}\mathbf{1}^T(\mathbf{I} - \varrho\mathbf{C})^{-1}\mathbf{1}]$, and particular values of this ratio are as follows, for selected regular, square lattice sizes:

ϱ	3-by-3	5-by-5	7-by-7	9-by-9	11-by-11
0.225	0.9429	0.8657	0.8351	0.8271	0.8289
0.150	0.9832	0.9733	0.9739	0.9764	0.9788
0.075	0.9969	0.9960	0.9965	0.9969	0.9973
−0.075	0.9979	0.9980	0.9984	0.9986	0.9988
−0.150	0.9927	0.9940	0.9950	0.9959	0.9965
−0.225	0.9851	0.9886	0.9910	0.9927	0.9938

These numerical results imply several interesting efficiency features regarding the GLS estimate as opposed to the estimate that overlooks spatial autocorrelation. First, the GLS estimate is always more efficient. Second, its efficiency improves as $|\varrho| \to 0.25$. Third, its efficiency is relatively less impressive for the case of negative spatial autocorrelation as compared with the case of positive spatial autocorrelation. And, fourth, this relative efficiency seems to diminish as $n \to \infty$.

Given the variable spatial mean conditional autoregressive model, for statistic T

$$\begin{aligned} &E[\{[\mathbf{X}^T(\mathbf{I} - \varrho\mathbf{C})\mathbf{X}]^{-1}\mathbf{X}^T(\mathbf{I} - \varrho\mathbf{C})\mathbf{Y} - \boldsymbol{\beta}\}^2] \\ &\quad = [\mathbf{X}^T(\mathbf{I} - \varrho\mathbf{C})\mathbf{X}]^{-1}\mathbf{X}^T(\mathbf{I} - \varrho\mathbf{C})E(\mathbf{Y}\mathbf{Y}^T) \\ &\qquad \times (\mathbf{I} - \varrho\mathbf{C})\mathbf{X}[\mathbf{X}^T(\mathbf{I} - \varrho\mathbf{C})\mathbf{X}]^{-1} - \boldsymbol{\beta}\boldsymbol{\beta}^T \\ &\quad = \sigma^2[\mathbf{X}^T(\mathbf{I} - \varrho\mathbf{C})\mathbf{X}]^{-1}. \end{aligned} \tag{4.36}$$

Meanwhile for T^*,

$$\begin{aligned} &E\{[(\mathbf{X}^T\mathbf{X})^{-1}\mathbf{X}^T\mathbf{Y} - \boldsymbol{\beta}]^2\} = (\mathbf{X}^T\mathbf{X})^{-1}\mathbf{X}^T E(\mathbf{Y}\mathbf{Y}^T)\mathbf{X}(\mathbf{X}^T\mathbf{X})^{-1} - \boldsymbol{\beta}\boldsymbol{\beta}^T \\ &\quad = \sigma^2(\mathbf{X}^T\mathbf{X})^{-1}\mathbf{X}^T(\mathbf{I} - \varrho\mathbf{C})^{-1}\mathbf{X}(\mathbf{X}^T\mathbf{X})^{-1}. \end{aligned} \tag{4.37}$$

Again, the ratio of expression (4.36) to (4.37) equals unity if $\varrho = 0$.

Another efficiency measure works with the information contained in a sample (Lindgren, 1976). Letting $I(\theta)$ denote this information

$$I(\theta) = E[\{\partial \ln [f(\mathbf{X}, \theta)]/\partial\theta\}^2].$$

In the class of unbiased estimators, this information value provides a meaningful bound, independent of the estimator, which simply is the reciprocal of the information. Thus the absolute efficiency of a statistic T that is an unbiased estimate of parameter θ is defined as the ratio

$$e(T) = 1/[I(\theta)\ \mathrm{VAR}(T)].$$

Accordingly an efficient estimator has an absolute efficiency value of 1.

Once more consider the constant spatial mean conditional autoregressive model. From the derivatives presented in Section 4.5 one can see that

$$\begin{aligned} I(\hat{\mu}) &= E[\{-[-\mathbf{1}^T(\mathbf{I} - \varrho\mathbf{C})\mathbf{Y} + \mu\mathbf{1}^T(\mathbf{I} - \varrho\mathbf{C})\mathbf{1}]/\sigma^2\}^2] \\ &= [\mathbf{1}^T(\mathbf{I} - \varrho\mathbf{C})E(\mathbf{Y}\mathbf{Y}^T)\ (\mathbf{I} - \varrho\mathbf{C})^T\mathbf{1} \\ &\quad - \mu^2\mathbf{1}^T(\mathbf{I} - \varrho\mathbf{C})\mathbf{1}\mathbf{1}^T(\mathbf{I} - \varrho\mathbf{C})\mathbf{1}]/\sigma^4 \\ &= \mathbf{1}^T(\mathbf{I} - \varrho\mathbf{C})\mathbf{1}/\sigma^2. \end{aligned}$$

The asymptotic variance of $\hat{\mu}$ also is given in Section 4.6. Using this variance value,

$$e(\hat{\mu}) = 1/\{[\mathbf{1}^T(\mathbf{I} - \varrho\mathbf{C})\mathbf{1}/\sigma^2][\sigma^2/\mathbf{1}^T(\mathbf{I} - \varrho\mathbf{C})\mathbf{1}]\} = 1.$$

Hence, asymptotically $e(\hat{\mu}) = 1$ for GLS estimation.

With respect to the spatial autoregressive parameter ϱ, if the Jacobian constraint is removed from the likelihood function, then expressions such as (4.5) reduce to OLS estimation equations. Ord (1975) notes that the efficiency of the OLS estimator relative to an ML estimator such as that given by expression (4.5) declines drastically as $|\varrho| \rightarrow 0.25$. The rate of change here will vary with the structure of matrix **W** or matrix **C**, whichever is used in the spatial linear operator term.

Consequently, the spatial statistics derived in this chapter can be shown to be relatively efficient compared to their counterparts that ignore the presence of spatial autocorrelation. And in an absolute sense these statistics tend to be asymptotically efficient. But the behavior of this absolute efficiency for small lattice situations remains unknown.

4.7. CONSISTENCY OF SPATIAL STATISTICS ESTIMATORS

The last salient property of statistical estimators to be discussed in this chapter is that of consistency, which pertains to the mean squared error of the estimator decreasing to zero as more and more observations are incorporated into its computation (i.e., as $n \to \infty$). This property will hold if and only if both the variance of the estimator statistic and the bias for the estimator statistic tend to zero as the sample size n becomes infinite. (Lindgren, 1976). In other words, the sampling distribution of a sample statistic T has a probability that increasingly concentrates about its corresponding parameter θ as n increases.

Mardia and Marshall (1984) have provided four conditions that must be met if regression parameter estimates are to be consistent for the spatially autocorrelated errors model. Consider the conditional spatial autoregressive model, where the constant spatial mean version simply is the special case of regressing variable Y on vector $\mathbf{1}$. The variance-covariance matrix here is $\mathbf{V} = (\mathbf{I} - \varrho\mathbf{C})^{-1}\sigma^2$. Now

$$\partial\mathbf{V}/\partial\varrho = \mathbf{C}(\mathbf{I} - \varrho\mathbf{C})^{-2}\sigma^2, \tag{4.38}$$

$$\partial^2\mathbf{V}/\partial\varrho^2 = 2\mathbf{C}^2(\mathbf{I} - \varrho\mathbf{C})^{-3}\sigma^2, \tag{4.39}$$

$$\partial\mathbf{V}^{-1}/\partial\varrho = -\mathbf{C}\sigma^{-2}, \tag{4.40}$$

and

$$\partial^2\mathbf{V}^{-1}/\partial\varrho^2 = 0.$$

The n eigenvalues of matrix $\mathbf{V}$ are $1/(1 - \varrho\lambda_i)$, where λ_i are the eigenvalues of matrix $\mathbf{C}$. The n eigenvalues of matrix (4.38) are $\sigma^2\lambda_i/(1 - \varrho\lambda_i)^2$. And the n eigenvalues of matrix (4.39) are $2\sigma^2\lambda_i^2/(1 - \varrho\lambda_i)^3$. The first condition Mardia and Marshall specify is that the limit as $n \to \infty$ of the principal eigenvalue of each of these three matrices must be finite. For a regular, square lattice, the principal eigenvalue is given by

$$\lim_{n\to\infty} \lambda_p = 4 \lim_{n\to\infty} (1/2)\,\{\cos[\pi/(n + 1)] + \cos[\pi/(n + 1)]\} = 4. \tag{4.41}$$

Thus,

$$\lim_{n\to\infty} 1/(1 - \varrho\lambda_p) = 1/(1 - 4\varrho) < \infty, \qquad \text{since } |\varrho| < 0.25,$$

$$\lim_{n\to\infty} \lambda_p/(1 - \varrho\lambda_p)^2 = 4/(1 - 4\varrho)^2 < \infty, \qquad \text{since } |\varrho| < 0.25,$$

and

$$\lim_{n\to\infty} \lambda_p^2/(1 - \varrho\lambda_p)^3 = 16/(1 - 4\varrho)^3 < \infty, \quad \text{since } |\varrho| < 0.25,$$

Hence the first condition is satisfied.

The second condition states that the Euclidean norm, denoted $\|\cdot\|$, of the matrix $\partial\mathbf{V}/\partial\varrho$ must be of order of magnitude $n^{-1/2-\delta}$, for some $\delta > 0$, or

$$\|\partial\mathbf{V}/\partial\varrho\|^{-2} = O(n^{-1/2-\delta}).$$

Now (see Fox, 1965)

$$\|\partial\mathbf{V}/\partial\varrho\|^{-2} = \|\mathbf{C}(\mathbf{I} - \varrho\mathbf{C})^{-2}\sigma^2\|^{-2} = \{[\lambda_p^2/(1 - \varrho\lambda_p)^4]^{1/2}\sigma^2\}^{-2}$$
$$= (1 - \varrho\lambda_p)^4/(\sigma^2\lambda_p)^2. \tag{4.42}$$

As was shown in equation (4.41), the principal eigenvalue λ_p is a function of n, so that one may write $\lambda_p(n)$. The order of magnitude of equation (4.42) is (see Ross, 1980)

$$\lim_{n\to 0} \{[1 - \varrho\lambda_p(n^{-1/2-\delta})]^4/[\sigma^2\lambda_p(n^{-1/2-\delta})]^2\}/n^{-1/2-\delta}$$
$$= [(1 - 4\varrho)^4/16\sigma^4]/\infty = 0,$$

since $\lambda_p(\infty) = 4$. Thus for $\delta = 1/2$ the order of magnitude of equation (4.42) is n^{-1}. Hence the second condition is satisfied.

The third condition states that

$$a_{1,1} = \lim_{n\to\infty} \{\text{tr}\,[\mathbf{V}^{-1}\,\partial\mathbf{V}/\partial\varrho\,\mathbf{V}^{-1}\,\partial\mathbf{V}/\partial\varrho]/\text{tr}\,[\mathbf{V}^{-1}\,\partial\mathbf{V}/\partial\varrho\,\mathbf{V}^{-1}\,\partial\mathbf{V}/\partial\varrho]\},$$

and that this scalar is other than zero. The result for this third condition reduces to a scalar in this case, and hence provides a somewhat trivial solution, because only one parameter is present. If many parameters are present, like in the case of some Kriging specifications of matrix $\mathbf{V}$, then this condition is far more complicated to evaluate. Nevertheless, this third condition is satisfied in the case of a conditional spatial autoregressive errors model.

The fourth and final condition comes directly from classical regression theory, and states that

$$\lim_{n\to\infty} (\mathbf{X}^T\mathbf{X})^{-1} = 0.$$

Clearly this condition is satisfied in the constant spatial mean case because $\mathbf{X} = \mathbf{1}$, and so $\mathbf{X}^T\mathbf{X} = \mathbf{1}^T\mathbf{1} = n$. Thus

$$\lim_{n\to\infty} 1/n = 0.$$

Hence this fourth condition is satisfied.

Therefore, the parameter estimates $\hat{\mu}$, for the constant spatial mean model, and $\hat{\boldsymbol{\beta}}$, for the variable spatial mean model, are consistent. Mardia and Marshall (1984) furnish a proof of these four aforementioned sufficient conditions of weak consistency. They also provide evidence from several lattice simulation exercises.

Meanwhile, Ord (1975) shows that the MLE $\hat{\varrho}$ of equation (4.5), for the simultaneous model, which has as its accompanying constraint $|\varrho| < 1$, is consistent if

$$\lim_{n\to\infty} \{\mathrm{tr}\,[\mathbf{W}^T(\mathbf{I} - \varrho\mathbf{W})^{-1}]/\mathrm{tr}\,[(\mathbf{I} - \varrho\mathbf{W}^T)^{-1}\mathbf{W}^T\mathbf{W} \times (\mathbf{I} - \varrho\mathbf{W})^{-1}]\} = 0. \tag{4.43}$$

As can be seen from the demonstrations that will be presented in Chapter 5,

$$\lim_{n\to\infty} (\mathbf{I} - \varrho\mathbf{W})^{-1} = \lim_{n\to\infty} [\mathbf{I} - (\varrho/4)\mathbf{C}]^{-1} = \boldsymbol{\Omega},$$

where matrix $\boldsymbol{\Omega}$ can be estimated from the spectral counterpart of the conditional spatial autoregressive model for a regular, square lattice. Thus

$$\lim_{n\to\infty} \mathrm{tr}\,[\mathbf{W}^T(\mathbf{I} - \varrho\mathbf{W})^{-1}] = n\varrho_{1,0},$$

$$\lim_{n\to\infty} \mathrm{tr}\,[(\mathbf{I} - \varrho\mathbf{W}^T)^{-1}\mathbf{W}^T\mathbf{W}(\mathbf{I} - \varrho\mathbf{W})^{-1}] = knf(\varrho_{i,j}),$$

where $\varrho_{i,j}$ is the first-order (i, j) lag spatial autocorrelation, and as matrix $\boldsymbol{\Omega}$ indicates, $f(\varrho_{i,j})$ is some constant, and so equation (4.43) is satisfied as $n \to \infty$.

Limit (4.43) also is equivalent to stating that

$$E[\boldsymbol{\xi}^T\mathbf{W}^T(\mathbf{I} - \varrho\mathbf{W})\mathbf{Y}] = E[\boldsymbol{\xi}^T\mathbf{W}^T(\mathbf{I} - \varrho\mathbf{W})(\mathbf{I} - \varrho\mathbf{W})^{-1}\boldsymbol{\xi}] = E[\boldsymbol{\xi}^T\mathbf{W}^T\boldsymbol{\xi}] = 0,$$

since $w_{ii} = 0$, and so only the uncorrelated random error cross-product $\xi_i \xi_j$ $(i \neq j)$ terms appear in $\boldsymbol{\xi}^T \mathbf{W}^T \boldsymbol{\xi}$.

Consequently the parameter estimates of the conditional and simultaneous spatial autoregressive models are consistent.

4.8. CONCLUSIONS

To date the regression model with spatially autocorrelated errors has been extensively investigated. Model specification and estimation problems have been solved for it. And, the recasting of other statistical models as a linear regression model facilitates an understanding of relationships between them and spatial autocorrelation. Furthermore, asymptotic variance results allow hypothesis testing to be carried out for this sort of regression model. Estimates obtained are efficient and consistent. While the findings summarized here are encouraging, one must realize that only a very limited class of statistical models has been evaluated.

4.9. REFERENCES

Bivand, R., 1980, A Monte Carlo study of correlation coefficient estimation with spatially autocorrelated observations, *Quaestiones Geographicae*, Vol. 6: 5–10.

Cliff, A., and J. Ord, 1975, The comparison of means when samples consist of spatially autocorrelated observations, *Environment and Planning* A, Vol. 7: 725–734.

Dorien, P., 1980, Linear models with spatially distributed data, *Sociological Methods and Research*, Vol. 9: 29–60.

Dorien, P., 1981, Estimating linear models with spatially distributed data, in *Sociological Methodology*, edited by S. Leinhardt. San Francisco: Jossey-Bass, pp. 359–388.

Dorien, P., 1982, Maximum likelihood methods for linear models, *Sociological Methods and Research*, Vol. 10: 243–269.

Fox, L., 1965, *An Introduction to Numerical Linear Algebra*. New York: Oxford University Press.

Griffith, D., 1978, A spatially adjusted ANOVA model, *Geographical Analysis*, Vol. 10: 296–301.

Griffith, D., 1979, Urban dominance, spatial structure and spatial dynamics: some theoretical conjectures and empirical implications, *Economic Geography*, Vol. 55: 95–113.

Griffith, D., 1980, Towards a theory of spatial statistics, *Geographical Analyis*, Vol. 12: 325–339.

Haining, R., 1978, A spatial model for high plains agriculture, *Annals*, Association of American Geographers, Vol. 68: 493–504.
Lindgren, B., 1976, *Statistical Theory* (3rd ed.). New York: MacMillan.
Mardia, K., and R. Marshall, 1984, Maximum likelihood estimation of models for residual covariance in spatial regression, *Biometrika*, Vol. 71: 135–146.
Neter, J., W. Wasserman, and M. Kutner, 1985, *Applied Linear Statistical Models* (2nd ed.). Homewood, Ill.: Richard D. Irwin.
Ord, K., 1975, Estimation methods for models of spatial interaction, *Journal of the American Statistical Association*, Vol. 70: 120–126.
Ross, S., 1980, *Introduction to Probability Models*. New York: Academic Press.
Tatsuoka, M., 1971, *Multivariate Analysis: Techniques for Education and Psychological Research*. New York: Wiley.
Upton, G., and B. Fingleton, 1985, *Spatial Data Analysis by Example: Point Pattern and Quantitative Data*, Vol. 1. New York: Wiley.

CHAPTER 5

Spatial Autocorrelation and Spectral Analysis

In earlier chapters the restrictions of $|\varrho| < 1/\lambda_p$ for the conditional model and $|\varrho| < 1$ for the simultaneous model employing matrix **W** (the stochastic version of matrix **C**) were employed to insure stationarity [i.e., $(\mathbf{I} - \varrho\mathbf{C})^{-1}$ and $(\mathbf{I} - \varrho\mathbf{W})^{-1}$ will be convergent series in their geometric series expansion forms], as well as invertibility between the autoregressive and moving average models. The restrictions also are necessary for writing the spectral density counterparts of spatial autoregressive models. The aim of this chapter is to outline spectral representations of spatial autoregressive models for both infinite and finite regular tessellation surfaces. In doing so, the selected tabular results presented in Bartlett (1975) will be extended. A second use of spectral analysis to be studied here has to do with the quantitative analysis of two-dimensional shapes. The dual axis Fourier analysis approach to constructing shape indices will be reviewed and evaluated.

5.1. A BRIEF BACKGROUND FOR SPECTRAL ANALYSIS

The primary purposes of this section are to briefly describe spectral analysis, and to introduce necessary terminology. Students interested in pursuing a more detailed study of this subject, especially in terms of geographical applications, should consult Rayner (1971).

The form of model discussed in the first four chapters of this text is linear, even though ϱ must be obtained through non-linear optimization. An alternative form of model is one that summarizes pattern with cycles. The presence of cyclical behavior implies that a periodic function is needed in order for phenomena to be described in an adequate fashion. Moreover, for some function $f(x)$ there exists a real number p such that

$$f(x + p) = f(x)$$

for all values of x contained in the domain of function f. The period of this function is the value $p_{\min} > 0$ for which this above equation holds. Examples of such functions are the sine and cosine trigonometric functions (see Figure 5.1).

Three properties of these two trigonometric functions are of interest. The amplitude, which equals one half the difference between the least upper bound and greatest lower bound of $f(x)$, identifies extremes of a cycle. The phase shift, which is the number of units that a graph of the function $f(x)$ must be shifted along the x-axis in order for $f(x)$ to coincide with $f(x + b)$, allows concomitant cyclical behavior to be uncovered, and furnishes a means for identifying lead and lag effects. Of most relevance here is the frequency, which is the number of cycles completed by a graph of $f(x)$ in a vertical band one unit wide (the width usually equals 2π). For example, for the function

$$y = A \cos (ax + b),$$

$|A|$ is the amplitude, $A \cos (ax)$ coincides with $A \cos (ax + b)$ if it is shifted along the x-axis $-b/a$ units (i.e., $ax = 0 \Rightarrow x = 0$, and so $ax + b = 0 \Rightarrow x = -b/a$), and $a/(2\pi)$ is the frequency.

The frequency domain is important because it allows cyclical behavior to be uncovered. If the frequency domain produces a uniform distribution, then no cyclical behavior is present. Deviations from the uniform distribution suggest the presence of cycles. This contention can be better seen by inspecting the form of the frequency domain function, which is

$$f(\theta) = (2\pi)^{-1} \sum_{\tau=-\infty}^{\infty} \gamma(\tau) \exp(-i\theta\tau), \quad |\theta| < \pi, \qquad (5.1)$$

where τ denotes the lag involved. But

$$\begin{aligned}\sqrt{i} &= (-1)^{1/2} = \exp[(2r + 1)\pi i/2] \\ &= \cos[(2r + 1)\pi/2] + i \sin[(2r + 1)\pi/2].\end{aligned}$$

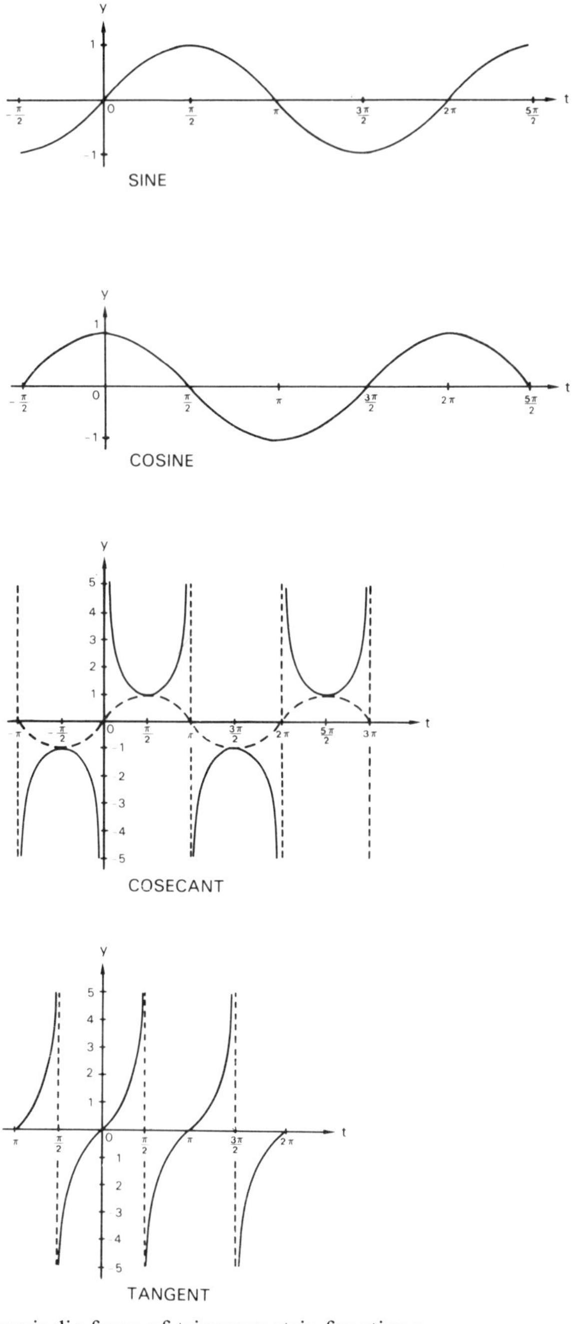

Fig. 5.1. The periodic form of trigonometric functions.

This decomposition into sine and cosine functions is the reason these two particular trigonometric functions have been mentioned in this section.

5.2. RELATIONSHIPS BETWEEN AUTOREGRESSIVE AND SPECTRAL MODELS

There exists an entire family of spatial autoregressive models that can be written in spectral terms. Dacey (1976) discusses some applications of certain of these models. Haining (1978) has assembled their spectral form for easy reference. The simplest of these will be analyzed here to illustrate the conversion steps involved in moving from an autoregressive to a spectral model.

Consider a one-dimensional, single directional dependency structure operating on an infinite, regular lattice (this specification is comparable to what often is posited for a time series). Then at any point on the surface, say t,

$$X_t = \varrho X_{t-1} + \xi_t, \quad |\varrho| < 1. \tag{5.2}$$

The theoretical autocovariance function at lag zero is defined to be

$$\begin{aligned}\gamma(0) &= E(X_t X_t)\\ &= E[(\varrho X_{t-1} + \xi_t)(\varrho X_{t-1} + \xi_t)]\\ &= E\{[\varrho(\varrho X_{t-2} + \xi_{t-1}) + \xi_t][\varrho(\varrho X_{t-2} + \xi_{t-1}) + \xi_t]\}\\ &= E\left[\left(\sum_{k=0}^{\infty} \varrho^k \xi_{t-k}\right)\left(\sum_{k=0}^{\infty} \varrho^k \xi_{t-k}\right)\right]\\ &= \sigma^2/(1 - \varrho^2).\end{aligned} \tag{5.3}$$

Here once again a restriction is placed upon ϱ to insure that convergence of a geometric series will occur. One should note that because the ξ_{t-k} are independent and identically distributed (iid), then $E(\xi_{t-k}\xi_{t-l}) = 0$ for $k \neq l$, and $E(\xi_{t-k}\xi_{t-k}) = \sigma^2$

The theoretical autocovariance function at lag one is defined as

$$\begin{aligned}\gamma(1) &= E(X_t X_{t-1})\\ &= E[(\varrho X_{t-1} + \xi_t)(X_{t-1})]\\ &= \varrho E(X_{t-1}^2) + E(X_{t-1}\xi_t)\\ &= \varrho\sigma^2/(1 - \varrho^2) + 0.\end{aligned} \tag{5.4}$$

The first term in Step 3 of equation (5.4) refers to an expectation at lag zero. The second term of this step, when expanded, always will have products of the form $\xi_t \xi_{t-k}$ for $k > 0$, all of whose expectations, as was seen previously, are zero.

The general form of the autocovariance function suggested by equations (5.3) and (5.4) is

$$\gamma(\tau) = \varrho^{\tau} \sigma^2/(1 - \varrho^2). \tag{5.5}$$

Equation (5.5) has been shown to hold for the cases of $\tau = 0,1$. Assume this equation is true for $k > 1$, such that

$$\gamma(k) = E(X_t X_{t-k}) = \varrho^k \sigma^2/(1 - \varrho^2).$$

Then for the case of $(k + 1)$,

$$\begin{aligned} \gamma(k + 1) &= E(X_t X_{t-(k+1)}) \\ &= \varrho^{k+1} \sigma^2/(1 - \varrho^2) \\ &= \varrho[\varrho^k \sigma^2/(1 - \varrho^2)] \\ &= \varrho\gamma(k). \end{aligned} \tag{5.6}$$

Therefore equation (5.5) holds for all k by mathematical induction.

Next the spectral frequency domain may be defined as in equation (5.1), where the angle is expressed in radians. Thus equation (5.1) may be rewritten as

$$f(\theta) = (2\pi)^{-1} \sum_{\tau=-\infty}^{\infty} [\gamma(\tau) [\cos(\theta\tau) - i \sin(\theta\tau)]\}. \tag{5.7}$$

Now the autocovariance function is symmetric, meaning $\gamma(\tau) = \gamma(-\tau)$. Further,

$$\cos(\theta\tau) = \cos(-\theta\tau),$$

and

$$\sin(\theta\tau) = -\sin(-\theta\tau).$$

Accordingly equation (5.7) may be rewritten as

$$f(\theta) = (2\pi)^{-1} \left[\sum_{\tau=-\infty}^{\infty} \gamma(\tau) \cos(\theta\tau) - i \sum_{\tau=-\infty}^{\infty} \gamma(\tau) \sin(\theta\tau) \right]$$

$$= (2\pi)^{-1}\Big\{\gamma(0)\cos(0) + 2\sum_{\tau=1}^{\infty}\gamma(\tau)\cos(\theta\tau)$$

$$- i\Big[\sum_{\tau=-\infty}^{-1}\gamma(\tau)\sin(\theta\tau) + \gamma(0)\sin(0) + \sum_{\tau=1}^{\infty}\gamma(\tau)\sin(\theta\tau)\Big]\Big\}$$

$$= (2\pi)^{-1}\Big[\gamma(0) + 2\sum_{\tau=1}^{\infty}\gamma(\tau)\cos(\theta\tau)\Big]. \qquad (5.8)$$

Substituting equation (5.5) into (5.8) yields

$$f(\theta) = (2\pi)^{-1}\Big[\sigma^2/(1-\varrho^2) + 2\sum_{\tau=1}^{\infty}\varrho^{\tau}\sigma^2\cos(\theta\tau)/(1-\varrho^2)\Big]$$

$$= \{\sigma^2/[2\pi(1-\varrho^2)]\}\Big[1 + 2\sum_{\tau=1}^{\infty}\varrho^{\tau}\cos(\theta\tau)\Big].$$

But $\cos(\theta\tau) = [\exp(i\theta\tau) + \exp(-i\theta\tau)]/2$, resulting in

$$f(\theta) = \{\sigma^2/[2\pi(1-\varrho^2)]\}\Big\{1 + 2\sum_{\tau=1}^{\infty}\varrho^{\tau}[\exp(i\theta\tau)$$

$$+ \exp(-i\theta\tau)]/2\Big\}$$

$$= \{\sigma^2/[2\pi(1-\varrho^2)]\}\Big\{\Big[1 + \sum_{\tau=1}^{\infty}\varrho^{\tau}\exp(i\theta\tau)\Big]$$

$$+ \Big[1 + \sum_{\tau=1}^{\infty}\varrho^{\tau}\exp(-i\theta\tau)\Big] - 1\Big\}$$

$$= \{\sigma^2/[2\pi(1-\varrho^2)]\}\{1/[1-\varrho\exp(i\theta)]$$

$$+ 1/[1-\varrho\exp(-i\theta)] - 1]$$

$$= \sigma^2/\{2\pi[1+\varrho^2 - 2\varrho\cos(\theta)]\}. \qquad (5.9)$$

Again one should note the importance of the restriction $|\varrho| < 1$. If $\varrho = 0$, meaning no spatial autocorrelation is present, then equation (5.9) reduces to $\sigma^2/(2\pi)$, which is the spectral density function for white noise. In contradistinction, if $\varrho = 1/4$, then equation (5.9) becomes

$$f(\lambda) = \sigma^2/\{[17/8 - \cos(\theta)]\pi\},$$

whose value at $\theta = 0$ is $8\sigma^2/(9\pi)$, and whose value at $\theta = \pi$ is $8\sigma^2/(25\pi)$. Clearly the spectral density function no longer is a constant.

Equations (5.3) thru (5.9) demonstrate how equation (5.2) can be converted to its spectral form. Parallel developments can be constructed for each of the four remaining spectral forms to be inspected here; this inversion will be left as an exercise for the reader to carry out. The next autoregressive model variate to be considered builds on equation (5.2), positing a one-dimensional, bidirectional dependency structure operating on an infinite, regular lattice (this situation is comparable to a linear spatial landscape). If the dependency structure that prevails is symmetric, then at any point on the surface, say i,

$$X_i = \varrho(X_{i-1} + X_{i+1}) + \xi_i, \quad |\varrho| < 1/2. \tag{5.10}$$

One approach to converting this equation to its spectral counterpart rewrites it as a second-order difference equation, such that

$$X_{i+1} = (1/\varrho)X_i - X_{i-1} - (1/\varrho)\xi_i.$$

The spectral counterpart to equation (5.10) is

$$f(\theta) = \sigma^2/\{2\pi[1 - 2\varrho \cos(\theta)]^2\}. \tag{5.11}$$

Again if $\varrho = 0$ then equation (5.11) reduces to $\sigma^2/(2\pi)$, the spectral density function for white noise. Suppose $\varrho = 1/4$, then equation (5.11) becomes

$$f(\theta) = \sigma^2/\{2\pi[1 - (1/2)\cos(\theta)]^2\},$$

whose value at $\theta = 0$ is $2\sigma^2/\pi$, whose value at $\theta = \pi$ is $2\sigma^2/(9\pi)$, and whose slope is $-\sigma^2 \sin(\theta)/\{2\pi[1 - \cos(\theta)/2]^3\}$ as compare with a slope of $-\sigma^2 \sin(\theta)/\{\pi[17/8 - \cos(\theta)]^2\}$ for the example application of equation (5.9). Obviously multidirectional dependencies cause the slope of the spectral density function to become steeper.

If the symmetry constraint attached to equation (5.10) is relaxed, then at any point in the surface, say i,

$$X_i = \varrho_1 X_{i-1} + \varrho_2 X_{i+1} + \xi_i, \quad |\varrho_1\varrho_2| < 1/4. \tag{5.12}$$

The spectral counterpart to equation (5.12) is

$$f(\theta) = \sigma^2/\{2\pi[1 + \varrho_1^2 + \varrho_2^2 - 2(\varrho_1 + \varrho_2)\cos(\theta) + 2\varrho_1\varrho_2 \cos(2\theta)]\}. \tag{5.13}$$

If $\varrho_1 = \varrho_2 = \varrho$ then equation (5.13) reduces to

$$\begin{aligned} f(\theta) &= \sigma^2/\{2\pi[1 + 2\varrho^2 - 4\varrho\cos(\theta) + 2\varrho^2\cos(2\theta)] \\ &= \sigma^2/\{2\pi[1 - 2\varrho\cos(\theta)]^2\}, \end{aligned}$$

since $\cos^2(\theta) = [2\cos(2\theta) - 1]$. Unfortunately asymmetric dependencies are difficult both to detect and to model in an efficient manner. Thus spatial autoregressive models that have been formulated to date fail to exploit this property. One attempt to provide guidance for this kind of endeavor is found in Haining (1977, 1979).

Equation (5.2), (5.10) and (5.12) refer to one-dimensional landscapes. Of more relevance to spatial data series analysis are models for two-dimensional processes. The simplest of these represents unidirectional dependence, and has the autoregressive form, for any point on the surface, say (i, j),

$$X_{i,j} = \varrho(X_{i-1,j} + X_{i,j-1}) + \xi_{i,j}, \quad |\varrho| < 1/2. \tag{5.14}$$

The spectral counterpart to equation (5.14) is

$$f(\theta_1, \theta_2) = \sigma^2/[4\pi^2\{1 - 2\varrho\,[\cos(\theta_1) + \cos(\theta_2)] + 2\varrho^2[1 + \cos(\theta_1 - \theta_2)]\}]. \tag{5.15}$$

Here the frequency domain is two-dimensional, requiring the two distinct arguments θ_1 and θ_2, since the surface is two-dimensional. And $4\pi^2 = (2\pi)(2\pi)$ results from this two-dimensionality, too. Now if $\varrho = 0$ the spectral function for white noise becomes $\sigma^2/(4\pi^2)$. If $\theta_1 = 0$ then equation (5.15) reduces to

$$\begin{aligned} f(0, \theta_2) &= \sigma^2/[4\pi^2\{1 - 2\varrho[1 + \cos(\theta_2)] + 2\varrho^2[1 + \cos(-\theta_2)]\}] \\ &= \sigma^2/[2\pi\{[1 - 2\varrho\cos(\theta_2)]^2\}]. \end{aligned}$$

Finally, the conditional spatial autoregressive model may be written, for any point on a two-dimensional lattice surface, say (i, j), as

$$X_{i,j} = \varrho(X_{i-1,j} + X_{i+1,j} + X_{i,j-1} + X_{i,j+1}) + \xi_{i,j}, \quad |\varrho| < 1/4. \tag{5.16}$$

The spectral counterpart to equation (5.16) is

$$f(\theta_1, \theta_2) = \sigma^2/[4\pi^2\{1 - 2\varrho[\cos(\theta_1) + \cos(\theta_2)]\}^2]. \tag{5.17}$$

Equation (5.17) will supply the basis for most of the subsequent analyses presented in this chapter.

5.3. DEFINING THE COVARIANCE MATRIX OF A CONDITIONAL SPATIAL MODEL USING THE SPECTRAL DENSITY FUNCTION

Bartlett (1975) shows that the autocorrelation value $\varrho_{k,l}$ can be obtained from the spectral density function given by equation (5.17) using numerical integration techniques, such that

$$\varrho_{k,l} = \int_0^\pi \int_0^\pi \cos(k, \theta_1) \cos(l, \theta_2)/[1 - 2\varrho \cos(\theta_1) - 2\varrho \cos(\theta_2)] \, d\theta_1 \, d\theta_2. \tag{5.18}$$

If ϱ is known, then, theoretical autocorrelation values can be calculated at various lags for a two-dimensional, regular lattice surface over which a conditional autoregressive mechanism operates. Because of symmetry, $\varrho_{k,l} = \varrho_{l,k}$. Bartlett (1975) has provided numerical results for selected values of ϱ and low order lags.

Equation (5.18) yields autocorrelation values for the infinite lattice surface. Its resulting numerical values are equivalent to those obtained from $(\mathbf{I} - \varrho\mathbf{C})^{-1}$, if matrix $\mathbf{C}$ is for a regular, infinite lattice. Hence equation (5.18) furnishes a means for comparing small, finite lattice spatial autocorrelation values with what can be considered to be asymptotic values. Further, the values are for a conditional spatial autoregressive model, since this particular model has $\sigma^2(\mathbf{I} - \varrho\mathbf{C})^{-1}$ as its variance–covariance matrix. Clearly, then, the theoretical values of a conditional spatial autoregressive model can be determined when ϱ is known.

The numerical results presented in Bartlett exhibit several conspicuous trends. First, as $\varrho \to 0.25$, $\varrho_{k,l}$ increases, but at a decreasing rate. This tendency can be seen in the following sequence for $\varrho_{1,0}$:

ϱ	$\varrho_{1,0}$	ϱ	$\varrho_{1,0}$
0.02	0.0200	0.16	0.1869
0.04	0.0403	0.18	0.2222
0.06	0.0611	0.20	0.2659
0.08	0.0827	0.22	0.3260
0.10	0.1055	0.24	0.4341
0.12	0.1299	0.249 998 0	0.7725
0.14	0.1567	0.25	1.0000

What these sorts of trends suggest is that the level of spatial dependencies most often observed in geographical data are quite weak, since ϱ usually lies relatively far from the boundary of the parameter spacc. For only values of ϱ extremely close to 0.25 (e.g., 0.249 999 999 999) are accompanied by correlations that would be considered quite marked. Second, as $k \to \infty$ and $l \to \infty$, $\varrho_{k,l} \to 0$. Further, this convergence to zero follows a negative exponentially declining trajectory. Moreover, just as in time series analysis (see Vandaele, 1983), the spatial autocorrelation functions are well behaved, and should provide useful information for model identification and estimation. Third, $\varrho = 0.25$ has spatial autocorrelation values of unity at all lags. Accordingly, it is equivalent to a trend surface. This specific finding again links the autoregressive and trend surface models. In general these results can be seen in the sample spatial autocorrelation values appearing in Table 5.1.

Selected correlograms, whose entries have been calculated using equation (5.18), appear in Table 5.1. As can be seen from the results reported in this table, the stationary spatial random field expands in size as $\varrho \to 0.25$. For instance, if $\varrho = 0.02$ then the dependency structure cannot be detected beyond a second spatial lag, whereas for $\varrho = 0.249$ the dependency structure extends to a spatial lag of seven. In addition, as was mentioned above, the strength of this spatial random field increases as $\varrho \to 0.25$. For example, $\varrho_{1,0}(0.02) = 0.0200$ whereas $\varrho_{1,0}(0.249) = 0.5898$. Meanwhile the negative exponential decline can be detected in either of the following two series, for $\varrho = 0.249$:

k, l	$\varrho_{k,l}$	k, l	$\varrho_{k,l}$
0, 0	1.0000	0, 0	1.0000
1, 0	0.5898	1, 1	0.4795
2, 0	0.4094	2, 2	0.3163
3, 0	0.3083	3, 3	0.2266
4, 0	0.2420	4, 4	0.1687
5, 0	0.1945	5, 5	0.1285
6, 0	0.1588	6, 6	0.0994
7, 0	0.1311		

The left-hand series may be described by the equation

$$\varrho_{k,l} = 0.80121 \exp(-0.27668k),$$
$$(t = -2.58) \quad (t = -13.47)$$

TABLE 5.1

Spatial Autocorrelation Values at Selected Lags for Various Values of ρ

		$\rho_{k,l}$					
ρ	k	l= 0	1	2	3	4	5
0.000	0	1.0000					
	1	.0000	.0000				
	2	.0000	.0000	.0000			
	3	.0000	.0000	.0000	.0000		
	4	.0000	.0000	.0000	.0000	.0000	
	5	.0000	.0000	.0000	.0000	.0000	.0000
0.020	0	1.0000					
	1	.0200	.0008				
	2	.0004	.0000	.0000			
	3	.0000	.0000	.0000	.0000		
	4	.0000	.0000	.0000	.0000	.0000	
	5	.0000	.0000	.0000	.0000	.0000	.0000
0.100	0	1.0000					
	1	.1055	.0218				
	2	.0114	.0034	.0007			
	3	.0012	.0005	.0001	.0000		
	4	.0001	.0001	.0000	.0000	.0000	
	5	.0000	.0000	.0000	.0000	.0000	.0000
0.200	0	1.0000					
	1	.2659	.1260				
	2	.0777	.0491	.0237			
	3	.0243	.0180	.0101	.0049		
	4	.0080	.0065	.0041	.0022	.0011	
	5	.0027	.0023	.0016	.0009	.0005	.0002
0.249	0	1.0000					
	1	.5898	.4795				
	2	.4094	.3732	.3163			
	3	.3083	.2933	.2621	.2266		
	4	.2420	.2345	.2162	.1929	.1687	
	5	.1945	.1902	.1789	.1632	.1458	.1285
0.250	0	1.0000					
	1	1.0000	1.0000				
	2	1.0000	1.0000	1.0000			
	3	1.0000	1.0000	1.0000	1.0000		
	4	1.0000	1.0000	1.0000	1.0000	1.0000	
	5	1.0000	1.0000	1.0000	1.0000	1.0000	1.0000

NOTE: These values were calculated using IMSL subroutine DBLIN.

TABLE 5.2

Spatial Autocorrelation Values for Selected Finite Lattice Sizes at Selected Lags for $\rho = 0.249$

Lattice Size	k l=			$\rho_{k,l}$			
		0	1	2	3	4	5
3-by-3	0	1.0000					
	1	.3311	.1649				
5-by-5	0	1.0000					
	1	.4295	.2272				
	2	.1703	.1272	.0633			
7-by-7	0	1.0000					
	1	.4797	.3397				
	2	.2472	.2024	.1328			
	3	.1080	.0933	.0643	.0320		
9-by-9	0	1.0000					
	1	.5106	.3788				
	2	.2932	.2500	.1820			
	3	.1667	.1500	.1155	.0761		
	4	.0765	.0702	.0556	.0374	.0186	
11-by-11	0	1.0000					
	1	.5314	.4052				
	2	.3238	.2822	.2165			
	3	.2046	.1878	.1526	.1122		
	4	.1225	.1148	.0962	.0727	.0481	
	5	.0576	.0545	.0464	.0356	.0238	.0118

which has a squared multiple correlation coefficient value of $R^2 = 0.968$, and a Durbin–Watson statistic value of 0.83. The left-hand series may be described by the equation

$$\varrho_{k,k} = 0.76560 \exp(-0.36386k),$$
$$(t = -2.29) \quad (t = -11.23)$$

which has a squared multiple correlation coefficient value of $R^2 = 0.962$, and a Durbin–Watson statistic value of 1.13. Although the sample sizes for these two series are too small to draw sound inferences here, and the diagnostic statistics are not completely encouraging, these numerical results do corroborate the general negative exponential decline description.

While equation (5.18) furnishes asymptotic spatial autocorrelation values, differences do occur between these theoretical limits and their small lattice counterparts. Table 5.2 presents results for selected small, bounded lattice cases. Complications partially uncovered here that are attributed to boundedness of the lattice surface will be discussed at some length in Chapter 7. Entries in Table 5.2 have been calculated using $(\mathbf{I} - \varrho\mathbf{C})^{-1}$, where $\varrho = 0.249$, and are for the central lattice areal unit (in order to minimize edge effects). Comparing these results with those reported in Table 5.1, for selected lags, yields [the selected lags are the same as those focused on in Bartlett (1975)]

n	$\varrho_{1,0}$	$\varrho_{1,1}$	$\varrho_{2,0}$
9	0.3311	0.1649	*
25	0.4295	0.2272	0.1703
49	0.4797	0.3397	0.2472
81	0.5106	0.3788	0.2932
121	0.5314	0.4052	0.3238
∞	0.5898	0.4795	0.4094

TABLE 5.3

Variance Estimates From $(\mathbf{I} - 0.249\mathbf{C})^{-1}$ for Selected Finite Lattice Sizes

Lattice Size	Lattice Dimension	Standardized Lattice Configuration Variances 1	2	3	4	5	6
3-by-3	1	.80009					
	2	.88258	1.00000				
5-by-5	1	.68882					
	2	.77870	.93193				
	3	.79274	.96302	1.00000			
7-by-7	1	.62931					
	2	.71369	.86221				
	3	.73189	.90680	.96730			
	4	.73567	.91742	.98283	1.00000		
9-by-9	1	.59239					
	2	.67232	.81413				
	3	.69044	.85969	.92389			
	4	.69598	.87625	.94992	.98157		
	5	.69733	.88055	.95705	.99049	1.00000	
11-by-11	1	.56751					
	2	.64424	.78070				
	3	.66187	.82541	.88921			
	4	.66766	.84300	.91760	.95335		
	5	.66982	.85016	.93003	.96980	.98863	
	6	.67040	.85214	.93356	.97459	.99419	1.00000

NOTE: Because of the dual four quadrant symmetry present in regular lattice results, only the fundamental information is displayed here. A complete lattice set of values can be obtained by reflecting the above results along their respective diagonals, and then reflecting these results into each of the remaining three quadrants of the lattice surface.

Clearly although the global patterns exhibited in Table 5.2 mirror those appearing in Table 5.1 (e.g., negative exponential declining correlograms), these small lattice values differ substantially from their asymptotic counterparts (apparently convergence is occurring, though). Even for 121 areal units, a size that most geographic samples do not exceed, the differences are not negligible. This small lattice behavior raises questions about the use of $(\mathbf{I} - \varrho\mathbf{C})^{-1}$ for the finite lattice variance–covariance matrix.

Another conspicuous property of the variance–covariance matrix for the conditional model, which once again reflects edge effects, has to do with the diagonal entries of this inverted matrix. The asymptotic counterpart to these diagonal values is unity, as can be seen from equation (5.18) and the numerical results reported in Table 5.1. The boundedness of the small lattice

TABLE 5.4

$\underline{\mathbf{E}}_{max}$ From Matrix $\underline{\mathbf{C}}$ for Selected Finite Lattice Sizes

Lattice Size	Lattice Dimension	Standardized $\underline{\mathbf{E}}_{max}$ for Lattice Configuration 1	2	3	4	5	6
3-by-3	1	.50000					
	2	.70711	1.00000				
5-by-5	1	.25000					
	2	.43301	.75000				
	3	.50000	.86603	1.00000			
7-by-7	1	.14645					
	2	.27060	.50000				
	3	.35355	.65328	.85355			
	4	.38268	.70711	.92388	1.00000		
9-by-9	1	.09549					
	2	.18164	.34549				
	3	.25000	.47553	.65451			
	4	.29389	.55902	.76942	.90451		
	5	.30902	.58779	.80902	.95106	1.00000	
11-by-11	1	.06699					
	2	.12941	.25000				
	3	.18301	.35355	.50000			
	4	.22414	.43301	.61237	.75000		
	5	.25000	.48296	.68301	.83652	.93301	
	6	.25882	.50000	.70711	.86603	.96593	1.00000

NOTE: Because of the dual four quadrant symmetry present in regular lattice results, only the fundamental information is displayed here. A complete lattice set of values can be obtained by reflecting the above results along their respective diagonals, and then reflecting these results into each of the remaining three quadrants of the lattice surface.

suppresses many of these diagonal values, with this distortion increasing as one moves from the center toward the edge of the bounded lattice. This pattern can be viewed in those results for selected lattice sizes that are reported in Table 5.3. One trend latent in these tabular results is that as $n \to \infty$, the central values converge upon unity, whereas the single band of edge lattice areal units converge upon zero, with this convergence being most rapid at the four corners of the finite lattice.

One additional property of the variance–covariance matrix that is noteworthy refers to the geographic pattern of diagonal values appearing in Table 5.3. Close inspection reveals that this spatial distribution is highly correlated with that for the principal eigenvector $\mathbf{E}_{\max}$ of the connectivity matrix $\mathbf{C}$ (see Table 5.4). Moreover, there is a very strong correspondence between these two variables, a relation that permits the diagonal values to be written as a function of $\mathbf{E}_{\max}$ (see Griffith, 1985). In fact the same tendencies uncovered in Table 5.3 hold in Table 5.4, such as the central values converging upon unity.

5.4. SPECTRAL ANALYSIS AND TWO-DIMENSIONAL SHAPE MEASUREMENT

Spectral analysis also has been employed to measure two-dimensional shapes. Various indices have been developed in an attempt to quantify shape, but none possess the property of a continuous one-to-one mapping of shape onto the set of real numbers (Lee and Sallee, 1970). Many of these indices compare a given shape with a circle (e.g., elongation ratio, form ratio, circularity ratio, compactness ratio). Some of the indices that have been developed actually are redundant [e.g., ellipticity index = $\pi/(4$ form ratio)]. One of the most appealing indices is the one developed by Boyce and Clark, namely the radial shape index, which again uses the circle as its benchmark. A useful review of shape indices has been furnished by Austin (1984).

The battery of shape indices mentioned above has proven to be inadequate, primarily because of the absence of the one-to-one mapping property mentioned above. As a result, Moellering and Rayner (1981) have explored the use of harmonic analysis for measuring shape. These two researchers point out that as a shape descriptor harmonic analysis uses the variance spectrum and the modulus ratio spectrum, both of which are made independent of scale, translation and rotation, and both of which decompose a shape into orthogonal components. The resulting shape frequency components, variances and modulus ratios can be directly interpreted in terms of importance, ellipticity, and rotation of each frequency. More important, harmonic analysis is information preserving in that an original shape can be reconstructed from the spectra.

This approach requires that a closed set of digitized points be determined for a shape, and the center of the shape be calculated. Olson (1976) has provided a simple and efficient procedure for determining the physical center of a two-dimensional shape. Let

$$\{(x_i, y_i), i = 1, 2, \ldots, n - 1, (x_1, y_1) = (x_n, y_n)\}$$

be the closed set of digitized points uniformly spaced at angle θ around the center of the shape in question, and positioned along the boundary of this shape. Then

$$\text{Area} = \left| \sum_{i=1}^{n-1} (x_i y_{i+1} - y_i x_{i+1}) \right| / 2$$

$$x\text{-center} = \left| \sum_{i=1}^{n-1} (y_i - y_{i+1})(x_i^2 + 2x_i x_{i+1} + x_{i+1}^2) \right| / (6 \times \text{Area}),$$

and

$$y\text{-center} = \left| \sum_{i=1}^{n-1} (x_i - x_{i+1})(y_i^2 + 2y_i y_{i+1} + y_{i+1}^2) \right| / (6 \times \text{Area}).$$

Now, parallelling the notation employed by Moellering and Rayner,

$$AX_j = \sum_{i=0}^{n-1} (x_i - x\text{-center}) \{\cos [2\pi(1 - ji/n]\}/n,$$

$$j = 1, 2, \ldots, n,$$

$$AY_j = \sum_{i=0}^{n-1} (y_i - y\text{-center}) \{\cos [2\pi(1 - ji/n]\}/n,$$

$$j = 1, 2, \ldots, n,$$

$$BX_j = \sum_{i=0}^{n-1} (x_i - x\text{-center}) \{\sin [2\pi(1 - ji/n]\}/n,$$
$$j = 1, 2, \ldots, n,$$
$$BY_j = \sum_{i=0}^{n-1} (y_i - y\text{-center}) \{\sin [2\pi(1 - ji/n]\}/n,$$
$$j = 1, 2, \ldots, n,$$

and

$$VXY_j = (AX_j + BY_j)^2 + (BX_j - AY_j)^2,$$
$$j = 1, 2, \ldots, n.$$

The spectral index is given by

$$(VXY_j/VXY_{n-j})^2,$$

and accounts for the following amount of variance (again in terms of a circular benchmark):

$$VXY_j \Big/ \sum_{j=1}^{n} VXY_j, \quad j = 1, 2, \ldots, n.$$

Usually a small number of the n indices account for nearly all of the variation displayed by the set of digitized points.

Griffith (1982) and Griffith *et al.* (1986) studied the performance of the aforementioned spectral shape index on two different sets of cities. In both cases several conspicuous tendencies were uncovered. First, the battery of traditional shape indices reduced to three basic dimensions, namely one measuring ellipticity, one measuring compactness, and one measuring concaveness. Second, in an overwhelming number of cases a single spectral index, based upon the first frequency, accounted for well over 90% of the variance in question. Third, the spectral index accounted for only about 50% of the variance exhibited by the three dimensions of shape, indicating that it does not capture all the information available about a two-dimensional shape. Finally, there was a marked correspondence between the Boyce–Clark radial shape index and the spectral index, so much so that these two indices seem to represent the same information.

This last relation is not surprising, since the Boyce-Clark index is based upon the same set of shape radii as is the spectral index, and is calculated with the following formula:

$$100 \sum_{i=1}^{n} \left| d_i \Big/ \sum_{i=1}^{n} d_i - 1/n \right|$$

Again distances are being assessed.

As an aside, Griffith (1982) found that a relationship exists between the level of spatial autocorrelation measured for some geographic distribution, and the boundary shape of the region under study. The general problem of edge effects of this sort will be addressed in Chapter 7.

5.5. CONCLUDING COMMENTS

This chapter has outlined links between autoregressive models and spectral density functions, and then used these links to explore properties of the variance–covariance matrix of the conditional spatial autoregressive model. Finite lattice results have been compared with asymptotic, infinite lattice results. Descriptive features of spatial correlograms have been studied. And several complications due to edge effects have been noted.

The importance of findings summarized in this chapter is threefold. First, one needs to be able to determine theoretical spatial autocorrelation values. One method of achieving this goal is reviewed here, in terms of spectral representations of autoregressive models. Second, in order to better understand the behavior of finite lattice statistics, one needs to know properties of the corresponding variance–covariance matrix. This second point has special implications regarding Monte Carlo simulation experiments. Finally, the stage is set now for exploring the problem of edge effects, which has been alluded to in this chapter, and will be treated in Chapter 7.

5.6. REFERENCES

Austin, R., 1984, Measuring and comparing two-dimensional shapes, in *Spatial Statistics and Models*, edited by G. Gaile and C. Willmott. Boston: Reidel, pp. 293–312.

Bartlett, M., 1975, *The Statistical Analysis of Spatial Pattern*. London: Chapman and Hall.

Dacey, M., 1976, Solutions to 135 simple recurrences that generate spatial trends, unpublished manuscript, Department of Geography, Northwestern University.

Griffith, D., 1982, Geometry and spatial interaction, *Annals*, Association of American Geographers, Vol. 72: 332–346.

Griffith, D., 1985, *Correcting for Edge Effects in Spatial Statistical Analyses*, unpublished master's thesis, Department of Statistics, The Pennsylvania State University.

Griffith, D., M. O'Neill, W. O'Neill, L. Leifer, and R. Mooney, 1986, Shape indices: useful measures of red herrings?, *The Professional Geographer*, Vol. 38: 263–270.

Haining, R., 1977, Model specification in stationary random fields, *Geographical Analysis*, Vol. 9: 107–129.

Haining, R., 1978, *Specification and Estimation Problems in Models of Spatial Dependence*. Evanston, Ill.: Department of Geography, Northwestern University.

Haining, R., 1979, Statistical tests and process generators for random field models, *Geographical Analysis*, Vol. 11: 45–64.

Lee, D., and T. Sallee, 1970, A method of measuring shape, *The Geographical Review*, Vol. 60: 555–563.

Moellering, H., and J. Rayner, 1981, The harmonic analysis of spatial shapes using dual axis Fourier shape analysis (DAFSA), *Geographical Analysis*, Vol. 13: 64–77.

Olson, J., 1976, Noncontiguous area cartograms, *The Professional Geographer*, Vol. 28: 371–380.

Rayner, J., 1971, *Introduction to Spectral Analysis*. London: Pion.

Vandaele, W., 1983, *Applied Time Series and Box–Jenkins Models*. New York: Academic Press.

CHAPTER 6

The Missing Data Problem for a Two-Dimensional Surface

The topics, themes and subject matter of Chapters 1 thru 5 are of a reference nature. Besides the complexities discussed in these preceding chapters, the field of spatial statistics also embraces several prominent difficult, bothersome, and unresolved problems. One concern has to do with incomplete data, which is the topic of this chapter. Geographical data sets sometimes contain missing observations that need to be estimated. An exact maximum likelihood solution for this problem is discussed, both in terms of parameter and missing value estimation, for multivariate normal spatial data sets satisfying the first-order spatial Markov property with constant mean. Moreover, information at neighboring or contiguous observed sites is used to estimate the missing values, and then the complete spatial distribution is used to estimate model parameters. The solution procedure is iterative, and is akin to the Orchard and Woodbury missing information principle. Results are reported for extensions to a second-order, simultaneous model, and from an empirical example used to explore the behavior of these estimates. Also, tentative simulation experiment findings are reviewed.

6.1. THE INCOMPLETE DATA PROBLEM STATEMENT

The major impetus for research on the problem of incomplete data comes from the general nuisance that is encountered when spatial analysts are faced with studying such data sets. Specific examples that sparked interest in research on this topic include missing (possibly due to equipment malfunctioning) or obstructed (probably due to cloud cover and reflectance glare) pixel information in LANDSAT images, suppressed areal unit data in census publications (because of confidentiality legislation), and the loss of rainfall measures (due to a gauge malfunctioning). These three instances exemplify a broader class of different but related problems, though. A considerable amount of work has been published in the literature about censored and truncated data. These two forms of incomplete data are non-random in nature, with truncation meaning either or both tails of a distribution have been eliminated, more than likely because sampling in these regions is not possible (e.g., a rain gauge overflow). Generally speaking, censoring refers to right-sided truncation, where usually a conditional experiment is of interest. In contradistinction, the concern in this research is with random incompleteness of data. For example, equipment malfunctioning often can be viewed as the outcome of a random, Poisson process.

The purpose of this chapter is to report findings regarding the problem of statistically analyzing incomplete data, with special emphasis being placed on geographical data sets, where the goal is to estimate values for observations having missing ones, and not simply parameter estimation. The motivation for studying this particular problem arises from, for instance, the need to have some 'best estimate' of a missing value in order to calibrate mathematical optimization models on data sets. Moreover, it is one approach whose aim is to minimize the amount of sampled data that might end up being discarded.

The problem of handling missing data has two components. First, one could ask about how parameters should be estimated, given an incomplete data set, for some model that has been posited. Most of the effort reported in the literature has addressed this issue. Second, one could ask how the missing values themselves could be estimated. While these might appear to be two separate questions, the approach outlined here actually treats both of them simultaneously. Moreover, the procedure is to set up a model for the data, and then simultaneously estimate the parameters of this model together with the missing values.

TABLE 6.1

A Possible Taxonomy of the Missing Data Literature

General

Hartley and Hocking (1971)	Koul and Susarla (1980)
Sundberg (1974)	U. S. National Academy of Sciences (1980)
Rubin (1974, 1976)	Hill and Dixon (1981)
Gleason and Staelin (1975)	Anderson, Basilevsky and Hun (1983)
John and Prescott (1975)	Little and Rubin (1983)
Singh (1977)	Titterington and Jiang (1983)
Woolson, Leeper and Clarke (1978)	Titterington and Mill (1983)

Parameter Estimation

1.	General:	Titterington (1984)
2.	Univariate μ and σ:	Akaike and Ishiguro (1980) Eubank and La Riccia (1982) Cheng and Ling (1983)
	Difference of proportions:	Campbell (1984) Gokhale and Sirtonik (1984)
	Difference of means:	Lin (1971) Bouza (1983) Lin and Stivers (1974) Bhoj (1984)
	Difference of variances:	Bhoj (1984)
	Normality:	Hocking and Smith (1972)
3.	General bivariate analysis:	Ratkowsky (1974) Hamdan, Pirie and Khun (1976) Dahiya and Korwar (1980) Korwar and Dahiya (1982)
	Analysis of variance:	Rubin (1972)
	Correlation:	Papaioannou and Speevak (1977) Brouwer and Vijn (1980) Weier and Basu (1980) Jones and Facer (1982) Iwase and Seto (1984) Papaioannou and Loukas (1984)
4.	General multivariate:	Afifi and Elashoff (1966) Beale and Little (1975) Titterington (1977) Hosking (1981) Smith and Riggs (1984)
	$\underline{\mu}$ for MVN:	Morrison and Bhoj (1973) Little (1976)
	$\underline{\mu}$ and $\underline{\Sigma}$ for MVN:	Hocking and Smith (1968) Morrison (1971) Murry (1979) Huseby, Schwertman and Allen (1980)

TABLE 6.1 (continued)

		Andrade and Helms (1984)
	$\underline{\Sigma}$	Engleman (1982) Rubin and Szatrowksi (1982)
	Regression:	Press and Scott (1976) Donner (1982) Donner and Rosner (1982) Miller and Halpern (1982) del Pino (1984)
	Canonical correlation:	Campbell (1984)
	Discriminant function analysis:	Chan and Dunn (1974)
	Factor analysis:	Hinkins (1980) de Ligny (1981)
	General linear model:	Drygas (1976) Hosking (1981)
5.	Time series:	Limonard (1978) Jones (1980) Marshall (1980) Robinson (1980) Vo-Dai (1980) Abraham (1981) Dunsmuir and Robinson (1981) Harvey (1981) Richardson and White (1981) Hoffman and Anderson (1982) Honohan and McCarthy (1982) Ljung (1982) Shumway and Stoffer (1982) Dagenais and Dufour (1984) Hinich and Weber (1984) Milhoej (1984)
6.	EM algorithm:	Dempster, Laird and Rubin (1977) Boyles (1983) Redner and Walker (1984)
7.	Inadmissibility of MLE:	Radhakrishnan (1982)
	Posterior Distributions:	Laird and Louis (1982)
	Discrete Data:	Hocking and Oxspring (1971) Selvin (1980) Pettitt (1983)
	Algorithms:	Morgan and Titterington (1977) Titterington (1977) Hocking and Marx (1979) van Guilder and Azen (1981)

TABLE 6.1 (continued)

	Autocorrelation:	Feingold (1982) Martin (1984)
	Exponential family:	Sundberg (1974)
Missing Values Estimation		
1.	General:	Preece (1971) Rubin (1972) John and Prescott (1975) Vacek and Ashikaga (1980) Chapman (1982) Greenlees, et. al. (1982) Tabony (1982) Kemp, et. al. (1983)
2.	Time series:	Chow and Lin (1976) Damsleth (1980) Harvey and McKenzie (1984) Harvey and Pierse (1984)
3.	Exchangeability:	Wei (1983)

6.2. BACKGROUND

The statistical literature on missing data is quite sparse through the beginning of the 1970s. At that point in time it began growing, seemingly at an exponential rate. Table 6.1 represents a summary of this literature. Most of the articles appearing since 1970 have been problem-oriented. And, an overwhelming number of the articles that appear in Table 6.1 are concerned solely with parameter estimation in the presence of an incomplete data set. Far fewer papers have addressed the problem of estimating the missing values themselves.

Table 6.1 furnishes a brief, and by no means comprehensive, survey of the relevant literature. A good, recent review can be found in Anderson, Basilevsky and Hum (1983). One should note that probably the first significant contribution was made by Dempster, Rubin and Laird (1977), who have provided a two-step iterative procedure that converges in a finite number of steps. At least four papers about the algorithm itself have followed the publication of this seminal piece of work (see Boyles, 1983; Engleman, 1982; Rubin and Szatrowski, 1982; van Guilder and Azen, 1981).

With regard to the geographical literature, Bennett, Griffith and Haining (1984) have supplied an overview of the problem in geography. Tobler (1979), Kennedy and Tobler (1983), and Tobler and Kennedy (1985) have discussed the interpolation of estimates in two dimensions. Unfortunately their work lacks sufficient analytical rigor, and appears to be based mostly on intuition. Martin (1984) discusses the problem in terms of correlated Gaussian processes, and derives the accompanying Jacobian of the transformation from an autocorrelated space to an unautocorrelated space, for incomplete geographic data. Haining, Griffith and Bennett (1984) outline the MLEs for a first-order Gaussian Markov model. Their results are consistent with the findings of Besag (1974). Finally, Upton (1985) has discussed the problem of geographic interpolation using distance weighting.

6.3. SOLUTIONS AVAILABLE IN COMMERCIAL STATISTICAL PACKAGES

The point of departure here is twofold. The embarkation for this section will be those methods available in the major commercial statistical packages for handling missing data. The next section will extend the work by Martin (1984), and by Haining, Griffith and Bennett (1984).

Commercial statistical packages tend to offer three options for dealing with missing values. The default in MINITAB, which is an option in SPSS, is pairwise deletion. Here, if a variable contains missing values, then the corresponding observations are ignored only when that particular variable is treated. In contrast, by invoking the MANOVA proc in SAS, an observation is ignored if it has missing values for any of its variables. Accordingly a considerable amount of sampled information could be lost. Meanwhile, SPSS offers the option of substituting the mean of the observed values for missing values contained in a given variable.

An example was contrived in order to illustrate these different options. Results for this example appear in Table 6.2, and this example consisted of a random sample of 15 observations made on 5 independent variables, each drawn from a population having $\mu = 0$ and $\sigma = 1$. Next, 15 observed values were selected at

TABLE 6.2

Selected Eigenstructures for the Contrived Example

Deletion Type	Eigenvalues 1	2	3	4	5	Example
pairwise	1.8	1.5	1.1	0.6	0.0	MINITAB default
pairwise (>15)	2.7	1.6	0.5	0.2	-0.1	MINITAB default
observation	2.2	1.4	0.9	0.5	0.0	SAS (MANOVA proc)
$\overline{x}$ substitute	1.7	1.4	1.1	0.7	0.1	SPSS option
limiting cases						
single column						
best	1.4	1.2	0.8	0.4	***	variable 3
nearest to average	1.6	1.2	0.8	0.4	***	variable 2
worst	1.5	1.3	0.9	0.2	***	variable 4
three rows						
best	1.5	1.4	1.1	0.8	0.2	rows 8, 10, 12
nearest to average	1.8	1.4	1.0	0.6	0.2	rows 1, 5, 15
worst	2.3	1.7	0.6	0.3	0.1	rows 1, 4, 7
population	1.0	1.0	1.0	1.0	1.0	*****
sample	1.6	1.4	1.1	0.7	0.2	*****

This table is based upon a random sample of 15 observations made on 5 independent variables, which were drawn from a population whose mean is $\underline{\mathbf{0}}$ and whose variance-covariance matrix is $\underline{\mathbf{I}}$. Next, 15 observed values were selected at random for deletion following a Poisson selection process, where 15 numbers were randomly selected between 1 and 5 (denoting the variable selection), and 15 numbers were selected between 1 and 15 (denoting the observation selection). This sampling was done without replacement.

random for deletion following a Poission selection process, without replacement. For comparative purposes, the eigenvalues of the associated variance-covariance matrices were calculated.

Since $\mathbf{\Sigma} = \mathbf{I}$, then the population eigenvalues are all unity. The complete sample has eigenvalues that range from 0.2 to 1.6. The major problem with pairwise deletion is demonstrated with the deletion of additional observations – one can end up with a variance–covariance matrix that no longer is positive definite (e.g., negative eigenvalues are present), because latent transitive relations may be destroyed. The primary problem with deleting any observation having at least one missing value is that much sampled information is thrown away, and the sample size may be diminished substantially. Accordingly, in the general case especially the principal eigenvalue becomes inflated. This can be minimized to some extent as the limiting case is achieved where all missing values are concentrated in a few observations. For the other limiting case, where all missing values are concentrated in a single variable, clearly omission of that variable would be preferable to deletion of all observations. But bias can be introduced here; as shown in Table 6.2, only four eigenvalues can be calculated for the contrived example.

Let $\bar{x}_0$ be the mean of the observed values. Substitution of $\bar{x}_0$ has a sound statistical basis, and important drawbacks, too. One must assume that unknown data values can be treated as additional parameters to be estimated from the sample data (see Charernkavanich and Cohen, 1984). Hence, for the univariate $N(\mu, \sigma^2)$ model, the missing value terms can be factored out so that

$$
\begin{aligned}
L(\mu\ \sigma^2) &= (2\pi)^{-n/2}\sigma^{-n} \exp\left[-\sum_{m=1}^{n_m}(x_m - \mu)^2/(2\sigma^2)\right] \\
&\quad \times \exp\left[-\sum_{i=1}^{n_0}(x_i - \mu)/(2\sigma^2)\right], \qquad (6.1)
\end{aligned}
$$

where n_0 denotes the number of observed values, and n_m denotes the number of missing values. Now $\partial \ln (L)/\partial x_m = 0$ yields as the MLE $\hat{x}_m = \mu$, which when μ is unknown renders $\hat{x}_m = \bar{x}_0$. This estimate has the following two properties:

$$E(\hat{x}_m) = \mu \quad \text{and} \quad \mathrm{VAR}(\hat{x}_m) = \sigma^2/n_0.$$

Clearly, and intuitively, this substitution will deflate the observed sample variance. In fact

$$
\begin{aligned}
\lim_{n_m \to n} s^2 &= 0, \\
E(s^2) &= [(n_m + n_0^2 - n)/(nn_0)]\sigma^2, \\
\lim_{n_m \to 0} E(s^2) &= [(n - 1)/n]\sigma^2,
\end{aligned}
$$

and

$$\lim_{n_m \to n} E(s^2) = 0.$$

One possible way to preserve this sample variance is to draw n_m random samples from the conditional normal distribution, setting

$$\mu = \bar{x}_0 \quad \text{and} \quad \sigma^2 = [n_0/(n_0 - 1)]s_0^2.$$

Regardless, the conditional mean appears to be preferable to the preceding two options.

Now consider an unbounded planar surface that has been partitioned into n mutually exclusive and collectively exhaustive areal units that form a regular square tessellation, or lattice. Suppose the following stationary, isotropic first-order Gaussian spatial Markov process generates some n-by-1 vector random variable $\mathbf{X}$.

$$X_{i,j} = \mu + \varrho[(X_{i,j-1} - \mu) + (X_{i,j+1} - \mu) + (X_{i-1,j} - \mu) + (X_{i+1,j} - \mu)] + \xi_{i,j}, \tag{6.2}$$

with $|\varrho| < 0.25$, $E(X_{i,j}|\mathbf{CX}) = \mu$, and $\mathrm{VAR}(X_{i,j}|\mathbf{CX}) = \sigma^2$, where $\mathbf{C}$ is the aforementioned binary 0–1 matrix denoting the juxtaposition of observations [it functions as a lag operator in converting vector $\mathbf{X}$ into equation (6.2)]. The accompanying MLEs are (Ord, 1975)

$$\hat{\mu} = \mathbf{1}^T(\mathbf{I} - \varrho\mathbf{C})\mathbf{X}[\mathbf{1}^T(\mathbf{I} - \varrho\mathbf{C})\mathbf{1}]^{-1}, \tag{6.3}$$

$$\hat{\sigma}^2 = (\mathbf{X} - \mu\mathbf{1})^T(\mathbf{I} - \varrho\mathbf{C})\,(\mathbf{X} - \mu\mathbf{1})/n, \tag{6.4}$$

and

$$\underset{|\varrho| < 1/\lambda_{\max}}{\mathrm{MIN:}} \left[\prod_{i=1}^{n} (1 - \varrho\lambda_i)\right]^{-1/n} n\hat{\sigma}^2, \tag{6.5}$$

where, as before, $\mathbf{1}$ is an n-by-1 vector of ones, and λ_i are the eigenvalues of matrix $\mathbf{C}$. Equations (6.3) thru (6.5) are the conditional spatial autoregressive counterparts to equations (4.3) thru (4.5), which were given for the simultaneous model.

Next, let $\mathbf{X}^T = (\mathbf{X}_0^T \vdots \mathbf{X}_m^T)$, and

$$(\mathbf{I} - \varrho\mathbf{C}) = \begin{pmatrix} (\mathbf{I}_0 - \varrho\mathbf{C}_{00}) & \vdots & -\varrho\mathbf{C}_{0m} \\ \hline -\varrho\mathbf{C}_{m0} & \vdots & (\mathbf{I}_m - \varrho\mathbf{C}_{mm}) \end{pmatrix}, \tag{6.6}$$

where $\mathbf{C}_{00}$ is the partitioned portion of matrix $\mathbf{C}$ affiliated with the covariation of observed values, $\mathbf{C}_{mm}$ is the part of matrix $\mathbf{C}$ associated with the covariation of missing values, and $\mathbf{C}_{m0}$ and $\mathbf{C}_{0m}$ are those sections of matrix $\mathbf{C}$ connected with the covariation of observed with missing values. Now the Jacobian becomes (Martin, 1984)

$$\left\{\left[\prod_{k=1}^{n_m} (1 - \varrho\eta_k)\right] \Big/ \left[\prod_{i=1}^{n} (1 - \varrho\lambda_i)\right]\right\}^{1/(n-n_m)},$$

where η_k are the eigenvalues of matrix $\mathbf{C}_{mm}$.

Haining, Griffith and Bennett (1984) have shown that the MLE of $\mathbf{X}_m$ for this model is

$$\hat{\mathbf{X}}_m = E(\mathbf{X}_m|\mathbf{X}_0) = \mu\mathbf{1}_m + \varrho(\mathbf{I} - \varrho\mathbf{C}_{mm})^{-1}\mathbf{C}_{m0}(\mathbf{X}_0 - \mu\mathbf{1}_0), \tag{6.7}$$

which is a conditional mean. Equation (6.7) can reduce in certain cases in that the matrix inversion disappears when $n_m = 1$ ($\mathbf{C}_{mm} = 0$), or when none of the missing observations are juxtaposed ($\mathbf{C}_{mm} = \mathbf{0}$). If $n_m = 1$, then $\mathbf{C}_{mm}$ is a 1-by-1 matrix (i.e., a scalar) whose single entry is zero. Hence equation (6.7) reduces to

$$\hat{x}_m = \mu + \varrho \sum_{j=1}^{n} c_{mj}(x_j - \mu), \tag{6.8}$$

which is the conditional forecasting equation derived in Besag (1974). Moreover, the single missing geographical value is a weighted average of the surrounding areal unit values. This is the type of solution that Tobler and Kennedy have proposed (1983, 1985), and is a direct result of stochastic processes theory. Meanwhile, if the set of missing values is completely dispersed over a two-dimensional surface, such that no two missing values are adjacent, then although $\mathbf{C}_{mm}$ is not a scalar it is a null matrix. Therefore the missing value estimation equation reduces to a set of n_m equations, each being of the form displayed by equation (6.8).

The inverse term $(\mathbf{I} - \varrho\mathbf{C}_{mm})^{-1}$ does not disappear from equation (6.7) only when missing values are clustered on a two-dimensional surface. This particular result is not a straightforward consequence of stochastic processes theory. Operationally the matrix inversion term results from iteratively and repeatedly substituting equation (6.8) for each missing value. The sequence of substitutions produces a convergent geometric series in terms of ϱ that is summarized by the matrix inversion. To exemplify this point, consider the case of a linear geographic landscape with two non-border, adjacent missing values at locations i and $i + 1$. From equation (6.7) one gets

$$\begin{pmatrix} x_i \\ x_{i+1} \end{pmatrix} = \mu \begin{pmatrix} 1 \\ 1 \end{pmatrix} + \varrho \begin{pmatrix} 1 & -\varrho \\ -\varrho & 1 \end{pmatrix}^{-1} \begin{pmatrix} x_{i-1} - \mu \\ x_{i+2} - \mu \end{pmatrix},$$

which implies that

$$\hat{x}_i - \mu = (1 - \varrho^2)^{-1}[\varrho(x_{i-1} - \mu) + \varrho^2(x_{i+2} - \mu)],$$

and

$$\hat{x}_{i+1} - \mu = (1 - \varrho^2)^{-1}[\varrho^2(x_{i-1} - \mu) + \varrho(x_{i+2} - \mu)].$$

The geometric series is captured here by the term

$$(1 - \varrho^2)^{-1} = \sum_{i=0}^{\infty} (\varrho^2)^i.$$

Now using equation (6.8),

$$\begin{aligned}
\hat{x}_i &= \mu + \varrho[(x_{i-1} - \mu) + (x_{i+1} - \mu)] \\
&= \mu + \varrho(x_{i-1} - \mu) + \varrho\{[(x_i - \mu) + (x_{i+2} - \mu)] - \mu\} \\
&= \mu + \varrho(x_{i-1} - \mu) + \varrho^2(x_{i+2} - \mu) + \varrho\{\mu + \varrho[(x_{i-1} - \mu) \\
&\quad + (x_{i+1} - \mu)] - \mu\} \\
&= \mu + \varrho(x_{i-1} - \mu) + \varrho^2(x_{i+2} - \mu) + \varrho^2(x_{i-1} - \mu) \\
&\quad + \varrho^2\{\mu + \varrho[(x_i - \mu) + (x_{i+2} - \mu)] - \mu\} \\
&\;\vdots \\
&= \mu + \varrho \sum_{k=0}^{\infty} \varrho^k(x_{i-1} - \mu) + \varrho^2 \sum_{k=0}^{\infty} \varrho^k(x_{i+2} - \mu).
\end{aligned}$$

Consequently, the missing value estimator, in the case of two adjacent missing values, iterates between estimates of the two missing values using equation (6.8). The series expansion summarized by the matrix inversion term appearing in equation (6.7) describes this iterative process.

Now equations (6.3), (6.4) and (6.5) are replaced by

$$\begin{aligned}
\hat{\mu} &= \mathbf{1}_0^T[(\mathbf{I} - \varrho\mathbf{C}_{00}) - \varrho^2\mathbf{C}_{0m}(\mathbf{I} - \varrho\mathbf{C}_{mm})^{-1}\mathbf{C}_{m0}]\mathbf{X}_0 \\
&\quad \times \{\mathbf{1}_0^T[(\mathbf{I} - \varrho\mathbf{C}_{00}) - \varrho^2\mathbf{C}_{0m}(\mathbf{I} - \varrho\mathbf{C}_{mm})^{-1}\mathbf{C}_{m0}]\mathbf{1}_0\}^{-1}
\end{aligned}$$

$$\hat{\sigma}^2 = \{(\mathbf{X}_0 - \mu\mathbf{1}_0)^T(\mathbf{I} - \varrho\mathbf{C}_{00})(\mathbf{X}_0 - \mu\mathbf{1}) - \varrho^2[(\mathbf{X}_0 - \mu\mathbf{1}_0)^T \times \mathbf{C}_{0m}(\mathbf{I} - \varrho\mathbf{C}_{mm})^{-1}\mathbf{C}_{m0}(\mathbf{X}_0 - \mu\mathbf{1}_0)]\}/(n - n_m),$$

and

$$\underset{|\varrho|<1/\lambda_{\max}}{\text{MIN:}} \left\{\left[\prod_{k=1}^{n_m}(1 - \varrho\eta_k)\right]\Big/\left[\prod_{i=1}^{n}(1 - \varrho\lambda_i)\right]\right\}^{1/(n-n_m)}(n - n_m)\hat{\sigma}^2. \tag{6.9}$$

Further,

$$E(\hat{\mathbf{X}}_m) = \mu\mathbf{1}_m,$$

and

$$\text{COV}(\hat{\mathbf{X}}_m) = \varrho^2\sigma^2(\mathbf{I} - \varrho\mathbf{C}_{mm})^{-1}\mathbf{C}_{m0}(\mathbf{I} - \varrho\mathbf{C}_{00})^{-1} \times \mathbf{C}_{0m}(\mathbf{I} - \varrho\mathbf{C}_{mm})^{-1}.$$

Besides the foregoing conditional model, one could posit a simultaneous model for the Gaussian spatial Markov process that generates the vector $\mathbf{X}$. As such,

$$\mathbf{X} = (\mathbf{I} - \varrho\mathbf{C})^{-1}\xi,$$

and the variance–covariance matrix becomes

$$\sigma^2[(\mathbf{I} - \varrho\mathbf{C})^T(\mathbf{I} - \varrho\mathbf{C})]^{-1}.$$

One should note that matrix $\mathbf{C}$ no longer must be symmetric. Employing the same partitioning as in equation (6.6), the MLE of $\mathbf{X}_m$ for equation this second variance–covariance matrix is

$$\hat{\mathbf{X}}_m = \mu\mathbf{1}_m + \varrho[(\mathbf{I} - \varrho\mathbf{C}_{mm})^T(\mathbf{I} - \varrho\mathbf{C}_{mm}) + \varrho^2\mathbf{C}_{0m}^T\mathbf{C}_{0m}]^{-1}[\mathbf{C}_{0m}^T(\mathbf{I} - \varrho\mathbf{C}_{00}) + (\mathbf{I} - \varrho\mathbf{C}_{mm})^T\mathbf{C}_{m0}](\mathbf{X}_0 - \mu\mathbf{1}_0). \tag{6.10}$$

In contrast with equation (6.7), for equation (6.10) considerable simplification does not occur for dispersed missing values. Here when $\mathbf{C}_{mm} = \mathbf{0}$, the inversion term becomes

$$(\mathbf{I} + \varrho^2\mathbf{C}_{0m}^T\mathbf{C}_{0m})^{-1}.$$

Now equations (6.3), (6.4) and (6.5) are replaced by

$$\hat{\mu} = [\mathbf{1}_0^T\{[(\mathbf{I} - \varrho\mathbf{C}_{00})^T(\mathbf{I} - \varrho\mathbf{C}_{00}) + \varrho^2\mathbf{C}_{m0}^T\mathbf{C}_{m0}] - \varrho^2[(\mathbf{I} - \varrho\mathbf{C}_{00})^T\mathbf{C}_{0m} + \mathbf{C}_{m0}^T(\mathbf{I} - \varrho\mathbf{C}_{mm})] \times [(\mathbf{I} - \varrho\mathbf{C}_{mm})^T(\mathbf{I} - \varrho\mathbf{C}_{mm}) + \varrho^2\mathbf{C}_{0m}^T\mathbf{C}_{0m}]^{-1}$$

$$
\begin{aligned}
&\times [\mathbf{C}_{0m}^T(\mathbf{I} - \varrho\mathbf{C}_{00}) + (\mathbf{I} - \varrho\mathbf{C}_{mm}^T)\mathbf{C}_{m0}]\}\mathbf{X}_0] \\
&\times [\mathbf{1}_0^T\{[(\mathbf{I} - \varrho\mathbf{C}_{00})^T(\mathbf{I} - \varrho\mathbf{C}_{00}) + \varrho^2\mathbf{C}_{m0}^T\mathbf{C}_{m0}] \\
&- \varrho^2[(\mathbf{I} - \varrho\mathbf{C}_{00})^T\mathbf{C}_{0m} + \mathbf{C}_{m0}^T(\mathbf{I} - \varrho\mathbf{C}_{mm})] \\
&\times [(\mathbf{I} - \varrho\mathbf{C}_{mm})^T(\mathbf{I} - \varrho\mathbf{C}_{mm}) + \varrho^2\mathbf{C}_{0m}^T\mathbf{C}_{0m}]^{-1} \\
&\times [\mathbf{C}_{0m}^T(\mathbf{I} - \varrho\mathbf{C}_{00}) + (\mathbf{I} - \varrho\mathbf{C}_{mm}^T)\mathbf{C}_{m0}]\}\mathbf{1}_0]^{-1}
\end{aligned}
$$

$$
\begin{aligned}
\hat{\sigma}^2 = &\{(\mathbf{X}_0 - \mu\mathbf{1}_0)^T[(\mathbf{I} - \varrho\mathbf{C}_{00})^T(\mathbf{I} - \varrho\mathbf{C}_{00}) + \varrho^2\mathbf{C}_{m0}^T\mathbf{C}_{m0}] \\
&\times (\mathbf{X}_0 - \mu\mathbf{1}_0) - \varrho^2(\mathbf{X}_0 - \mu\mathbf{1}_0)^T[(\mathbf{I} - \varrho\mathbf{C}_{00})^T\mathbf{C}_{0m} \\
&+ \mathbf{C}_{m0}^T(\mathbf{I} - \varrho\mathbf{C}_{mm})]\,[(\mathbf{I} - \varrho\mathbf{C}_{mm})^T(\mathbf{I} - \varrho\mathbf{C}_{mm}) \\
&+ \varrho^2\mathbf{C}_{0m}^T\mathbf{C}_{0m}]^{-1}[\mathbf{C}_{0m}^T(\mathbf{I} - \varrho\mathbf{C}_{00}) + (\mathbf{I} - \varrho\mathbf{C}_{mm}^T)\mathbf{C}_{m0}] \\
&\times (\mathbf{X}_0 - \mu\mathbf{1}_0)\}/(n - n_m),
\end{aligned}
$$

and

$$
\underset{|\varrho| < 1/\lambda_{\max}}{\text{MIN:}} \left\{\left[\prod_{k=1}^{n_m}(1 - \varrho\eta_k)\right]\bigg/\left[\prod_{i=1}^{n}(1 - \varrho\lambda_i)\right]\right\}^{2/(n-n_m)} (n - n_m)\hat{\sigma}^2 .
$$

Furthermore, $E(\hat{\mathbf{X}}_m) = \mu\mathbf{1}_m$, so that $\hat{\mathbf{X}}_m$ still is unbiased.

Only the conditional and simultaneous models are presented here because, as was mentioned earlier, rarely in geographic research have other higher-order models been found to provide better data descriptions. Consequently, once an appropriate spatial model has been identified, conditional expectations in the form of MLEs of missing values may be obtained, but possibly with considerable numerical effort. A problem that remains, though, is one of incorporating additional exogenous information, such as preservation of the sample variation that was mentioned earlier.

6.5. PROPERTIES OF THE CONDITIONAL MODEL WHEN DATA ARE INCOMPLETE

Properties of equation (6.7) and the MLEs associated with it have been explored through the analysis of an empirical example (see Haining, Griffith and Bennett, 1984). They also will be inspected here through the use of asymptotic variances and the conducting of simulation experiments. For the empirical example, rainfall

Fig. 6.1. The spatial structure of the weather station surface.

data for 41 weather stations were analyzed. The study area came from Kansas–Nebraska, and the region was divided into areal units for the purpose of constructing matrix **C** with the use of Thiessen polygons (see Figure 6.1). A jackknife type of procedure was used to assess the distribution of $\hat{\varrho}$, with $n_m = 1$. The value of ϱ for the complete data surface was found to be 0.1813, with an associated maximum value of $1/\lambda_{\max} = 0.1851$. As was seen in Chapter 5, the overall spatial level of spatial dependency measured here is moderate (relatively speaking, $\varrho \simeq 0.245$, indicating that the first-order correlation is approximately 0.5).

TABLE 6.3

Jackknife Results for a Single Missing Valuc of the Rainfall Data

Areal Unit	$\hat{\rho}$	$\overline{x}$	x_i	$\hat{x}_i$
Imperial	.18183	.11252	-1.35272	-.35014
Hayes Center	.18147	.04706	-1.02584	-.28255
Curtis	.18094	-.03730	.60138	.16390
Holdregc	.18108	-.03419	1.84180	.44871
Minden	.18108	-.13832	2.24311	-.18263
Hastings	.18229	.19888	-.86617	.35339
Benkelman	.18117	.00788	-1.21052	-.97307
Trenton	.18145	-.01588	.02938	-1.17188
McCook	.18106	-.01584	-.90613	-.52425
Beaver City	.18156	.03230	1.05293	-.17852
Alma	.18133	.01454	-.62570	.10279
Franklin	.18090	-.05919	-.77588	.07523
Rosemont	.18175	.08058	-.99772	.14722
St. Francis	.18133	.04289	-2.10446	-1.33406
Atwood	.18093	-.02281	-1.95509	-1.45284
Oberlin	.18098	-.01925	-1.60582	-1.18932
Norton	.18159	.02841	-.27288	-1.20731
Phillipsburg	.18112	-.00685	-1.28658	-1.29219
Smith Center	.18119	.00475	-.93411	-.44570
Burr Oak	.18098	-.04634	.08793	-.54735
Goodland	.18165	.08146	-2.01308	-.77082
Colby	.18113	-.00626	-2.24331	-2.26506
Hoxie	.18121	.00198	-2.25613	-2.35584
Hill City	.18107	.00015	-1.91315	-1.42173
Plainville	.18084	-.02785	-2.38120	-1.77449
Alton	.18181	.04313	-.24184	-1.47415
Beloit	.18109	-.01099	-.24595	-.30065
Sharon Springs	.18191	-.15301	2.55526	-.69690
Oakley	.18039	-.08534	-2.57365	-1.55261
Gove	.18098	.00330	-2.73343	-1.88992
Wakeeney	.18164	.02702	-1.65506	-2.40129
Hays	.18108	-.00609	-2.04583	-1.90361
Russell	.18034	-.10105	-2.68333	-1.38455
Lincoln	.18109	-.01488	-.55545	-.68285
Tribune	.18048	-.09633	.94289	.40718
Leoti	.18103	-.02424	.04188	-.21628
Healy	.18135	.04513	-2.08228	-1.28284
Ness City	.18111	-.00764	-1.87703	-1.88961
Bison	.18131	.03852	-1.91299	-1.20720
Ellinwood	.18079	-.09596	-.11037	-.97921
Ellsworth	.18110	-.01089	-.57696	-.61150
$\mu_{\overline{x}}$ =	.18121	-.0055	- .893	- .890
σ^2 =	.00040	.0645	1.323	.788

NOTE 1: The results reported in this table differ from those reported in Haining, Griffith and Bennett (1984) in the fourth and fifth digits to the right of the decimal point because these results were calculated with single precision on a VAX computer, whereas the other results were calculated with single precision on a CDC Cyber computer.

NOTE 2: The IMSL subroutine ZXMIN was used to calculate $\hat{\rho}$.

Results from this empirical example are reported in Table 6.3. The mean squared error for x_i with respect to $\hat{x}_i$ is 0.97235. As was demonstrated in Section 6.3, the variance for the complete data, 1.323^2, is greater than that for the estimated data, 0.788^2. But the mean for the two sets of numbers is nearly identical (-0.893 versus -0.890). The mean for the ξ_i's is zero, whereas the GLS mean from equation (6.3) is -0.0055. The mean of $\hat{\varrho}_m$ is 0.1812, as opposed to the original value of 0.1813. Hence there appears to be a slight bias in these results.

Figure 6.2 presents frequency distributions for the variates appearing in Table 6.3. These distributions compare favorably with a normal distribution, as can be seen from their calculated modified Wilk–Shapiro statistics (see Ryan, Joiner and Ryan, 1982):

Variate	Correlation coefficient
$\hat{\varrho}$	0.978
$\bar{x}$	0.969**
x_i	0.968**
$\hat{x}_i$	0.989
$x_i - \hat{x}_i$	0.953***

where the double and triple asterisks respectively denote significant departure from normality at the 5% and 1% levels (one should note that none of these values are significant at the 10% level). These statistics imply that the distribution of $\hat{\varrho}$ and $\hat{x}_i$, and to some extent $\bar{x}$ and x_i, conform to a normal distribution, when a single missing value is present. In addition, the distribution of error does not seem to conform to a normal distribution.

Certain relationships are latent in the data appearing in Table 6.3. First, as Figure 6.3 illustrates, there is a reasonably strong relationship between $\hat{\varrho}$ and $\bar{x}$ ($r = 0.683$). These two quantities are not independent. The likelihood function that results from substituting equation (6.6) and the aforementioned partitioned form of vector **X** into equation (4.33) is

$$\begin{aligned}
\ln (L_{\text{inc}}) = {} & -(n - n_m) \ln (2\pi)/2 - (n - n_m) \ln (\sigma^2)/2 \\
& + \ln \left[\prod_{i=1}^{n_m} (1 - \varrho\eta_i) \Big/ \prod_{i=1}^{n} (1 - \varrho\lambda_i) \right] /2 \\
& - \{\mathbf{X}_0 - \mu\mathbf{1}_0)^T[(\mathbf{I} - \varrho\mathbf{C}_{00}) - \varrho^2\mathbf{C}_{0m} \\
& \times (\mathbf{I} - \varrho\mathbf{C}_{mm})^{1}\mathbf{C}_{m0}] (\mathbf{X}_0 - \mu\mathbf{1}_0)\}/(2\sigma^2). \qquad (6.11)
\end{aligned}$$

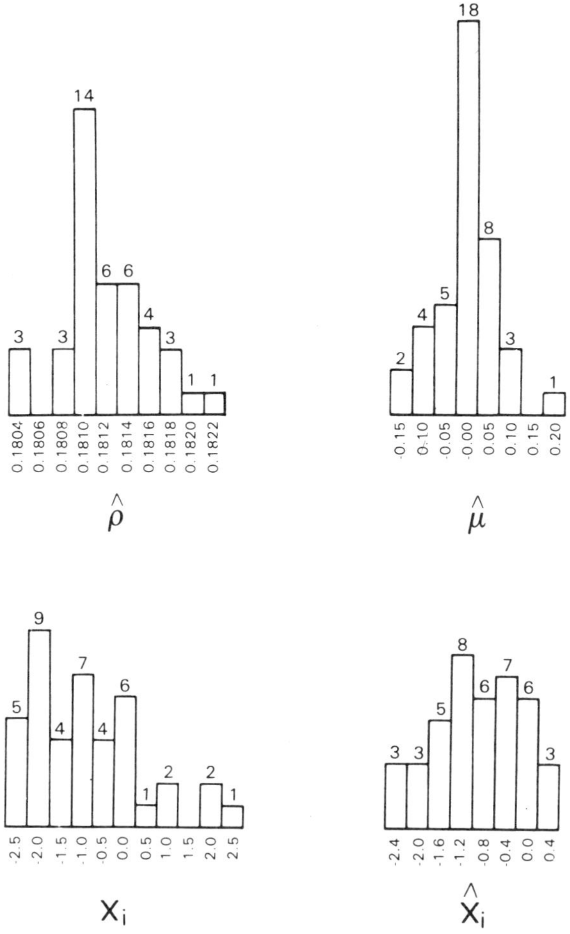

Fig. 6.2. Jackknife sampling distribution for $(n - 1)$ observations based upon the weather stations data.

If $n_m = 1$, then equation (6.11) reduces to

$$\begin{aligned}\ln (L_{\text{inc}}) = & - (n - 1) \ln (2\pi)/2 - (n - 1) \ln (\sigma^2)/2 \\ & - \ln \left[\prod_{i=1}^{n} (1 - \varrho\lambda_i) \right] /2 \\ & - \{\mathbf{X}_0 - \mu\mathbf{1}_0)^T[(\mathbf{I} - \varrho\mathbf{C}_{00}) \\ & - \varrho^2\mathbf{C}_{0m}\mathbf{C}_{m0}]\,(\mathbf{X}_0 - \mu\mathbf{1}_0)\}/(2\sigma^2). \qquad (6.12)\end{aligned}$$

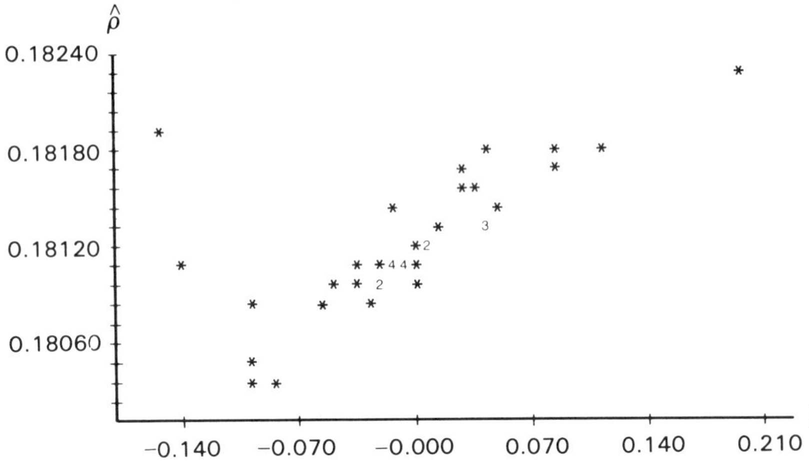

Fig. 6.3. The relationship between $\hat{\varrho}$ and $\bar{x}$ for the rainfall data.

The asymptotic covariance entry into matrix $\mathbf{V}^{-1}$ discussed in Chapter 4 for $\hat{\varrho}$ and $\hat{\mu}$ can be obtained from equation (6.12), following the mathematical and statistical arguments presented in this previous chapter, as

$$\begin{aligned}
\partial \ln (L_{\text{inc}})/\partial\mu &= -2\mathbf{1}_0^T[(\mathbf{I} - \varrho\mathbf{C}_{00}) - \varrho^2\mathbf{C}_{0m}\mathbf{C}_{m0}] \\
&\quad \times (\mathbf{X}_0 - \mu\mathbf{1}_0)/(2\sigma^2) \\
\partial^2 \ln (L_{\text{inc}})/\partial\mu\partial\varrho &= -2\mathbf{1}_0^T(-\mathbf{C}_{00} - 2\varrho\mathbf{C}_{0m}\mathbf{C}_{m0}) \\
&\quad \times (\mathbf{X}_0 - \mu\mathbf{1}_0)/(2\sigma^2) \\
- E[\partial^2 \ln (L_{\text{inc}})/\partial\mu\partial\varrho] &= 0.
\end{aligned}$$

Thus the strong relationship observed in this empirical example is attributable to the relatively small sample size of $n = 41$, and the use of the jackknife procedure.

Another somewhat obvious strong relationship exhibited by the data appearing in Table 6.3 is the one between x_i and $\hat{x}_i$. Here the correlation coefficient is 0.659. As Figure 6.3 confirms, the regression of vector $\mathbf{X}$ on vector $\hat{\mathbf{X}}$ produces an intercept that is not significantly different from zero, and a regression coefficient

TABLE 6.4

Statistics Calculated from 400 Simulated Surfaces for Dispersed Missing Values

ρ	n_m	$\bar{x}_{\hat{\rho}_m}$	$s_{\hat{\rho}_m}$	MSE($\hat{\rho}_m$)	$\bar{x}_{\hat{\rho}_m}$	$s_{\hat{\mu}_m}$	MSE($\hat{\mu}_m$)	MSE($\hat{x}_{31}$)
$\lvert\rho_m\rvert < 1/3.80423$								
0.075	0	.04991	.07291		-.00143	.08628		
	1	.04921	.07346	.00605	.00653	.13032	.01698	.9933
	4	.04676	.07713	.00673	.00733	.13109	.01720	1.0012
	9	.04383	.08235	.00774	.00641	.13492	.01820	1.0254
	16	.03931	.08706	.00884	.00943	.14179	.02014	1.0277
0.150	0	.11990	.06599		.00070	.28290		
	1	.11898	.06651	.00538	.00939	.15965	.02551	.8864
	4	.11607	.07093	.00617	.01092	.16045	.02580	.8949
	9	.11118	.07681	.00739	.01020	.16337	.02673	.9155
	16	.10561	.08146	.00859	.01256	.16947	.02881	.9145
0.225	0	.20125	.04569		-.00150	1.20140		
	1	.20069	.04603	.00270	.01520	.25010	.06263	.6506
	4	.19873	.04964	.00315	.01740	.25060	.06297	.6559
	9	.19472	.05364	.00379	.01790	.25170	.06350	.6622
	16	.19100	.05676	.00437	.01900	.25310	.06424	.6621
$\lvert\rho_m\rvert < 0.25$								
0.225	16	.19057	.05628	.00435	.01520	.24330	.05930	.6619

n_m	1	4	9	16
$r_{\hat{\rho}_m,\hat{\mu}_m}(0.075)$	.033	.040	.032	.025
$r_{\hat{\rho}_m,\hat{\mu}_m}(0.150)$	.065	.073	.068	.045
$r_{\hat{\rho}_m,\hat{\mu}_m}(0.225)$	.069	.075	.071	.049

that is not significantly different from unity:

$$x_i = \underset{(t\,=\,0.38)}{0.0904} + \underset{(t\,=\,0.52)}{1.1052}\,\hat{x}_m, \quad R^2 = 0.434.$$

Unfortunately, although the regression parameter estimates are appealing values, the R^2 value and the scatter of points (see Figure 6.3) are disappointing.

A set of simulation experiments also have been conducted in order to explore properties of equation (6.7). These experiments consisted of 400 geographical surfaces, each being a sample from a multivariate normal distribution with $\boldsymbol{\mu} = \mathbf{0}$ and $\boldsymbol{\Sigma} = \mathbf{I}$, and each having a 9-by-9 lattice configuration of areal units. Procedures for setting up these sorts of simulation experiments will be discussed in Chapter 9. Summary statistics for these experiments appear in Table 6.4. Two conspicuous trends can be identified with these results. First, as $|\varrho| \to 0.25$, the missing value estimates improve. Second, as n_m increases, the missing value estimates deteriorate. The single inconsistency here occurs with $n_m = 16$. Presumably the problem here is that $\lambda_{\max} = 1/3.80423$ rather than 0.25. Of the 400 simulated surfaces analyzed, 126 had a value of $\hat{\varrho}$ exceeding 0.225 whereas 29 had a value of $\hat{\varrho}$ exceeding 0.25. By imposing the more binding constraint of 0.25, the results for $n_m = 16$ become further suppressed (see Table 6.4).

Several additional noteworthy tendencies are displayed in Table 6.4. As was mentioned earlier, the estimate of ϱ is biased. This bias is attributable to the employment of nonlinear estimation techniques, and is transferred to the estimate of ϱ in the presence of incomplete data. Further, the standard errors and mean squared errors appear to increase as n_m increases, while the bias present seems to be negligible for the GLS estimate of μ. Finally, as was shown with the asymptotic covariance of $\hat{\varrho}_m$ and $\hat{\mu}_m$, the correlation of these two estimates essentially is zero.

6.6. AN ALGORITHM FOR THE CONDITIONAL SPATIAL CASE

An iterative algorithm has been developed for the case of a conditional model having a constant spatial mean. This algorithm is the one that has been used for those numerical calibrations described in the preceding section. The steps of this algorithm are as follows:

Step 1: calculate initial estimates of $\mathbf{X}_m$ and μ assuming $\varrho = 0$,
Step 2: maximize equation (6.5) for ϱ,
Step 3: sequentially calculate new estimates of $\mathbf{X}_m$ and μ based upon the estimate of ϱ obtained in the previous step,
Step 4: maximize the likelihood function

$$\left[\prod_{k=1}^{n_m}(1-\varrho\eta_k)\Big/\prod_{i=1}^{n}(1-\varrho\lambda_i)\right]^{1/(n-n_m)}(\mathbf{X}-\mu\mathbf{1})^T(\mathbf{I}-\varrho\mathbf{C})(\mathbf{X}-\mu\mathbf{1}),$$

Step 5: repeat Steps 3–4 until $\hat{\varrho}$ converges.

Theoretically convergence can be accelerated by directly substituting the expression for $\hat{\mu}$ in the presence of incomplete data, namely

$$\mathbf{1}_0^T[(\mathbf{I} - \varrho\mathbf{C}_{00}) - \varrho^2\mathbf{C}_{0m}(\mathbf{I} - \varrho\mathbf{C}_{mm})^{-1}\mathbf{C}_{m0}]\mathbf{X}_0$$
$$\times \{\mathbf{1}_0^T[(\mathbf{I} - \varrho\mathbf{C}_{00}) - \varrho^2\mathbf{C}_{0m}(\mathbf{I} - \varrho\mathbf{C}_{mm})^{-1}\mathbf{C}_{m0}]\mathbf{1}_0\}^{-1}$$

directly into equation (6.11). This form of the likelihood function is completely void of any explicit references to missing data, but is a considerably complex function in one unknown. The consolidated likelihood function that results from this substitution also can be optimized with respect to ϱ.

The algorithm presented here, whose FORTRAN computer code is given in Appendix 6A, was designed to use as little CPU core as possible. Accordingly, the eigenvalues of matrix **C** must be calculated outside this algorithm, and are an input to it. Fúrther, the matrix **C** is input in a condensed form, where only the non-zero row entries are noted. This input takes the form of one card image for each areal unit, and on this card image is listed the number of neighboring areal units, followed by the sequence of areal unit neighbor numbers. These numbers refer to the location of these areal units in the input data file. In addition, COMMON blocks are used extensively. Not only do these blocks save core, but they also permit information to be transferred into the external functions.

6.6.1. *COMMON Block Arguments*

All arguments are single precision with type in accordance with the usual FORTRAN naming convention.

N	the total number of areal units
NMISS	the number of areal units having missing data values
IOBSN	the number of areal units having observed data values
EIGEN(100)	the vector of eigenvalues for matrix **C**
DNUM(100)	a floating point version of the number of neighboring areal units for each areal unit
EMAX	the maximum eigenvalue in EIGEN(100)
SUMNGH	$\mathbf{1}^T\mathbf{C}\mathbf{1}$

SUMMNGH	$\mathbf{1}_0^T \mathbf{C}_{0m} \mathbf{1}_m$
SUMMNG2	$\mathbf{1}_0^T \mathbf{C}_{0m} \mathbf{C}_{m0} \mathbf{1}_0$
XLAG	$(\mathbf{X} - \mu\mathbf{1})^T \mathbf{C}(\mathbf{X} - \mu\mathbf{1})$
SQUARE	$(\mathbf{X} - \mu\mathbf{1})^T (\mathbf{X} - \mu\mathbf{1})$
DATA(100)	the data set, with 0.0 entered into the positions of the missing values
RINT	the starting value for ϱ in the non-linear optimization routine
NEIGH(100, 10)	the matrix of neighbor areal unit numbers
NUM(100)	the number of neighboring areal units for each areal unit
MISS(25)	the list of areal unit numbers for those areal units having missing values [these numbers designate the corresponding areal unit positions in DATA (100)]
WK(3)	work space needed by the non-linear optimization routine
SUMLOX	$\mathbf{X}_0^T \mathbf{C}_{00} \mathbf{X}_0$
SUMOX	$\mathbf{1}_0^T \mathbf{X}_0$
RSUMOX	$\mathbf{1}_0^T \mathbf{C}_{00} \mathbf{X}_0$
SUMOX2	$\mathbf{X}_0^T \mathbf{X}_0$
SMLMX	$\mathbf{1}_0^T \mathbf{C}_{0m} \mathbf{C}_{m0} \mathbf{X}_0$
SMLMX2	$\mathbf{X}_0^T \mathbf{C}_{0m} \mathbf{C}_{m0} \mathbf{X}_0$
CMISS(25, 25)	matrix $(\mathbf{I} - \varrho\mathbf{C}_{mm})$
EIGEN2(25)	the eigenvalues of matrix $\mathbf{C}_{mm}$
WORK(700)	work space needed by the matrix inversion subroutine
WMO10(25)	$\mathbf{C}_{m0} \mathbf{1}_0$
CTINV(25, 25)	the inverted matrix $(\mathbf{I} - \varrho\mathbf{C}_{mm})^{-1}$
WOOSUM	$\mathbf{1}_0^T \mathbf{C}_{00} \mathbf{1}_0$
WMOSUM	$\mathbf{1}_m^T \mathbf{C}_{m0} \mathbf{1}_0$
WMMSUM	$\mathbf{1}_m^T \mathbf{C}_{mm} \mathbf{1}_m$
WMOXO(25)	$\mathbf{C}_{m0} \mathbf{X}_0$

6.6.2. *Input*

CHOICE1 and CHOICE2 are declared as integers.

CHOICE1	= 1 if **C** is a stochastic matrix (row sums equal to unity)
	= 0 if **C** is a binary connectivity matrix

CHOICE2	= 1 if an iterative E–M solution is sought for equation (6.8) = 2 if a consolidated likelihood function solution is sought for equation (6.8) = 3 if an iterative E–M solution is sought for equation (6.7) = 4 if a consolidated likelihood function solution is sought for equation (6.7)
N	$\leqslant 100$
RINT	= 0.0 (this is a suggested value)
NMISS	$\leqslant 25$
(MISS(I),I = 1,NMISS)	the sequence of missing value areal unit numbers
(EIGEN(I),I = 1,N)	the sequence of eigenvalues for matrix **C**
NUM(I)	the number of neighboring areal units for areal unit *I*
(NEIGH(I,J),J = 1,NUM(I))	the sequence of neighboring areal unit numbers [these numbers refer to the areal unit positions in DATA(I)]
(DATA(I),I = 1,N)	the sequence of data values, with zeros entered into the missing value positions

6.6.3. *Subroutines*

ITERATE	the E–M type algorithm for equation (6.8)
FUNCTI	the likelihood function for equation (6.9) [note: a constraint is included to ensure that $\lvert\varrho\lambda_{\max}\rvert < 1$]
NONITER	the consolidated likelihood function algorithm for equation (6.8)
FUNCT2	the likelihood function for expression (6.9) when $\mathbf{C}_{mm} = 0$ [note: a constraint is included to ensure that $\lvert\varrho\lambda_{\max}\rvert < 1$]
WMM	constructs CMISS(25,25) and calculates WMMSUM and EIGEN2(25)
ITRATEC	the E–M type algorithm for equation (6.7)

FUNCT3 the likelihood function for equation (6.9) [note: a constraint is included to ensure that $|\varrho\lambda_{max}| < 1$]

NONITRC the consolidated likelihood function algorithm for equation (6.7)

FUNCT4 the likelihood function for expression (6.9) [note: a constraint is included to ensure that $|\varrho\lambda_{max}| < 1$]

ESTIMT1 calculates $\mathbf{X}_m$ with equation (6.8)

ESTIMT2 calculates $\mathbf{X}_m$ with equation (6.7)

6.6.4. *Output*

Error codes for all IMSL9 subroutines are written to the normal OUTPUT medium. Parameter and missing value estimates are written to a tape of the user's choice, designated by 7.

RHO ϱ

SIGDT the number of significant digits for ϱ

MISS(I) the missing value areal unit number [i.e., its position in DATA(100)]

ICOUNT the number of E–M iterations that occurred during the estimation of ϱ (this variable is set to zero when a consolidated likelihood function algorithm is used)

EXPECT $\hat{\mathbf{X}}_m$

6.6.5. *Working Space and Library Subroutines*

Working CPU core is provided by the arrays CMISS2(25,25), TEMPEIG(25), G(1), H(1), WK(3), WORK(700), Z(1,1), and MNEIGH(25).

The algorithm presented here calls three external routines, namely EIGRS, ZXMIN and LINV2F, from the IMSL9 library. The first of these extracts only the eigenvalues of a real symmetric matrix, the second performs non-linear optimization, and the third inverts an NMISS-by-NMISS matrix. Users can substitute their own subroutines here, in accordance with what they have available. Comment cards precede each CALL statement to these three subroutines, and these comment cards explain the parameters of each IMSL9 routine.

One way of incorporating additional sample information into the estimation of missing values is to constrain the MLEs. For instance, from equation (6.1), if $n_m = 1$ then $\hat{x}_m = \bar{x}_0$, which could be viewed as a constrained MLE in terms of the first sample moment. If $n_m = 2$, then a second constraint of $s_m^2 = s_0^2$ could be added to that of $\Sigma_{k=1}^{2} x_k/2 = \bar{x}_0$. Now the MLE set is $\{\bar{x}_0 - s_0, \bar{x}_0 + s_0\}$. If $n_m = 3$, then both of these constraints combined with that of third sample moment (the one associated with skewness) would yield an MLE set. Results for this third, as well as higher moment constraints, would need to be obtained numerically by solving a system of nonlinear equations. Nevertheless, the results become increasingly sample specific, for n_m missing values require the first n_m sample moments as constraints. Furthermore, an MLE set of estimates is obtained that then must be randomly allocated to the n_m missing values positions in the data, resulting in $n_m!$ different solutions. Hence the advantages of this approach over that of sampling from a conditional distribution remain elusive.

Of more promise is the case where some total is known for the sampled data, such that $\Sigma_{i=1}^{n} x_i = T$. Let $c = T - \Sigma_{i=1}^{n_0} x_i$. Then the MLEs from equation (6.1) become $\hat{x}_m = c/n_m = \bar{x}_m$. Clearly this sort of constrained optimization is more meaningful than the preceding one, and is directly relevant to the census data problem, where census figures are suppressed for areal units, but included in regional aggregates for a given variate. Since $E(\bar{x}_m) = \mu$, then $E(\hat{x}_m) = \mu$. Similarly, $\mathrm{VAR}(\hat{x}_m) = \sigma^2/n_m$.

Extending this last result to the conditional spatial autoregressive model yields

$$\begin{aligned}\hat{\mathbf{X}}_{m-1} = {} & \mu\mathbf{1}_{m-1} + (\mathbf{I} - \varrho\mathbf{C}_{m-1,m-1} + 2\varrho\mathbf{1}_{m-1}\mathbf{C}_{1,m-1} \\ & + \mathbf{1}_{m-1}\mathbf{1}_{m-1}^{T})^{-1}[c\mathbf{1}_{m-1} + \varrho c\mathbf{C}_{m-1,1} \\ & + \varrho(\mathbf{C}_{m-1,0} - \mathbf{1}_{m-1}\mathbf{C}_{1,0})(\mathbf{X}_0 - \mu\mathbf{1}_0)], \qquad (6.13)\end{aligned}$$

to which equation $\mathbf{1}_m^T\hat{\mathbf{X}}_m = c$ must be attached. Again $E(\hat{\mathbf{X}}_m) = \mu\mathbf{1}_m$.

Sometimes rather than just a single constraint c, each of L collections of areal units can be constrained, because the c_l ($l = 1, 2, \ldots, L$) totals are available. For example, data for U.S.

counties may be studied, with totals for each of the 50 states being available. Introduction of a vector of constraints, rather than a single constraint, produces a useful counterpart to equation (6.13). Let **c** be an L-by-1 vector of constraints, and let $\mathbf{\Omega}$ be an $(n_m - L)$-by-L binary matrix of indicator variables, whose ω_{ij} entry is unity if areal unit i belongs to group j, and zero otherwise. The MLE of $\mathbf{X}_{m-L}$ is

$$\hat{\mathbf{X}}_{m-L} = \mu\mathbf{1}_{m-L} + [(\mathbf{I} - \varrho\mathbf{C}_{m-L,m-L}) + \mathbf{\Omega}(\mathbf{I} - \varrho\mathbf{C}_{LL})\mathbf{\Omega}^T + \varrho(\mathbf{\Omega}\mathbf{C}_{L,m-L} + \mathbf{C}_{m-L,L}\mathbf{\Omega}^T)]^{-1}\{\varrho(\mathbf{C}_{m-L,0} - \mathbf{\Omega}\mathbf{C}_{L,0}) \times (\mathbf{X}_0 - \mu\mathbf{1}_0) + [\mathbf{\Omega}(\mathbf{I} - \varrho\mathbf{C}_{L,L}) + \varrho\mathbf{C}_{m-L,L}]\mathbf{c}\}. \tag{6.14}$$

Equation (6.14) is a set of $(n_m - L)$ equations in n_m unknowns. A solution is obtained by attaching the L equations $\mathbf{\Omega}^T\mathbf{X}_{m-L} = \mathbf{c}$ to (6.14).

6.8. CONCLUDING COMMENTS

This chapter extends the work of Haining, Griffith and Bennett (1984), especially to the case of constrained MLEs. Frequently the results reported here are shown to be unbiased. It remains to be shown which are MVUE, especially since $\hat{\varrho}$ must be calculated using nonlinear optimization procedures. One conjecture to keep in mind is the finding by Radhakrishnan (1982) that MLEs for parameters are inadmissible when some observations are missing.

One of the results reported in Table 6.4 also helps furnish a transition into the next chapter. Here, again, boundary effects can be detected, in terms of the constraint that is placed upon the optimization that occurs. The general problem of edge effects in spatial statistical analysis will be discussed now.

6.9. REFERENCES

Abraham, B., 1981, Missing observations in time series, *Communications in Statistics* A, Vol. 10: 1643–1653.

Afifi, A., and R. Elashoff, 1966, Missing observations in multivariate statistics, I: review of the literature, *Journal of the American Statistical Association*, Vol. 61: 595–604.

Akaike, H., and M. Ishiguro, 1980, Trend estimation with missing observations, *Annals of the Institute of Statistical Mathematics*, Vol. 32: 481–488.

Anderson, A., A. Basilevsky, and D. Hum, 1983, Missing data: a review of the literature, *Handbook of Survey Research*, Vol. 4: 415–494.

Beale, E., and R. Little, 1975, Missing values in multivariate analysis, *Journal of the Royal Statistical Society* B. Vol. 37: 129–146.

Bennett, R., D. Griffith, and R. Haining, 1984, The problem of missing data on spatial surfaces, *Annals*, Association of American Geographers, Vol. 74: 138–156.

Besag, J., 1974, Spatial interaction and the statistical analysis of lattice systems, *Journal of the Royal Statistical Society* B, Vol. 36: 192–236.

Bhoj, D., 1984, On difference of means of correlated variates with incomplete data on both responses, *Journal of Statistical Computation and Simulation*, Vol. 19: 275–290.

Bhoj, D., 1984, On testing equality of variances of correlated variates with incomplete data, *Biometrika*, Vol. 71: 639–641.

Boyles, R., 1983, On the convergence of the EM algorithm, *Journal of the Royal Statistical Society* B, Vol. 45: 47–50.

Bouza, C., 1983, Estimation of a difference in finite populations with missing observations, *Biometrical Journal*, Vol. 25: 123–128.

Box, M., 1971, A parameter estimation criterion for multiresponse models applicable when some observations are missing, *Applied Statistics*, Vol. 20: 1–7.

Brooks, R., 1982, On the loss of information through censoring, *Biometrika*, Vol. 69: 137–144.

Brouwer, U., and P. Vijn, 1980, A program to estimate the correlation coefficient in incomplete datasets, *COMPSTAT*, Vol. 4, Proceedings in Computational Statistics, edited by M. Barritt and D. Wishart, Vienna: Physica, pp. 194–200.

Campbell, N., 1984, Canonical variate analysis – a general model formulation, *Australian Journal of Statistics*, Vol. 26: 86–96.

Campbell, G., 1984, Testing equality of proportions with incomplete correlated data, *Journal of Statistical Planning and Inference*, Vol. 10: 311–321.

Chan, L., and O. Dunn, 1974, A note on the asymptotic aspects of the treatment of missing values in discriminant analysis, *Journal of the American Statistical Association*, Vol. 69: 672–673.

Chapman, D., 1982, Substitution for missing units, *Proceedings of Survey Research Methods Section*, American Statistical Association, 76–84.

Cheng, S., and K. Ling, 1983, On the BLUE's of location and scale parameters based on incomplete samples, *Soochow Journal of Mathematics*, Vol. 9: 35–45.

Chow, G., and A. Lin, 1976, Best linear unbiased estimation of missing observations in an economic time series, *Journal of the American Statistical Association*, Vol. 71: 719–721.

Dagenais, M., and J. Dufour, 1984, Durbin–Watson tests with missing observations: applications and comparisons, *Proceedings*, American Statistical Association Business and Economic Statistics Section, pp. 525–530.

Dahiya, R., and R. Korwar, 1980, Maximum likelihood estimates for a bivariate normal distribution with missing data, *Annals of Statistics*, Vol. 8: 687–692.

Damsleth, E., 1980, Interpolating missing values in a time series, *Scandinavian Journal of Statistics*, Vol. 7: 33–39.
de Ligny, C. *et al.*, 1981, An application of factor analysis with missing data, *Technometrics*, Vol. 23: 91–95.
del Pino, G., 1984, Linear restrictions and two step least squares with applications, *Statistics and Probability Letters*, Vol. 2: 245–248.
Dempster, A., N. Laird, and D. Rubin, 1977, Maximum likelihood from incomplete data via the EM algorithm, *Journal of the Royal Statistical Society* B, Vol. 39: 1–38.
Donner, A., 1982, The relative effectiveness of procedures commonly used in multiple regression analysis for dealing with missing values, *American Statistician*, Vol. 36: 378–381.
Donner, A., and B. Rosner, 1982, Missing value problems in multiple linear regression with two independent variables, *Communications in Statistics* A, Vol. 11: 127–140.
Drygas, H., 1976, Gauss-Markov estimation for multivariate linear models with missing observations, *Annals of Statistics*, Vol. 4: 779–787.
Dunsmuir, W., and P. Robinson, 1981, Estimation of time series models in the presence of missing data, *Journal of the American Statistical Association*, Vol. 76: 560–568.
Englman, L., 1982, An efficient algorithm for computing covariance matrices from data with missing values, *Communications in Statistics* B, Vol. 11: 113–121.
Eubank, R., and V. LaRiccia, 1982, Location and scale parameter estimation from randomly censored data, *Communications in Statistics* A, Vol. 11: 2869–2888.
Feingold, M., 1982, Missing data in linear models with correlated errors, *Communications in Statistics* A, Vol. 11: 2831–2843.
Gleason, T., and R. Staelin, 1975, A proposal for handling missing data, *Psychometrika*, Vol. 40: 229–252.
Gokhale, D., and B. Sirtonik, 1984, On tests for correlated proportions in the presence of incomplete data, *Psychometrika*, Vol. 49: 147–152.
Greenlees, J. *et al.*, 1982, Imputation of missing values when the probability of response depends on the variable being imputed, *Journal of the American Statistical Association*, Vol. 77: 251–261.
Haining, R., D. Griffith, and R. Bennett, 1984, A statistical approach to the problem of missing spatial data using a first-order Markov model, *The Professional Geographer*, Vol. 36: 338–345.
Hamdan, M., W. Pirie, and A. Khuri, 1976, Unbiased estimation of the common mean based on incomplete bivariate normal samples, *Biometrische Zeitschrift*, Vol. 18: 245–249.
Hartley, H., and R. Hocking, 1971, An analysis of incomplete data, *Biometrics*, Vol. 27: 783–823.
Harvey, A., 1981, The Kalman filter and its applications in econometrics and time series analysis, *Methods of Operations Research*, Vol. 44: 3–18.
Harvey, A., and C. McKenzie, 1984, Missing observations in dynamic econometric models, in *Time Series Analysis of Irregularly Observed Data*, edited by E. Parzen. New York: Springer-Verlag, pp. 108–133.

Harvey, A., and R. Pierse, 1984, Estimating missing observations in economic time series, *Journal of the American Statistical Association*, Vol. 79: 125–131.

Hill, M., and W. Dixon, 1981, Missing data: search for patterns, *Proceedings of the Statistical Computing Section*, American Statistical Association, pp. 57–60.

Hinich, M., and W. Weber, 1984, A method for estimating distributed lags when observations are randomly missing, *Journal of the American Statistical Association*, Vol. 79: 368–373.

Hinkins, S., 1980, RFACTOR – a program to create Rubin's factorization when there are incomplete multivariate data, *American Statistician*, Vol. 34: 182–183.

Hocking, R., and D. Marx, 1979, Estimation with incomplete data: an improved computational method and the analysis of nested data, *Communications in Statistics – Theory and Methods* A, Vol. 8: 1155–1181.

Hocking, R., and H. Oxspring, 1971, Maximum likelihood estimation with incomplete multinomial data, *Journal of the American Statistical Association*, Vol. 66: 65–70.

Hocking, R., and W. Smith, 1968, Estimation of parameters in the multivariate normal distribution with missing observations, *Journal of the American Statistical Association*, Vol. 63: 159–173.

Hocking, R., and W. Smith, 1972, Optimum incomplete normal samples, *Technometrics*, Vol. 14: 299–307.

Hoffmann, R., and J. Anderson, 1982, The effect of missing values on estimators for very short AR(1) time series, *Proceedings of the Statistical Computing Section*, American Statistical Association, pp. 224–227.

Honohan, P., and C. McCarthy, 1982, On the use of Durbin–Watson type statistics where there are missing observations, *The Statistician*, Vol. 31: 149–152.

Hosking, J., 1981, Missing data in multivariate linear models: a comparison of several estimation techniques, *Proceedings* of the SAS Users Group International Conference, Vol. 6: 46–51.

Huseby, J., N. Schwertman, and D. Allen, 1980, Computation of the mean vector and dispersion matrix for incomplete multivariate data, *Communications in Statistics* B, Vol. 9: 301–309.

Iwase, K., and N. Seto, 1984, A construction of incomplete sufficient unbiased estimators of the normal correlation coefficient, *Journal of the Japan Statistical Society*, Vol. 14: 49–61.

John, J., and P. Prescott, 1975, Estimating missing values in experiments, *Applied Statistics*, Vol. 24: 190–192.

Jones, B., and R. Facer, 1982, CORRMAT/PROB, a program to create and test a correlation coefficient matrix from data with missing values, *Computers and Geosciences*, Vol. 8: 191–198.

Jones, R., 1980, Maximum likelihood fitting of ARMA models to time series with missing observations, *Technometrics*, Vol. 22: 389–395.

Kemp, W., D. Burnell, D. Eberson, and A. Thomson, 1983, Estimating missing daily maximum and minimum temperatures, *Journal of Climate and Applied Meteorology*, Vol. 22: 1587–1593.

Kennedy, S., and W. Tobler, 1983, Geographic interpolation, *Geographical Analysis*, Vol. 15: 151–156.

Korwar, R., and R. Dahiya, 1982, Estimation of a bivariate distribution function from incomplete observations, *Communications in Statistics* A, Vol. 11: 887–897.

Koul, H., and V. Susarla, 1980, Testing for new better than used in expectation with incomplete data, *Journal of the American Statistical Association*, Vol. 75: 952–956.

Koziol, J., 1980, Goodness-of-fit tests for randomly censored data, *Biometrika*, Vol. 67: 693–696.

Laird, N., and T. Louis, 1982, Approximate posterior distributions for incomplete data problems, *Journal of the Royal Statistical Society* B, Vol. 44: 190–200.

Limonard, C., 1978, Missing values in time series and the implications on autocorrelation analysis, *Analytica Chimica Acta*, Vol. 103: 133–140.

Lin, P., and L. Stivers, 1974, On difference of means with incomplete data, *Biometrika*, Vol. 61: 325–334.

Lin, P., 1971, Estimation procedures for differences of means with missing data, *Journal of the American Statistical Association*, Vol. 66: 634–636.

Little, R., 1976, Inference about means from incomplete multivariate data, *Biometrika*, Vol. 63: 593–604.

Little, R., and D. Rubin, 1983, Missing data in large data sets, in *Statistical Methods and the Improvement of Data Quality*, edited by T. Wright. New York: Academic Press, pp. 215–243.

Liung, G., 1982, The likelihood function for a stationary Gaussian autoregressive-moving average process with missing observations, *Biometrika*, Vol. 69: 265–268.

Marshall, R., 1980, Autocorrelation estimation of time series with randomly missing observations, *Biometrika*, Vol. 67: 567–570.

Martin, R., 1984, Exact maximum likelihood for incomplete data from a correlated Gaussian process, *Communications in Statistics* A, Vol. 13: 1275–1288.

Milhoej, A., 1984, Bias correction in the frequency domain estimation of time series models, *Biometrika*, Vol. 71: 91–99.

Miller, R., and J. Halpern, 1982, Regression with censored data, *Biometrika*, Vol. 69: 521–531.

Morgan, B., and D. Titterington, 1977, A comparison of iterative methods for obtaining maximum likelihood estimates in contingency tables with a missing diagonal, *Biometrika*, Vol. 64: 265–270.

Morrison, D., 1971, Expectations and variances of maximum likelihood estimates of the multivariate normal distribution parameters with missing data, *Journal of the American Statistical Association*, Vol. 66: 602–604.

Morrison, D., and D. Bhoj, 1973, Power of the likelihood ratio test on the mean vector of the multivariate normal distribution with missing observations, *Biometrika*, Vol. 60: 365–368.

Murry, G., 1979, The estimation of multivariate normal density functions using incomplete data, *Biometrika*, Vol. 66: 375–380.

Nelson, F., 1981, A test for misspecification in the censored normal model *Econometrica*, Vol. 49: 1317–1330.

Ord, K., 1975, Estimation methods for models of spatial interaction, *Journal of the American Statistical Association*, Vol. 70: 120–126.

Papaioannou, T., and S. Loukas, 1984, Inequalities of rank correlation with missing data, *Journal of the Royal Statistical Society* B, Vol. 46: 68–71.

Papaioannou, T., and T. Speevak, 1977, Rank correlation inequalities with missing data, *Communications in Statistics* A, Vol. 6: 67–72.

Preece, D., 1971, Iterative procedures for missing values in experiments, *Technometrics*, Vol. 13: 743–754.

Press, S., and A. Scott, 1976, Missing variables in Bayesian regression II, *Journal of the American Statistical Association*, Vol. 71: 366–369.

Radhakrishnan, R., 1982, Inadmissibility of the maximum likelihood estimator for a multivariate normal distribution when some observations are missing, *Communications in Statistics* A, Vol. 11: 941–955.

Ratkowsky, D., 1974, Maximum likelihood estimation in small incomplete samples from the bivariate normal distribution, *Applied Statistics*, Vol. 23: 180–189.

Redner, R., and H. Walker, 1984, Mixture densities, maximum likelihood and the EM algorithm, *SIAM Review*, Vol. 26: 195–202.

Richardson, S., and K. White, 1981, The power of tests for autocorrelation with missing observations, *Econometrica*, Vol. 47: 785–788.

Robinson, P., 1980, Estimation and forecasting for time series containing censored or missing observations, in *Time Series*, edited by O. Anderson, Amsterdam: North-Holland, pp. 167–182.

Rubin, D., 1974, Characterizing the estimation of parameters in incomplete-data problems, *Journal of the American Statistical Association*, Vol. 69: 467–474.

Rubin, D., 1972, A noniterative algorithm for least squares estimation of missing values in any analysis of variance design, *Applied Statistics*, Vol. 21: 136–141.

Rubin, D., 1976, Inference and missing data, *Biometrika*, Vol. 63: 581–592.

Rubin, D., and T. Szatrowski, 1982, Finding maximum likelihood estimates of patterned covariance matrices by the EM algorithm, *Biometrika*, Vol. 69: 657–660.

Ryan, T., B. Joiner, and B. Ryan, 1982, *Minitab Reference Manual*. State College, Pa.: Minitab, Inc.

Selvin, S., 1980, Maximum likelihood estimation for complete or incomplete discrete data, *Computer Programs in Biomedicine*, Vol. 11: 83–87.

Shumway, R., and D. Stoffer, 1982, An approach to time series smoothing and forecasting using the EM algorithm, *Journal of Time Series Analysis*, Vol. 3: 253–264.

Singh, R., 1977, A note on the use of incomplete multi-auxiliary information in sample surveys, *Australian Journal of Statistics*, Vol. 19: 105–107.

Smith, W., and M. Riggs, 1984, Likelihood ratio testing on partial multinormal data, *Statistics and Probability Letters*, Vol. 2: 337–343.

Sundberg, R., 1974, Maximum likelihood theory for incomplete data from an exponential family, *Scandinavian Journal of Statistics*, Vol. 1: 49–58.

Susarla, V., and J. Van Ryzin, 1976, Nonparametric Bayesian estimation of survival curves from incomplete observations, *Journal of the American Statistical Association*, Vol. 71: 897–902.

Tabony, R., 1982, The estimation of missing values in highly correlated data, *COMPSTAT*, Vol. 5, Proceedings in Computational Statistics, edited by H. Caussinus, P. Ettinger and R. Tomassone. Vienna: Physica, pp. 425–430.

Titterington, D., 1977, Analysis of incomplete multivariate binary data by the kernel method, *Biometrika*, Vol. 64: 455–460.

Titterington, D., 1984, Recursive parameter estimation using incomplete data, *Journal of the Royal Statistical Society* B, Vol. 46: 257–267.

Titterington, D., and J. Jiang, 1983, Recursive estimation procedures for missing-data problems, *Biometrika*, Vol. 70: 613–624.

Titterington, D., and G. Mill, 1983, Kernel-based density estimates from incomplete data, *Journal of the Royal Statistical Society* B, Vol. 45: 258–266.

Tobler, W., 1979, Smooth pycnophlyactic interpolation for geographical regions, *Journal of the American Statistical Association*, Vol. 74: 519–530.

Tobler, W., and S. Kennedy, 1985, Smooth multidimensional interpolation, *Geographical Analysis*, Vol. 17: 251–257.

Upton, G., 1985, Distance-weighted geographic interpolation, *Environment and Planning* A, Vol. 17: 667–671.

U.S. National Academy of Sciences, 1980, *Panel on Incomplete Data*, Washington, D.C.: NAS.

Vacek, P., and T. Ashikaga, 1980, An examination of the nearest neighbor rule for imputing missing values, *Proceedings of the Statistical Computing Section*, American Statistical Association, pp. 326–331.

van Guilder, M., and S. Azen, 1981, Conclusions regarding algorithms for handling incomplete data, *Proceedings of the Statistical Computing Section*, American Statistical Association, pp. 53–56.

Vo-Dai, T., 1980, Time series analysis with missing or aberrant data, *COMPSTAT*, Vol. 4, Proceedings in Computational Statistics, ed. by M. Barritt and D. Wishart. Vienna: Physica, pp. 594–601.

Wei, L., 1983, Tests for interchangeability with incomplete paired observations, *Journal of the American Statistical Association*, Vol. 78: 725–729.

Weier, D., and A. Basu, 1980, An investigation of Kendall's τ modified for censored data with applications, *Journal of Statistical Planning and Inference*, Vol. 4: 381–390.

Wingo, D., 1982, Unimodality of the Pareto distribution likelihood function for multicensored samples and implications for estimation, *Communications in Statistics* A, Vol. 11: 1129–1138.

Woolson, R., J. Leeper, and W. Clarke, 1978, Analysis of incomplete data from longitudinal and mixed longitudinal studies, *Journal of the Royal Statistical Society* A, Vol. 141: 242–252.

APPENDIX 6A

FORTRAN Subroutine Using the IMSL Library for Estimating Missing Spatial Data

```
      SUBROUTINE MISSING(CHOICE1,CHOICE2)
C
C     THIS PROGRAM CALCULATES ESTIMATES OF MISSING SPATIAL DATA
C     USING A FIRST-ORDER SPATIAL AUTOREGRESSIVE MODEL.
C
C     THE INPUT OF THIS PROGRAM CONSISTS OF THE PROBLEM
C     PARAMETERS, THE EIGENVALUES OF THE SPATIAL CONFIGURATION
C     MATRIX (THIS MATRIX MUST BE SYMMETRIC), A COUNT OF THE
C     NUMBER OF NEIGHBORS FOR EACH AREAL UNIT TOGETHER WITH A LIST OF
C     THE NEIGHBORING AREAL UNIT NUMBERS, AND THE KNOWN SPATIAL DATA
C
C     EXCEPT FOR CHOICE1 AND CHOICE2, ALL OF THIS INFORMATION MUST
C     BE PASSED INTO THE SUBROUTINE MISSING FROM THE MAIN PROGRAM
C     THROUGH THE FIVE COMMON BLOCKS APPEARING IN THE SUBROUTINE
C     MISSING.
C
C     PROBLEM PARAMETERS ENTERED HAVE THE FOLLOWING ORGANIZATION:
C
C         NAME        DEFINITION
C
C         CHOICE1     IF 0, THEN THE CONNECTIVITY MATRIX IS USED;
C                     IF 1, THEN THE STOCHASTIC VERSION OF THE
C                           CONNECTIVITY MATRIX IS USED.
C         CHOICE2     IF 1, THEN AN ITERATIVE SOLUTION IS
C                           CALCULATED FOR UNCLUSTERED MISSING VALUES;
C                     IF 2, THEN A NON-ITERATIVE SOLUTION IS
C                           CALCULATED FOR UNCLUSTERED MISSING VALUES;
C                     IF 3, THEN AN ITERATIVE SOLUTION IS
C                           CALCULATED FOR CLUSTERED MISSING VALUES;
C                     IF 4, THEN A NON-ITERATIVE SOLUTION IS
C                           CALCULATED FOR CLUSTERED MISSING VALUES.
C         N           THE TOTAL NUMBER OF AREAL UNITS (MAX=100)
C         RINT        A STARTING ESTIMATE OF THE SPATIAL AUTO-
C                     REGRESSIVE PARAMETER REQUIRED BY IMSL9.1
C                     (SUGGESTION: SET TO 0.0)
C         NMISS       THE NUMBER OF MISSING VALUES (MAX = 25)
C         (MISS(I),     A LIST OF THE AREAL UNIT MISSING VALUES
C           I=1,NMISS) IN THE ORDER THAT THESE AREAL UNITS APPEAR
C                       IN THE DATA FILE.
C
C     THE EIGENVALUES ARE TRANSFERRED INTO EIGEN(I); THESE MUST BE
C     READ AS REALS.
C
C     THE NEIGHBORS STRUCTURE FOR THE DATA MUST BE READ AS
C     INTEGER, AND IS ORGANIZED AS FOLLOWS:
C
C         NUM(I) SPECIFIES THE NUMBER OF AREAL UNIT NEIGHBORS IN
C              JUXTAPOSITION WITH AREAL UNIT I;
C         (NEIGH(I,J),J=1,NUM(I)) IS THE LIST OF JUXTAPOSED AREAL
C                                   UNIT #S FOR AREAL UNIT I.
```

```
C
C     THE DATA MUST BE READ AS REALS. A ZERO SHOULD BE ENTERED IN
C     ALL LOCATIONS OF THIS FILE FOR WHICH DATA ARE MISSING
C
C     THE OUTPUT CONSISTS OF THE ESTIMATED AUTOREGRESSIVE
C     PARAMETER VALUE, THE LIST OF AREAL UNIT NUMBERS FOR THOSE
C     AREAL UNITS HAVING MISSING VALUES, THE NUMBER OF ITERATIONS
C     REQUIRED FOR CONVERGENCE (THIS IS SET TO ZERO FOR THE NON-
C     ITERATIVE SOLUTIONS), AND AN ESTIMATE FOR EACH OF THE
C     MISSING VALUES (THESE RESULTS ARE SENT TO UNIT 7).
C
C     THIS PROGRAM HAS TO BE COMPILED BY A FORTRAN 77 COMPILER;
C     IT WAS DEVELOPED ON A CDC 730 UNDER NOS 2.1. ONE SHOULD NOTE
C     THAT SINGLE PRECISION ON A CDC OFTEN IS EQUIVALENT TO DOUBLE
C     PRECISION ON OTHER MACHINES. THIS LATEST VERSION WAS RUN ON A VAX.
C
C     PROGRAM DEVELOPMENT INFORMATION:
C
C         COMPLETED UNDER THE AUSPICES OF THE NATO SCIENTIFIC
C         AFFAIRS DIVISION (GRANT # 059.81), ON FEBRUARY 11, 1984,
C         BY DR. DANIEL A. GRIFFITH, DEPARTMENT OF GEOGRAPHY,
C         STATE UNIVERSITY OF NEW YORK AT BUFFALO, AMHERST, NEW
C         YORK 14260 UNITED STATES
C
      COMMON/BLOCK1/N,NMISS,IOBSN,EIGEN(100),DNUM(100),EMAX,
     CSUMNGH,SUMMNGH,SUMMNG2
      COMMON/BLOCK2/XLAG,SQUARE
      COMMON/BLOCK3/DATA(100),RINT,NEIGH(100,10),NUM(100),
     CMISS(25),WK(3)
      COMMON/BLOCK4/SUMLOX,SUMOX,RSUMOX,SUMOX2,SUMLMX,SUMLMX2
      COMMON/BLOCK5/CMISS(25,25),EIGEN2(25),WORK(700),WMO1O(25),
     CCTINV(25,25),WOOSUM,WMOSUM,WMMSUM,WMOXO(25)
      REAL CMISS2(25,25), TEMPEIG(25)
      INTEGER CHOICE1, CHOICE2
      DO 5 I=1,N
      DO 4 J=I,N
      IF(EIGEN(I).GE.EIGEN(J)) GO TO 4
      T = EIGEN(I)
      EIGEN(I) = EIGEN(J)
      EIGEN(J) = T
4     CONTINUE
5     CONTINUE
      EMAX = EIGEN(1)
      IF (CHOICE1.EQ.1) GO TO 15
      SUMNGH = 0.0
      DO 12 I=1,N
      DNUM(I) = 1.0
12    SUMNGH = SUMNGH + FLOAT(NUM(I))
      SUMMNGH = 0.0
      SUMMNG2 = 0.0
      DO 14 I=1,NMISS
      SUMMNGH = SUMMNGH + FLOAT(NUM(MISS(I)))
4     SUMMNG2 = SUMMNG2 + FLOAT(NUM(MISS(I)))**2
      GO TO 17
15    DO 16 I = 1, N
```

```
16     DNUM(I) = FLOAT(NUM(I))
       SUMNGH = FLOAT(N)
       SUMMNGH = FLOAT(NMISS)
       SUMMNG2 = FLOAT(NMISS)
17     IF(CHOICE2.GE.3) GO TO 30
       IF(CHOICE2.EQ.2) GO TO 20
       CALL ITERATE(ICOUNT,RHO,SIGDT,XBAR)
       GO TO 50
20     CALL NONITER(RHO,SIGDT,XBAR)
       ICOUNT = 0
       GO TO 50
30     CALL WMM(CMISS2,TEMPEIG)
       IF(CHOICE2.EQ.4) GO TO 40
       CALL ITRATEC(ICOUNT,RHO,SIGDT,XBAR)
       GO TO 50
40     CALL NONITRC(RHO,SIGDT,XBAR)
       ICOUNT = 0
50     WRITE(7,6000) RHO,SIGDT
       IF(CHOICE2.GE.3) GO TO 60
       CALL ESTIMT1(ICOUNT,RHO,XBAR)
       GO TO 70
60     CALL ESTIMT2(ICOUNT,RHO,XBAR)
70     RETURN
6000   FORMAT(1X,'ESTIMATED RHO =',F8.5,5X,'SIG.DIGITS =',F6.2)
       END
C
       SUBROUTINE ITERATE(ICOUNT,RITER,SIGDT,XBAR)
C
C      THIS SUBROUTINE CALCULATES ITERATIVE ESTIMATES OF UNCLUSTERED
C      MISSING SPATIAL DATA USING AN E-M TYPE ALGORITHM BASED UPON
C      THE ORCHARD-WOODBURY MISSING INFORMATION PRINCIPLE.
C
       EXTERNAL FUNCT1
       COMMON/BLOCK1/N,NMISS,IOBSN,EIGEN(100),DNUM(100),EMAX,
      CSUMNGH,SUMMNGH,SUMMNG2
       COMMON/BLOCK2/XLAG,SQUARE
       COMMON/BLOCK3/DATA(100),RINT,NEIGH(100,10),NUM(100),
      CMISS(25),WK(3)
       REAL F,G(1),H(1),RHO(1),NUMR
       M = 1
       ICOUNT = 0
       K = 0
       SUM = 0.0
       DO 10 I=1,N
       DO 9 J=1,NMISS
9      IF(I.EQ.MISS(J)) GO TO 10
       SUM = SUM + DATA(I)
       K = K + 1
10     CONTINUE
       XBAR = SUM/FLOAT(K)
       DO 20 I = 1,NMISS
20     DATA(MISS(I)) = XBAR
       TRHO = 0.0
25     XLAG = 0.0
       SQUARE = 0.0
```

```
      DO 30 I=1,N
      RNEIGH = 0.0
      DO 29 J=1,NUM(I)
29    RNEIGH = RNEIGH + (DATA(NEIGH(I,J))-XBAR)/DNUM(I)
      SQUARE = SQUARE + (DATA(I)-XBAR)**2
30    XLAG = XLAG + (DATA(I)-XBAR)*RNEIGH
      RHO(1) = RINT
C
C     EXPLANATION OF PARAMETERS FOR THE IMSL9.1 SUBROUNTINE ZXMIN:
C
C     FUNCT1 - A USER SUPPLIED SUBROUTINE THAT CALCULATES THE
C              LIKELIHOOD FUNCTION F FOR SPATIAL AUTOREGRESSIVE
C              PAREMETER RHO(1).
C     M - THE NUMBER OF PARAMETERS OF THE LIKELIHOOD FUNCTION (M = 1).
C     10 - THE NUMBER OF DIGITS OF ACCURACY REQUIRED IN THE
C          CONVERGENCE OF THE ESTIMATE OF RHO(1).
C     500 - THE MAXIMUM NUMBER OF ITERATIONS (I. E., NUMBER OF
C           CALLS TO SUBROUTINE FUNCT1) ALLOWED.
C     0 - CAUSES ZXMIN TO INITIALIZE THE HESSIAN MATRIX H TO THE
C         IDENTITY MATRIX.
C     RHO - A VECTOR OF LENGTH M CONTAINING THE SPATIAL AUTO-
C           REGRESSIVE PARAMETER VALIE.
C     H - A VECTOR OF LENGTH M(M+1)/2 CONTAINING AN ESTIMATE OF
C         THE HESSIAN MATRIX.
C     G - A VECTOR OF LENGTH M CONTAINING AN ESTIMATE OF THE
C         GRADIENT AT THE FINAL PARAMETER ESTIMATE.
C     F - A SCALAR CONTAINING THE VALUE OF THE LIKELIHOOD
C         FUNCTION AT THE FINAL PARAMETER ESTIMATE.
C     WK - A VECTOR OF LENGTH 3*M USED AS WORKING SPACE. ON
C          OUTPUT, WK(1) CONTAINS THE NORM OF THE GRADIENT, WK(2)
C          CONTAINS THE NUMBER OF FUNCTION EVALUATIONS PERFORMED,
C          AND WK(3) CONTAINS AN ESTIMATE OF THE NUMBER OF
C          SIGNIFICANT DIGITS IN THE FINAL PARAMETER ESTIMATE.
C     IER - ERROR PARAMETER. 0 IMPLIES SUCCESSFUL CONVERGENCE WAS
C           ACHIEVED, 130 IMPLIES THAT THE ITERATION WAS TERMINATED
C           DUE TO ROUNDING ERRORS BECOMING DOMINANT, AND 131 IMPLIES
C           THAT THE MAXIMUM NUMBER OF ITERATIONS WAS EXCEEDED.
C
      CALL ZXMIN(FUNCT1,M,10,500,0,RHO,H,G,F,WK,IER)
      ICOUNT = ICOUNT + 1
      WRITE(6,6000) RHO(1),ICOUNT,IER
      DO 50 J = 1,NMISS
      DATA(MISS(J)) = 0.0
      DO 49 K=1,NUM(MISS(J))
49    DATA(MISS(J)) = DATA(MISS(J)) +
     CDATA(NEIGH(MISS(J),K))/DNUM(MISS(J))
50    DATA(MISS(J)) = RHO(1)*DATA(MISS(J))
      DEN = FLOAT(N-NMISS) - RHO(1)*SUMNGH + 2.0*RHO(1)*SUMMNGH -
     C(RHO(1)**2)*SUMMNG2
      SUM = 0.0
      RSUM = 0.0
      DO 60 I=1,N
      DO 59 J=1,NUM(I)
59    RSUM = RSUM + DATA(NEIGH(I,J))/DNUM(I)
60    SUM = SUM + DATA(I)
```

```
      NUMR = SUM - RHO(1)*RSUM
      XBAR = NUMR/DEN
      DO 70 I=1,NMISS
      DATA(MISS(I)) = 0.0
      DO 69 J=1,NUM(MISS(I))
69    DATA(MISS(I)) = DATA(MISS(I)) + RHO(1)*(DATA(NEIGH(MISS(I),
     CJ)) - XBAR)/DNUM(MISS(I))
70    DATA(MISS(I)) = XBAR + DATA(MISS(I))
      DIFF = ABS(RHO(1) - TRHO)
C
C     A CONVERGENCE CRITERION OF 0.00001 WAS SELECTED IN ORDER TO
C     AVOID DOMINANCE OF ROUNDING ERRORS IN CONVERGENCE.
C
      IF(DIFF.LE.0.00001) GO TO 80
      TRHO = RHO(1)
      GO TO 25
80    RITER = RHO(1)
      SIGDT = WK(3)
      RETURN
6000  FORMAT(1x,'RHO =',F8.5,5X,'ITERATION #',I5,5X,'ERROR CODE =',
     CI5)
      END
C
      SUBROUTINE FUNCT1(M,RHO,F)
C
C     THIS IS THE EXTERNAL FUNCTION REQUIRED BY IMSL9.1, WHICH DEFINES
C     THE EXACT LIKELIHOOD FUNCTION, AFTER MARTIN, NEEDED BY ZXMIN.
C
      COMMON/BLOCK1/N,NMISS,IOBSN,EIGEN(100),DNUM(100),EMAX,
     CSUMNGH,SUMMNGH,SUMMNG2
      COMMON/BLOCK2/XLAG,SQUARE
      REAL F,RHO(1)
      TEST = ABS(RHO(1)*EMAX)
      IF(TEST.LT.1.0) GO TO 5
      RHO(1) = 0.0
5     SUM = 0.0
      DO 10 I=1,N
10    SUM = SUM + ALOG((1.0 - RHO(1)*EIGEN(I))**2)
      F = - SUM/FLOAT(2*(N - NMISS)) +
     C(ALOG((SQUARE - RHO(1)*XLAG)**2))/2.0
      RETURN
      END
C
      SUBROUTINE NONITER(RNONIT,SIGDT,XBAR)
C
C     THIS SUBROUTINE CALCULATES NON-ITERATIVE ESTIMATES OF UNCLUSTERED
C     MISSING SPATIAL DATA USING A CONSOLIDATED LIKELIHOOD FUNCTION;
C     THE ALGORITHM IS BASED UPON THE ORCHARD-WOODBURY MISSING
C     INFORMATION PRINCIPLE.
C
      EXTERNAL FUNCT2
      COMMON/BLOCK1/N,NMISS,IOBSN,EIGEN(100),DNUM(100),EMAX,
     CSUMNGH,SUMMNGH,SUMMNG2
      COMMON/BLOCK3/DATA(100),RINT,NEIGH(100,10),NUM(100),
     CMISS(25),WK(3)
```

```
      COMMON/BLOCK4/SUMLOX,SUMOX,RSUMOX,SUMOX2,SUMLMX,SUMLMX2
      REAL F,G(1),H(1),RHO(1)
      M = 1
      RHO(1) = RINT
      SUMOX = 0.0
      SUMLOX = 0.0
      RSUMOX = 0.0
      SUMOX2 = 0.0
      IOBSN = 0
      DO 10 I=1,N
      DO 7 J=1,NMISS
7     IF(I.EQ.MISS(J)) GO TO 10
      SUMOX = SUMOX +DATA(I)
      SUMOX2 = SUMOX2 + DATA(I)**2
      IOBSN = IOBSN + 1
      RNEIGH = 0.0
      DO 9 J=1,NUM(I)
      DO 8 K=1,NMISS
8     IF(NEIGH(I,J).EQ.MISS(K)) GO TO 9
      RNEIGH = RNEIGH + DATA(NEIGH(I,J))/DNUM(I)
9     CONTINUE
      RSUMOX = RSUMOX + RNEIGH
      SUMLOX = SUMLOX + DATA(I)*RNEIGH
10    CONTINUE
      SUMLMX = 0.0
      SUMLMX2 = 0.0
      DO 20 I=1,NMISS
      RNEIGH = 0.0
      COEF = 0.0
      DO 19 J=1,NUM(MISS(I))
      COEF = COEF + 1.0/DNUM(MISS(I))
19    RNEIGH = RNEIGH + DATA(NEIGH(MISS(I),J))/DNUM(MISS(I))
      SUMLMX = SUMLMX + COEF*RNEIGH
20    SUMLMX2 = SUMLMX2 + RNEIGH**2
C
C     EXPLANATION OF PARAMETERS FOR THE IMSL9.1 SUBROUTINE ZXMIN:
C
C     FUNCT1 - A USER SUPPLIED SUBROUTINE THAT CALCULATES THE
C              LIKELIHOOD FUNCTION F FOR SPATIAL AUTOREGRESSIVE
C              PAREMETER RHO(1).
C     M - THE NUMBER OF PARAMETERS OF THE LIKELIHOOD FUNCTION (M = 1).
C     10 - THE NUMBER OF DIGITS OF ACCURACY REQUIRED IN THE
C          CONVERGENCE OF THE ESTIMATE OF RHO(1).
C     500 - THE MAXIMUM NUMBER OF ITERATIONS (I. E., NUMBER OF
C           CALLS TO SUBROUTINE FUNCT1) ALLOWED.
C     0 - CAUSES ZXMIN TO INITIALIZE THE HESSIAN MATRIX H TO THE
C         IDENTITY MATRIX.
C     RHO - A VECTOR OF LENGTH M CONTAINING THE SPATIAL AUTO-
C           REGRESSIVE PARAMETER VALIE.
C     H - A VECTOR OF LENGTH M(M+1)/2 CONTAINING AN ESTIMATE OF
C         THE HESSIAN MATRIX.
C     G - A VECTOR OF LENGTH M CONTAINING AN ESTIMATE OF THE
C         GRADIENT AT THE FINAL PARAMETER ESTIMATE.
C     F - A SCALAR CONTAINING THE VALUE OF THE LIKELIHOOD
C         FUNCTION AT THE FINAL PARAMETER ESTIMATE.
```

```
C     WK - A VECTOR OF LENGTH 3*M USED AS WORKING SPACE. ON
C          OUTPUT , WK(1) CONTAINS THE NORM OF THE GRADIENT, WK(2)
C          CONTAINS THE NUMBER OF FUNCTION EVALUATIONS PERFORMED,
C          AND WK(3) CONTAINS AN ESTIMATE OF THE NUMBER OF
C          SIGNIFICANT DIGITS IN THE FINAL PARAMETER ESTIMATE.
C     IER - ERROR PARAMETER. 0 IMPLIES SUCCESSFUL CONVERGENCE WAS
C           ACHIEVED, 130 IMPLIES THAT THE ITERATION WAS TERMINATED
C           DUE TO ROUNDING ERRORS BECOMING DOMINANT, AND 131 IMPLIES
C           THAT THE MAXIMUM NUMBER OF ITERATIONS WAS EXCEEDED.
C
      CALL ZXMIN(FUNCT2,M,10,500,0,RHO,H,G,F,WK,IER)
      WRITE(6,6000) RHO(1),IER
      RNONIT = RHO(1)
      SIGDT = WK(3)
      DEN = FLOAT(N - NMISS) - RHO(1)*SUMNGH + 2.0*RHO(1)*
     CSUMMNGH - (RHO(1)**2)*SUMMNG2
      XBAR = (SUMOX - RHO(1)*RSUMOX - (RHO(1)**2)*SUMLMX)/DEN
      RETURN
6000  FORMAT(1x,'RHO =',F8.5,5X,'ERROR CODE =',I5)
      END
C
      SUBROUTINE FUNCT2(M,RHO,F)
C
C     THIS IS THE EXTERNAL FUNCTION REQUIRED BY IMSL9.1, WHICH DEFINES
C     THE EXACT LIKELIHOOD FUNCTION, AFTER MARTIN, NEEDED BY ZXMIN.
C
      COMMON/BLOCK1/N,NMISS,IOBSN,EIGEN(100),DNUM(100),EMAX,
     CSUMNGH,SUMMNGH,SUMMNG2
      COMMON/BLOCK4/SUMLOX,SUMOX,RSUMOX,SUMOX2,SUMLMX,SUMLMX2
      REAL F,RHO(1)
      TEST = ABS(RHO(1)*EMAX)
      IF(TEST.LT.1.0) GO TO 5
      RHO(1) = 0.0
5     SUM = 0.0
      DO 10 I=1,N
10    SUM = SUM + ALOG((1.0 - RHO(1)*EIGEN(I))**2)
      DEN = FLOAT(N - NMISS) - RHO(1)*SUMNGH + 2.0*RHO(1)*SUMMNGH-
     C(RHO(1)**2)*SUMMNG2
      XBAR = (SUMOX - RHO(1)*RSUMOX - (RHO(1)**2)*SUMLMX)/DEN
      F = - SUM/FLOAT(2*(N - NMISS)) +
     CALOG((SUMOX2 - 2.0*XBAR*SUMOX + 2.0*RHO(1)*XBAR*RSUMOX
     C- RHO(1)*SUMLOX + FLOAT(IOBSN)*(XBAR**2) -
     CRHO(1)*(SUMNGH-2.0*SUMMNGH)*(XBAR**2) - (RHO(1)**2)*SUMLMX2
     C+ 2.0*XBAR*(RHO(1)**2)*SUMLMX -
     C((RHO(1)*XBAR)**2)*SUMMNG2)**2)/2.0
      RETURN
      END
C
      SUBROUTINE WMM(CMISS2,TEMPEIG)
C
C     THIS SUBROUTINE CONSTRUCTS THE CONFIGURATION MATRIX FOR CLUSTERED
C     MISSING VALUES, AND THEN CALCULATES THE EIGENVALUES OF THIS MATRIX
C
      COMMON/BLOCK1/N,NMISS,IOBSN,EIGEN(100),DNUM(100),EMAX,
     CSUMNGH,SUMMNGH,SUMMNG2
```

```
      COMMON/BLOCK3/DATA(100),RINT,NEIGH(100,10),NUM(100),
     CMISS(25),WK(3)
      COMMON/BLOCK5/CMISS(25,25),EIGEN2(25),WORK(700),WMO1O(25),
     CCTINV(25,25),WOOSUM,WMOSUM,WMMSUM,WMOXO(25)
      REAL CMISS2(NMISS,NMISS),TEMPEIG(NMISS),Z(1,1)
      WMMSUM = 0.0
      DO 10 I=1,NMISS
      DO 9 J=1,NMISS
      CMISS(I,J) = 0.0
      CMISS2(I,J) = CMISS(I,J)
      DO 8 K=1,NUM(MISS(J))
      IF(NEIGH(MISS(J),K).NE.MISS(I)) GO TO 8
      CMISS(I,J) = 1.0/DNUM(MISS(I))
      CMISS2(I,J) = CMISS(I,J)
8     CONTINUE
9     WMMSUM = WMMSUM + CMISS(I,J)
10    WRITE(6,6000) (CMISS(I,J),J=1,NMISS)
C
C     EXPLANATION OF PARAMETERS FOR THE IMSL9.1 SUBROUTINE EIGRS:
C
C     CMISS2   INPUT REAL SYMMETRIC MATRIX OF ORDER NMISS,
C              WHOSE EIGENVALUES ARE TO BE COMPUTED. CMISS2 IS
C              DESTROYED DURING THESE CALCULATIONS
C     NMISS - INPUT ORDER OF MATRIX CMISS2
C     10 - COMPUTE EIGENVALUES ONLY; CMISS2 IS IN FULL STORAGE MODE
C     TEMPEIG - OUTPUT VECTOR OF LENGTH NMISS, CONTAINING THE
C               EIGENVALUES OF MATRIX CMISS2
C     Z - A DUMMY, 1-BY-1 MATRIX FOR THE EIGENVECTORS THAT ARE NOT
C         CALCULATED
C     1 - INPUT ROW DIMENSION OF MATRIX Z EXACTLY AS SPECIFIED IN
C         THE DIMENSION STATEMENT IN THE SUBROUTINE WMM
C     WORK - WORK AREA
C     IER - ERROR PARAMETER. 0 IMPLIES THAT THE EIGENVALUES WERE
C           SUCCESSFULLY CALCULATED
C
      CALL EIGRS(CMISS2,NMISS,10,TEMPEIG,Z,1,WORK,IER)
      DO 20 I=1,NMISS
20    EIGEN2(I) = TEMPEIG(I)
      WRITE(6,6001) IER
      RETURN
6000  FORMAT(20F4.2,/,20F4.2,/,10F4.2)
6001  FORMAT(1x,'EIGENFUNCTION ANALYSIS ERROR CODE =',I5)
      END
C
      SUBROUTINE ITRATEC(ICOUNT,RITER,SIGDT,XBAR)
C
C     THIS SUBROUTINE CALCULATES ITERATIVE ESTIMATES OF CLUSTERED
C     MISSING SPATIAL DATA USING AN E-M TYPE ALGORITHM BASED UPON
C     THE ORCHARD-WOODBURY MISSING INFORMATION PRINCIPLE.
C
      EXTERNAL FUNCT3
      COMMON/BLOCK1/N,NMISS,IOBSN,EIGEN(100),DNUM(100),EMAX,
     CSUMNGH,SUMMNGH,SUMMNG2
      COMMON/BLOCK2/XLAG,SQUARE
      COMMON/BLOCK3/DATA(100),RINT,NEIGH(100,10),NUM(100),
```

```
     CMISS(25),WK(3)
      COMMON/BLOCK5/CMISS(25,25),EIGEN2(25),WORK(700),WMO1O(25),
     CCTINV(25,25),WOOSUM,WMOSUM,WMMSUM,WMOXO(25)
      REAL F,G(1),H(1),RHO(1),CMISS2(25,25),MNEIGH(25)
      M = 1
      ICOUNT = 0
      K = 0
      SUM = 0.0
      DO 10 I=1,N
      DO 9 J=1,NMISS
9     IF(I.EQ.MISS(J)) GO TO 10
      SUM = SUM + DATA(I)
      K = K + 1
10    CONTINUE
      SUMOX = SUM
      XBAR = SUM/FLOAT(K)
      RSUMOX = 0.0
      DO 15 I=1,N
      DO 12 J=1,NMISS
12    IF(I.EQ.MISS(J)) GO TO 15
      DO 14 J=1,NUM(I)
      DO 13 K=1,NMISS
13    IF(NEIGH(I,J).EQ.MISS(K)) GO TO 14
      RSUMOX = RSUMOX + DATA(NEIGH(I,J))/DNUM(I)
14    CONTINUE
15    CONTINUE
      WMOSUM = 0.0
      DO 18 I=1,NMISS
      WMO1O(I) = 0.0
      RNEIGH = 0.0
      DO 17 J=1,NUM(MISS(I))
      DO 16 K=1,NMISS
16    IF(MISS(K).EQ.NEIGH(MISS(I),J)) GO TO 17
      RNEIGH = RNEIGH + DATA(NEIGH(MISS(I),J))/DNUM(MISS(I))
      WMO1O(I) = WMO1O(I) + 1.0/DNUM(MISS(I))
17    CONTINUE
      WMOXO(I) = RNEIGH
      WMOSUM = WMOSUM + WMO1O(I)
18    CONTINUE
      DO 20 I=1,NMISS
20    DATA(MISS(I)) = XBAR
      TRHO = 0.0
25    XLAG = 0.0
      SQUARE = 0.0
      DO 30 I=1,N
      RNEIGH = 0.0
      DO 29 J=1,NUM(I)
29    RNEIGH = RNEIGH + (DATA(NEIGH(I,J)) - XBAR)/DNUM(I)
      SQUARE = SQUARE + (DATA(I) - XBAR)**2
30    XLAG = XLAG + (DATA(I) - XBAR)*RNEIGH
      RHO(1) = RINT
C
C     EXPLANATION OF PARAMETERS FOR THE IMSL9.1 SUBROUTINE ZXMIN:
C
C     FUNCT1 - A USER SUPPLIED SUBROUTINE THAT CALCULATES THE
```

```
C            LIKELIHOOD FUNCTION F FOR SPATIAL AUTOREGRESSIVE
C            PAREMETER RHO(1).
C     M - THE NUMBER OF PARAMETERS OF THE LIKELIHOOD FUNCTION (M = 1).
C     10 - THE NUMBER OF DIGITS OF ACCURACY REQUIRED IN THE
C          CONVERGENCE OF THE ESTIMATE OF RHO(1).
C     500 - THE MAXIMUM NUMBER OF ITERATIONS (I. E., NUMBER OF
C           CALLS TO SUBROUTINE FUNCT1) ALLOWED.
C     0 - CAUSES ZXMIN TO INITIALIZE THE HESSIAN MATRIX H TO THE
C         IDENTITY MATRIX.
C     RHO - A VECTOR OF LENGTH M CONTAINING THE SPATIAL AUTO-
C           REGRESSIVE PARAMETER VALIE.
C     H - A VECTOR OF LENGTH M(M+1)/2 CONTAINING AN ESTIMATE OF
C         THE HESSIAN MATRIX.
C     G - A VECTOR OF LENGTH M CONTAINING AN ESTIMATE OF THE
C         GRADIENT AT THE FINAL PARAMETER ESTIMATE.
C     F - A SCALAR CONTAINING THE VALUE OF THE LIKELIHOOD
C         FUNCTION AT THE FINAL PARAMETER ESTIMATE.
C     WK - A VECTOR OF LENGTH 3*M USED AS WORKING SPACE. ON
C          OUTPUT , WK(1) CONTAINS THE NORM OF THE GRADIENT, WK(2)
C          CONTAINS THE NUMBER OF FUNCTION EVALUATIONS PERFORMED,
C          AND WK(3) CONTAINS AN ESTIMATE OF THE NUMBER OF
C          SIGNIFICANT DIGITS IN THE FINAL PARAMETER ESTIMATE.
C     IER - ERROR PARAMETER. 0 IMPLIES SUCCESSFUL CONVERGENCE WAS
C           ACHIEVED, 130 IMPLIES THAT THE ITERATION WAS TERMINATED
C           DUE TO ROUNDING ERRORS BECOMING DOMINANT, AND 131 IMPLIES
C           THAT THE MAXIMUM NUMBER OF ITERATIONS WAS EXCEEDED.
C
      CALL ZXMIN(FUNCT3,M,10,500,0,RHO,H,G,F,WK,IER)
      WRITE(6,6000) IER
      DO 40 I=1,NMISS
      DO 39 J=1,NMISS
39    CMISS2(I,J) = - RHO(1)*CMISS(I,J)
40    CMISS2(I,I) = 1.0
C
C     EXPLANATION OF PARAMETERS FOR THE IMSL9.1 SUBROUTINE LINV2F:
C
C     CMISS2 - INPUT MATRIX OF DIMENSION NMISS-BY-NMISS CONTAINING
C              THE MATRIX TO BE INVERTED
C     NMISS - ORDER OF MATRIX CMISS2
C     25 - ROW DIMENSION OF MATRICES CMISS2 AND CTINV EXACTLY AS
C          SPECIFIED IN THE COMMON BLOCK5 STATEMENT
C     CTINV - OUTPUT MATRIX OF DIMENSION NMISS-BY-NMISS CONTAINING
C             THE INVERSE OF CMISS2 (CMISS2 AND CTINV MUST OCCUPY
C             SEPARATE CORE LOCATIONS)
C     5 - THE ELEMENTS OF CTINV ARE ESTIMATED TO 5 DECIMAL DIGITS
C     WORK - WORK AREA OF DIMENSION 25**2 + 3*25
C     IER - ERROR PARAMETER. 0 IMPLIES THAT THE INVERSION WAS
C           SUCCESSFUL, 129 IMPLIES THAT CMISS2 IS ALGORITHMICALLY
C           SUNGULAR, 131 IMPLIES THAT THE MATRIX IS TOO ILL-
C           CONDITIONED FOR ITERATIVE IMPROVEMENT TO BE EFFECTIVE
C
      CALL LINV2F(CMISS2,NMISS,25,CTINV,5,WORK,IER)
      WRITE(6,6001) IER
      TERM1 = 0.0
      TERM2 = 0.0
```

```
      TERM3 = 0.0
      DO 50 I=1,NMISS
      WMMWMOX = 0.0
      WMMWMO1 = 0.0
      DO 49 J=1,NMISS
      WMMWMO1 = WMMWMO1 + CTINV(I,J)*WMO1O(J)
49    WMMWMOX = WMMWMOX + CTINV(I,J)*WMOXO(J)
      TERM1 = TERM1 + WMO1O(I)*WMMWMOX
      TERM2 = TERM2 + WMO1O(I)*WMMWMO1
50    TERM3 = TERM3 + WMOXO(I)*WMMWMOX
      XBAR = (SUMOX - RHO(1)*RSUMOX - (RHO(1)**2)*TERM1)/(FLOAT
     C(N-NMISS) - RHO(1)*SUMNGH + RHO(1)*(WMMSUM+2.0*WMOSUM) -
     C(RHO(1)**2)*TERM2)
      DO 60 I=1,NMISS
      MNEIGH(I) = 0.0
      DO 59 J=1,NUM(MISS(I))
      DO 58 K=1,NMISS
58    IF(MISS(K).EQ.NEIGH(MISS(I),J)) GO TO 59
      MNEIGH(I) = MNEIGH(I) + (DATA(NEIGH(MISS(I),J)) -
     CXBAR)/DNUM(MISS(I))
59    CONTINUE
60    CONTINUE
      DO 70 I=1,NMISS
      DATA(MISS(I)) = 0.0
      DO 69 J=1,NMISS
69    DATA(MISS(I)) = DATA(MISS(I)) + CTINV(I,J)*MNEIGH(J)
70    DATA(MISS(I)) = XBAR + RHO(1)*DATA(MISS(I))
      ICOUNT = ICOUNT + 1
      WRITE(6,6002) RHO(1),ICOUNT,IER
      DIFF = ABS(RHO(1) - TRHO)
C
C     A CONVERGENCE CRITERION OF 0.00001 WAS SELECTED IN ORDER TO
C     AVOID DOMINANCE OF ROUNDING ERRORS IN CONVERGENCE.
C
      IF(DIFF.LE.0.00001) GO TO 80
      TRHO = RHO(1)
      GO TO 25
80    RITER = RHO(1)
      SIGDT =  WK(3)
      RETURN
6000  FORMAT(1x,'NONLINEAR MINIMIZATION ERROR CODE =',I5)
6001  FORMAT(1x,'MATRIX INVERSION ERROR CODE=',I5)
6002  FORMAT(1x,'RHO =',F8.5,5X,'ITERATION #',I5,5X,'ERROR CODE =',
     CI5)
      END
C
      SUBROUTINE FUNCT3(M,RHO,F)
C
C     THIS IS THE EXTERNAL FUNCTION REQUIRED BY IMSL9.1, WHICH DEFINES
C     THE EXACT LIKELIHOOD FUNCTION, AFTER MARTIN, NEEDED BY ZXMIN.
C
      COMMON/BLOCK1/N,NMISS,IOBSN,EIGEN(100),DNUM(100),EMAX,
     CSUMNGH,SUMMNGH,SUMMNG2
      COMMON/BLOCK2/XLAG,SQUARE
      COMMON/BLOCK5/CMISS(25,25),EIGEN2(25),WORK(700),WMO1O(25),
```

```
     CCTINV(25,25),WOOSUM,WMOSUM,WMMSUM,WMOXO(25)
      REAL F, RHO(1)
      TEST = ABS(RHO(1)*EMAX)
      IF(TEST.LT.1.0) GO TO 5
      RHO(1) = 0.0
5     SUM = 0.0
      SUM2 = 0.0
      DO 10 I=1,N
10    SUM = SUM + ALOG((1.0 - RHO(1)*EIGEN(I))**2)
      DO 20 I=1,NMISS
20    SUM2 = SUM2 + ALOG((1.0 - RHO(1)*EIGEN2(I))**2)
      F = (-SUM + SUM2)/FLOAT(2*(N - NMISS))
     C+ (ALOG((SQUARE - RHO(1)*XLAG)**2))/2.0
      RETURN
      END
C
      SUBROUTINE NONITRC(RNONITC,SIGDT,XBAR)
C
C     THIS SUBROUTINE CALCULATES NON-ITERATIVE ESTIMATES OF
C     CLUSTERED MISSING SPATIAL DATA USING A CONSOLIDATED
C     LIKELIHOOD FUNCTION; THE ALGORITHM IS BASED UPON THE ORCHARD
C     -WOODBURY MISSING INFORMATION PRINCIPLE.
C
      EXTERNAL FUNCT4
      COMMON/BLOCK1/N,NMISS,IOBSN,EIGEN(100),DNUM(100),EMAX,
     CSUMNGH,SUMMNGH,SUMMNG2
      COMMON/BLOCK3/DATA(100),RINT,NEIGH(100,10),NUM(100),
     CMISS(25),WK(3)
      COMMON/BLOCK4/SUMLOX,SUMOX,RSUMOX,SUMOX2,SUMLMX,SUMLMX2
      COMMON/BLOCK5/CMISS(25,25),EIGEN2(25),WORK(700),WMO1O(25),
     CCTINV(25,25),WOOSUM,WMOSUM,WMMSUM,WMOXO(25)
      REAL F,G(1),H(1),RHO(1)
      M = 1
      RHO(1) = RINT
      SUMOX = 0.0
      SUMLOX = 0.0
      RSUMOX = 0.0
      SUMOX2 = 0.0
      IOBSN = 0
      WOOSUM = 0.0
      DO 10 I=1,N
      DO 7 J=1,NMISS
7     IF(I.EQ.MISS(J)) GO TO 10
      SUMOX = SUMOX + DATA(I)
      SUMOX2 = SUMOX2 + DATA(I)**2
      IOBSN = IOBSN + 1
      RNEIGH = 0.0
      DO 9 J=1,NUM(I)
      DO 8 K=1,NMISS
8     IF(NEIGH(I,J).EQ.MISS(K)) GO TO 9
      RNEIGH = RNEIGH + DATA(NEIGH(I,J))/DNUM(I)
      WOOSUM = WOOSUM + 1.0/DNUM(I)
9     CONTINUE
      RSUMOX = RSUMOX + RNEIGH
      SUMLOX = SUMLOX + DATA(I)*RNEIGH
```

```
10    CONTINUE
      WMOSUM = 0.0
      SUMLMX = 0.0
      SUMLMX2 = 0.0
      DO 25 I=1,NMISS
      WMO1O(I) = 0.0
      RNEIGH = 0.0
      DO 24 J=1,NUM(MISS(I))
      DO 23 K=1,NMISS
23    IF(MISS(K).EQ.NEIGH(MISS(I),J)) GO TO 24
      RNEIGH = RNEIGH + DATA(NEIGH(MISS(I),J))/DNUM(MISS(I))
      WMO1O(I) = WMO1O(I) + 1.0/DNUM(MISS(I))
24    CONTINUE
      WMOXO(I) = RNEIGH
      WMOSUM = WMOSUM + WMO1O(I)
      SUMLMX = SUMLMX + RNEIGH
25    SUMLMX2 = SUMLMX2 + RNEIGH**2
C
C     EXPLANATION OF PARAMETERS FOR THE IMSL9.1 SUBROUNTINE ZXMIN:
C
C     FUNCT1 - A USER SUPPLIED SUBROUTINE THAT CALCULATES THE
C              LIKELIHOOD FUNCTION F FOR SPATIAL AUTOREGRESSIVE
C              PAREMETER RHO(1).
C     M - THE NUMBER OF PARAMETERS OF THE LIKELIHOOD FUNCTION (M = 1).
C     10 - THE NUMBER OF DIGITS OF ACCURACY REQUIRED IN THE
C          CONVERGENCE OF THE ESTIMATE OF RHO(1).
C     500 - THE MAXIMUM NUMBER OF ITERATIONS (I. E., NUMBER OF
C           CALLS TO SUBROUTINE FUNCT1) ALLOWED.
C     0 - CAUSES ZXMIN TO INITIALIZE THE HESSIAN MATRIX H TO THE
C         IDENTITY MATRIX.
C     RHO - A VECTOR OF LENGTH M CONTAINING THE SPATIAL AUTO-
C           REGRESSIVE PARAMETER VALIE.
C     H - A VECTOR OF LENGTH M(M+1)/2 CONTAINING AN ESTIMATE OF
C         THE HESSIAN MATRIX.
C     G - A VECTOR OF LENGTH M CONTAINING AN ESTIMATE OF THE
C         GRADIENT AT THE FINAL PARAMETER ESTIMATE.
C     F - A SCALAR CONTAINING THE VALUE OF THE LIKELIHOOD
C         FUNCTION AT THE FINAL PARAMETER ESTIMATE.
C     WK - A VECTOR OF LENGTH 3*M USED AS WORKING SPACE. ON
C          OUTPUT , WK(1) CONTAINS THE NORM OF THE GRADIENT, WK(2)
C          CONTAINS THE NUMBER OF FUNCTION EVALUATIONS PERFORMED,
C          AND WK(3) CONTAINS AN ESTIMATE OF THE NUMBER OF
C          SIGNIFICANT DIGITS IN THE FINAL PARAMETER ESTIMATE.
C     IER - ERROR PARAMETER. 0 IMPLIES SUCCESSFUL CONVERGENCE WAS
C           ACHIEVED, 130 IMPLIES THAT THE ITERATION WAS TERMINATED
C           DUE TO ROUNDING ERRORS BECOMING DOMINANT, AND 131 IMPLIES
C           THAT THE MAXIMUM NUMBER OF ITERATIONS WAS EXCEEDED.
C
      CALL ZXMIN(FUNCT4,M,10,500,0,RHO,H,G,F,WK,IER)
      WRITE(6,6000) RHO(1),IER
      RNONITC = RHO(1)
      SIGDT = WK(3)
      TERM1 = 0.0
      TERM2 = 0.0
      TERM3 = 0.0
```

```
      DO 20 I=1,NMISS
      WMMWMOX = 0.0
      WMMWMO1 = 0.0
      DO 19 J=1,NMISS
      WMMWMO1 = WMMWMO1 + CTINV(I,J)*WMO1O(J)
19    WMMWMOX = WMMWMOX + CTINV(I,J)*WMOXO(J)
      TERM1 = TERM1 + WMO1O(I)*WMMWMOX
      TERM2 = TERM2 + WMO1O(I)*WMMWMO1
20    TERM3 = TERM3 + WMOXO(I)*WMMWMOX
      XBAR = (SUMOX - RHO(1)*RSUMOX - (RHO(1)**2)*TERM1)/
     C(FLOAT(N-NMISS) - RHO(1)*SUMNGH + RHO(1)*(WMMSUM+2.0*WMOSUM)
     C- (RHO(1)**2)*TERM2)
      RETURN
6000  FORMAT(1x,'RHO =',F8.5,5X,'ERROR CODE =',I5)
      END
C
      SUBROUTINE FUNCT4(M,RHO,F)
C
C     THIS IS THE EXTERNAL FUNCTION REQUIRED BY IMSL9.1, WHICH DEFINES
C     THE EXACT LIKELIHOOD FUNCTION, AFTER MARTIN, NEEDED BY ZXMIN.
C
      COMMON/BLOCK1/N,NMISS,IOBSN,EIGEN(100),DNUM(100),EMAX,
     CSUMNGH,SUMMNGH,SUMMNG2
      COMMON/BLOCK4/SUMLOX,SUMOX,RSUMOX,SUMOX2,SUMLMX,SUMLMX2
      COMMON/BLOCK5/CMISS(25,25),EIGEN2(25),WORK(700),WMO1O(25),
     CCTINV(25,25),WOOSUM,WMOSUM,WMMSUM,WMOXO(25)
      REAL F,RHO(1),CMISS2(25,25)
      TEST = ABS(RHO(1)*EMAX)
      IF(TEST.LT.1.0) GO TO 5
      RHO(1) = 0.0
5     SUM = 0.0
      DO 10 I=1,N
10    SUM = SUM + ALOG((1.0 - RHO(1)*EIGEN(I))**2)
      DO 15 I=1,NMISS
      DO 14 J=1,NMISS
14    CMISS2(I,J) = - RHO(1)*CMISS(I,J)
15    CMISS2(I,I) = 1.0
C
C     EXPLANATION OF PARAMETERS FOR THE IMSL9.1 SUBROUTINE LINV2F:
C
C     CMISS2 - INPUT MATRIX OF DIMENSION NMISS-BY-NMISS CONTAINING
C              THE MATRIX TO BE INVERTED
C     NMISS - ORDER OF MATRIX CMISS2
C     25 - ROW DIMENSION OF MATRICES CMISS2 AND CTINV EXACTLY AS
C          SPECIFIED IN THE COMMON BLOCK5 STATEMENT
C     CTINV - OUTPUT MATRIX OF DIMENSION NMISS-BY-NMISS CONTAINING
C            THE INVERSE OF CMISS2 (CMISS2 AND CTINV MUST OCCUPY
C            SEPARATE CORE LOCATIONS)
C     5 - THE ELEMENTS OF CTINV ARE ESTIMATED TO 5 DECIMAL DIGITS
C     WORK - WORK AREA OF DIMENSION 25**2 + 3*25
C     IER - ERROR PARAMETER. 0 IMPLIES THAT THE INVERSION WAS
C           SUCCESSFUL, 129 IMPLIES THAT CMISS2 IS ALGORITHMICALLY
C           SUNGULAR, 131 IMPLIES THAT THE MATRIX IS TOO ILL-
C           CONDITIONED FOR ITERATIVE IMPROVEMENT TO BE EFFECTIVE
C
```

```
      CALL LINV2F(CMISS2,NMISS,25,CTINV,5,WORK,IER)
      WRITE(6,6000) IER
      SUM2 = 0.0
      DO 30 I=1,NMISS
30    SUM2 = SUM2 + ALOG((1.0 - RHO(1)*EIGEN2(I))*2)
      TERM1 = 0.0
      TERM2 = 0.0
      TERM3 = 0.0
      DO 20 I=1,NMISS
      WMMWMOX = 0.0
      WMMWMO1 = 0.0
      DO 19 J=1,NMISS
      WMMWMO1 = WMMWMO1 + CTINV(I,J)*WMO1O(J)
19    WMMWMOX = WMMWMOX + CTINV(I,J)*WMOXO(J)
      TERM1 = TERM1 + WMO1O(I)*WMMWMOX
      TERM2 = TERM2 + WMO1O(I)*WMMWMO1
20    TERM3 = TERM3 + WMOXO(I)*WMMWMOX
      XBAR = (SUMOX - RHO(1)*RSUMOX - (RHO(1)**2)*TERM1)/
     C(FLOAT(N-NMISS) - RHO(1)*SUMNGH + RHO(1)*(WMMSUM + 2.0*
     CWMOSUM) - (RHO(1)**2)*TERM2)
      F = (-SUM + SUM2)/FLOAT(2*(N - NMISS)) +
     CALOG((SUMOX2 - RHO(1)*SUMLOX - 2.0*XBAR*SUMOX +
     C2.0*XBAR*RHO(1)*RSUMOX + FLOAT(IOBSN)*(XBAR**2) -
     CRHO(1)*(XBAR**2)*WOOSUM - (RHO(1)**2)*TERM3 +
     C2.0*XBAR*(RHO(1)**2)*TERM1 -
     C((RHO(1)*XBAR)**2)*TERM2)**2)/2.0
      RETURN
6000  FORMAT(1x,'MATRIX INVERSION ERROR CODE =',I5)
      END
C
      SUBROUTINE ESTIMT1(ICOUNT,RHO,XBAR)
C
C     THIS SUBROUTINE CALCULATES THE MISSING VALUE ESTIMATES FOR
C     UNCLUSTERED MISSING SPATIAL DATA.
C
      COMMON/BLOCK1/N,NMISS,IOBSN,EIGEN(100),DNUM(100),EMAX,
     CSUMNGH,SUMMNGH,SUMMNG2
      COMMON/BLOCK3/DATA(100),RINT,NEIGH(100,10),NUM(100),
     CMISS(25),WK(3)
      DO 30 I=1,NMISS
      EXPECT = 0.0
      DO 29 J=1,NUM(MISS(I))
29    EXPECT = EXPECT + RHO*(DATA(NEIGH(MISS(I),J)) -
     CXBAR)/DNUM(MISS(I))
      EXPECT = XBAR + EXPECT
30    WRITE(7,6000) MISS(I),ICOUNT,EXPECT,XBAR
      RETURN
6000  FORMAT(1X,'OBSERVATION #',I5,2X,
     C'NUMBER OF ITERATIONS =',I5,2X,'MISSING VALUE ESTIMATE =',
     CE14.6,2X,'XBAR =',E14.6)
      END
C
      SUBROUTINE ESTIMT2(ICOUNT,RHO,XBAR)
C
C     THIS SUBROUTINE CALCULATES THE MISSING VALUE ESTIMATES FOR
```

```
C     CLUSTERED MISSING SPATIAL DATA.
C
      COMMON/BLOCK1/N,NMISS,IOBSN,EIGEN(100),DNUM(100),EMAX,
     CSUMNGH,SUMMNGH,SUMMNG2
      COMMON/BLOCK3/DATA(100),RINT,NEIGH(100,10),NUM(100),
     CMISS(25),WK(3)
      COMMON/BLOCK5/CMISS(25,25),EIGEN2(25),WORK(700),WMO1O(25),
     CCTINV(25,25),WOOSUM,WMOSUM,WMMSUM,WMOXO(25)
      REAL CMISS2(25,25),MNEIGH(25)
      DO 10 I=1,NMISS
      DO 9 J=1,NMISS
9     CMISS2(I,J) = - RHO*CMISS(I,J)
10    CMISS2(I,I) = 1.0
C
C     EXPLANATION OF PARAMETERS FOR THE IMSL9.1 SUBROUTINE LINV2F:
C
C     CMISS2 - INPUT MATRIX OF DIMENSION NMISS-BY-NMISS CONTAINING
C              THE MATRIX TO BE INVERTED
C     NMISS - ORDER OF MATRIX CMISS2
C     25 - ROW DIMENSION OF MATRICES CMISS2 AND CTINV EXACTLY AS
C          SPECIFIED IN THE COMMON BLOCK5 STATEMENT
C     CTINV - OUTPUT MATRIX OF DIMENSION NMISS-BY-NMISS CONTAINING
C             THE INVERSE OF CMISS2 (CMISS2 AND CTINV MUST OCCUPY
C             SEPARATE CORE LOCATIONS)
C     5 - THE ELEMENTS OF CTINV ARE ESTIMATED TO 5 DECIMAL DIGITS
C     WORK - WORK AREA OF DIMENSION 25**2 + 3*25
C     IER - ERROR PARAMETER. 0 IMPLIES THAT THE INVERSION WAS
C           SUCCESSFUL, 129 IMPLIES THAT CMISS2 IS ALGORITHMICALLY
C           SUNGULAR, 131 IMPLIES THAT THE MATRIX IS TOO ILL-
C           CONDITIONED FOR ITERATIVE IMPROVEMENT TO BE EFFECTIVE
C
      CALL LINV2F(CMISS2,NMISS,25,CTINV,5,WORK,IER)
      WRITE(6,6000) IER
      DO 20 I=1,NMISS
      MNEIGH(I) = 0.0
      DO 19 J=1,NUM(MISS(I))
      DO 18 K=1,NMISS
18    IF(MISS(K).EQ.NEIGH(MISS(I),J)) GO TO 19
      MNEIGH(I) = MNEIGH(I) + (DATA(NEIGH(MISS(I),J)) -
     CXBAR)/DNUM(MISS(I))
19    CONTINUE
20    CONTINUE
      DO 30 I=1,NMISS
      EXPECT = 0.0
      DO 29 J=1,NMISS
29    EXPECT = EXPECT + CTINV(I,J)*MNEIGH(J)
      EXPECT = XBAR + RHO*EXPECT
30    WRITE(7,6001) MISS(I),ICOUNT,EXPECT,XBAR
      RETURN
6000  FORMAT(1x,'MATRIX INVERSION ERROR CODE =',I5)
6001  FORMAT(1X,'OBSERVATION #',I5,2X,
     C'NUMBER OF ITERATIONS =',I5,2X,'MISSING VALUE ESTIMATE =',
     CE14.6,2X,'XBAR =',E14.6)
      END
*eot
```

CHAPTER 7

Correcting for Edge Effects in Spatial Statistical Analyses

Motivation for the research theme discussed in this chapter comes from the myriad of problems arising from dividing geographical landscapes into a finite number of areal units. Some of the more prominent of these problems already have been mentioned in earlier chapters. For example, in Chapter 2 the finite nature of a two-dimensional surface was shown to influence the answer to the question of whether or not data transformations should be employed. In Chapter 3 an empirical relationship was uncovered between the shape of a region and the level of spatial autocorrelation that was measured for its internal geographical distributions. In Chapter 5 the theoretical autocorrelation values for a conditional model, calculated with equation (5.16), were found to differ from those given by the spatial linear operator $(\mathbf{I} - \varrho\mathbf{C})^{-1}$. These differences were found to be most pronounced along the edges. And, in Chapter 6 edge effects were found to be linked to the estimation of values for ϱ that are close to the border of the parameter space. These are but some of the numerous complications introduced into a spatial analysis by edge effects.

Another edge effects issue has to do with inferential errors. Ways of overcoming inferential difficulties in the presence of non-zero spatial autocorrelation require estimating the nature and degree of that autocorrelation latent in a data set. In general, the best estimation procedure to employ with spatial data is that of maximum likelihood, especially if a conditional autoregressive model is posited. But if the method utilized to handle spatial autocorrelation fails to accommodate it fully, then inferential problems such as biased standard errors still persist, although probably at a less severe level.

7.1. PROBLEM STATEMENT

Consider an infinite planar surface that has been partitioned into mutually exclusive and collectively exhaustive cells (areal units) forming a regular, square tessellation over the surface. The sides of these squares constitute vertical and horizontal lines through the integral divisions of two real axes that form a grid, which determines the infinitude of points sometimes called a lattice. Let a set of independent and identically distributed (iid) random variables be distributed over this surface, where $\xi_{i,j}$ is the random variate for cell (i, j) of this lattice, $E(\xi_{i,j}) = 0$ and $\mathrm{VAR}(\xi_{i,j}) = \sigma_\xi^2$. In addition, let an autocorrelation mechanism operate over this two-dimensional surface in such a way that the first-order conditional autoregressive model describes it. Moreover, there exists a set of correlated variables $X_{i,j}$ based upon the $\xi_{i,j}$ such that

$$E[X_{i,j} | x_{i+k,j+l};\ \text{all } (k, l) \neq (0, 0)]$$
$$= \varrho(x_{i-1,j} + x_{i+1,j} + x_{i,j-1} + x_{i,j+1}),$$

with modifications at the edges of the study area where necessary, and with $|\varrho| < 1/4$ to insure stationarity, where ϱ is the autoregressive parameter, and

$$\mathrm{VAR}[X_{i,j} | x_{i+k,j+l};\ \text{all } (k, l) \neq (0, 0)] = \sigma^2.$$

[Cliff and Ord, 1981, p. 146]. Accordingly the variance–covariance matrix for these X's is $\sigma^2(\mathbf{I} - \varrho\mathbf{C})^{-1}$, where $\mathbf{C}$ is a binary matrix whose values in column positions $(i \pm 1, j)$ and $(i, j \pm 1)$ are unity for cell position (i, j), and zero elsewhere.

Now consider any finite set of coterminous cells comprising a sub-region, say Γ, of this lattice surface. Denoting the set of lattice cells in the complement of Γ as Γ^c, interdependencies between $x_{i,j} \in \Gamma$ and $x_{k,l} \in \Gamma^c$ are ignored in an analysis that is restricted solely to Γ. The importance of edge effects in this situation arises from the mode in which a spatial model is formulated or specified for this truncated spatial data series. There are two distinct, general viewpoints that may be utilized when specifying the spatial model. First, one could posit that some random vector **X** that is geographically distributed over the subregion Γ is statistically distributed MVN[$\boldsymbol{\mu}$, $\sigma^2(\mathbf{I} - \varrho\mathbf{C})^{-1}$]. This particular specification ignores edge effects altogether; moreover, these effects are being assumed away with this viewpoint. This specification may be considered satisfactory for regions that are well delineated. Usually, though, this is not the situation faced by a spatial analyst.

A second approach to model specification posits that the vector **X** is the result of a stationary spatial process, resulting in

$$E(X_{i,j}) = \mu,$$

$$\mathrm{VAR}(X_{i,j}) = \sigma^2,$$

and

$$\mathrm{COV}(X_{i,j}, X_{i',j'}) = \gamma(|i - i'|, |j - j'|) = w_{rs},$$

where w_{rs} is a function of the distance separating lattice cells (i, j) and (i', j'). Here the problem of edge effects arises because $(\mathbf{I} - \varrho\mathbf{C})^{-1}$ fails to preserve the equality of w_{rs} for lattice cell pairs (i, j) and (i', j'), whose relative positions are the same, but for which w_{rs} changes when one pair is located in the interior of the lattice sub-region Γ while the other pair is located on the arbitrarily demarcated regional boundary, as opposed to both pairs being located in the interior of Γ. For instance, in Chapter 5 results are reported for $(\mathbf{I} - 0.249\mathbf{C})^{-1}$ where matrix **C** is based upon a 3-by-3 lattice. In this instance the values of $\varrho_{i,j}$ vary as follows:

Areal unit location	$\varrho_{0,0}$	$\varrho_{1,0}$	$\varrho_{1,1}$	$\varrho_{2,0}$	$\varrho_{2,1}$	$\varrho_{2,2}$
Corner	0.80009	0.26081	0.16490	0.08250	0.07031	0.03501
Central border	0.88254	0.33112 0.26081	0.16490	0.11746	0.07031	
Lattice center	1.00000	0.33112	0.16490			

In other words, the existence of a finite region and the assumption of stationarity become incompatible.

Consequently, the question to be addressed here may be stated as follows:

> Can a correction factor be constructed that adjusts w_{rs} in such a way that the absolute locations in some finite region of cells (i, j) and (i', j') become irrelevant when their relative locations are held constant?

As an aside, such a factor is obtainable in time-series work, and takes on a very simple form (Johnston, 1972, p. 260).

7.2. MAJOR PROPOSED SOLUTIONS

A relatively simple correction technique for point patterns has been found, and is discussed extensively by Ripley (1981). With regard to quantitative spatial data, though, little progress has been made, although numerous attempts have been made to develop such a correction technique. Griffith (1983) has provided a survey of these techniques.

Traditionally edge effects have been coped with in terms of eight solutions, the first five of which will be referred to here as operational solutions. These nearly half dozen different procedures for handling boundary effects are ignoring them, undertaking a torus mapping, construction of an empirical buffer zone, construction of an artificial buffer zone, and extrapolation into a buffer zone. Each of these operational solutions will be discussed in turn in this section, with an emphasis on their shortcomings. Their communality lies in a quest for unbiased parameter estimation for spatial series. Hence, they are equivalent only in that they attempt to provide a medium for removing the bias introduced into spatial statistical analyses by the presence of edge effects. The first method actually yields a trivial solution, since it assumes an infinite surface in order to remove edge effects. Its principal drawback is that all empirical spatial analyses occur within a finite region, and an infinite surface never exists in the real world.

In an attempt to circumvent the constraint of finiteness, finite surfaces have been mapped onto a torus in order to emulate an infinite surface (see, for instance, Goldsmith, 1980; Ripley, 1981). But the spectral results for an infinite surface differ substantially from those for a torus mapping, as was shown in Chapter 5. In fact, in the limit the torus estimates failed to converge upon the infinite surface values. Mapping onto a torus is permissible because the process generating error term ξ is assumed to be stationary over the plane. Moreover, spatial interdependencies will be of the same type in all directions. Another major shortcoming of this solution is that no physical situation corresponds to it and, hence, many researchers find working with a donut shape distasteful. A third important weakness of this solution pertains to the latent configurational structure of the x_{ij} values. These values are realizations of random variables, and as such are subject to sampling error. The spurious spatial structure embedded in this sampling error is compounded as estimation techniques pull the circular series of information off a torus.

A third method of dealing with border effects is to construct a buffer zone along the regional boundary (Ripley, 1981). If a first-order Markov structure is assumed (i.e., the conditional autoregressive model), then this zone would need to be a single areal unit in width. Not only is this width a direct consequence of stochastic processes theory, but it also will be demonstrated numerically in the next section. Once a buffer zone has been identified, its areal units are used in the estimation of the prevailing spatial autocorrelation within the region, serving as juxtaposed units for this internal region. However, autoregressive equations for these buffer areal units never appear in the estimation expression, and any subsequent statistical analyses discard units belonging to this zone. The fundamental shortcoming of this third method is affiliated with information loss. Once again, consider the regular lattice described in the problem statement. The circumference of a coterminous n_1-by-n_2 rectangular sub-region will consist of $2n_1 + 2n_2 + 4$ areal units. Now a Diophantine equation of the form

$$n^2 + (a - 4)n - (2a + 4) \geqslant 0 \tag{7.1}$$

will yield as its roots n_1 and n_2, given that information loss is to be minimized. If the constraint of a rectangular region is relaxed, then for $n_1 = 3$ one has the solution

$$2n_2 + a \geqslant 2n_2 + 10, \tag{7.2}$$

where a, an element of the positive integers, is the number of contiguous areal units in the third row of the rectangular-type region. Consequently, as a region becomes increasingly linear in shape, the buffer zone will consume more and more of the information contained in a spatial distribution. Griffith (1980) has shown that as the region expands towards an infinite lattice, the proportion of information contained in the buffer zone of a compact region approaches zero, whereas this proportion approaches 2/3 for a linear region having $n_1 = 1$. Here it is shown that, coupled with this compactness property, at least 50 areal units are needed in order for more areal units to be contained within the internal region than are in the buffer zone. Moreover, since geographical analyses deal with small numbers of areal units that tend to be somewhat dispersed in their regional shape, the buffer zone solution often will result in a considerable loss of empirical information.

A fourth method that attempts to avoid this type of information loss constructs an artificial buffer zone (Haining, 1978). These imaginary areal units forming this zone would be assigned the mean value of the observed spatial distribution. Then spatial autocorrelation statistics are calculated in the same way as they were in the third method. Again, subsequent statistical analyses would ignore this buffer zone. Results obtained with this method are strongly affected by the values that are taken on by buffer zone areal units and will tend to depress the observed variance, as was seen for the missing values problem in Chapter 6. Furthermore, having a constant value for all real units is inconsistent with the observed spatial variation within the region.

One variant of this fourth method is to extrapolate values into the artificial buffer zone using a trend surface model. One undesirable characteristic of this method is the tendency for better fitting polynomial surfaces within the region to be poorer extrapolatory models for the buffer zone (Haggett, 1980, 1981). Krumbein (1966) has suggested that the buffer zone for which extrapolation is reasonable contracts as the order, and regional fit, of the model increases. Meanwhile, Ripley (1981) points out that the calibration of a trend surface model is severely affected by the presence of boundaries. Further, the estimated surface tends to pucker along the edges in order to fit points in the center of the region. This phenomenon has led some cartographers to adjust the boundary values by modifying them in terms of various derivatives of the trend surface equation in question.

Three statistical solutions provide alternatives to these preceding five operational solutions. The first of these is based upon generalized least squares theory. Let $\mathbf{\Omega}$ be the GLS variance–covariance matrix. In econometrics the autoregressive model is defined as

$$\xi_t = (\mathbf{I} - \varrho\mathbf{C})\mathbf{u}_t,$$

where $\mathbf{C}$ is a binary matrix with zeros everywhere except for the lower off-diagonal, which is unity. But

$$(\mathbf{I} - \varrho\mathbf{C})^T(\mathbf{I} - \varrho\mathbf{C}) \neq \Delta\mathbf{\Omega}^{-1},$$

for some constant Δ, so that some transformation matrix $\mathbf{\Gamma}$ is needed such that

$$\xi_t = \mathbf{\Gamma}(\mathbf{I} - \varrho\mathbf{C})\mathbf{u}_t \quad \text{or} \quad \xi_t = (\mathbf{I} - \varrho\mathbf{C})\mathbf{\Gamma}\mathbf{u}_t.$$

Now $(\mathbf{I} - \varrho\mathbf{C})^T\mathbf{\Gamma}^T\mathbf{\Gamma}(\mathbf{I} - \varrho\mathbf{C}) = \mathbf{\Omega}^{-1}$ only when $\mathbf{\Gamma}$ is a diagonal matrix whose first cell is $\gamma_{11} = (1 - \varrho^2)^{1/2}$ and whose remaining diagonal cells contain unity. As a result the boundary value adjustment for some time series is achieved by setting $\xi_1 = (1 - \varrho^2)^{1/2}u_1$ (see Wonnacott and Wonnacott, 1970). In contrast, $\mathbf{\Gamma}^T(\mathbf{I} - \varrho\mathbf{C})^T(\mathbf{I} - \varrho\mathbf{C})\mathbf{\Gamma} = \mathbf{\Omega}^{-1}$ produces a more complex form of matrix $\mathbf{\Gamma}$. For example, if the time series under study consists of the observation set $\{x_1, x_2, x_3\}$, then the matrix $\mathbf{\Gamma}$ for post-multiplying matrix $(\mathbf{I} - \varrho\mathbf{C})$ becomes

$$\begin{pmatrix} (1 - \varrho^2)^{1/2} & 0 & 0 \\ \varrho[(1 - \varrho^2)^{1/2} - 1] & 1 & 0 \\ \varrho^2[(1 - \varrho^2)^{1/2} - 1] & 0 & 1 \end{pmatrix}.$$

Clearly this form is more complex than its pre-multiplying matrix counterpart.

The same solution approach can be used in the case of a spatial data series. Accordingly, for the conditional autoregressive model, let n_I denote the number of interior lattice cells in the sub-region, and let n_B denote the number of edge lattice cells in the sub-region, where $n = n_I + n_B$. Now $\mathbf{\Omega}^{-1}$ is of the form

$$\begin{pmatrix} \mathbf{B}_{12} & \vdots & \mathbf{B}_1 \\ \cdots & \cdots & \cdots \\ \mathbf{B}_1^T & \vdots & \mathbf{I} - \varrho \mathbf{C}_I \end{pmatrix},$$

where $(\mathbf{I} - \varrho \mathbf{C}_I)$ is n_I-by-n_I, $\mathbf{B}_1$ is n_B-by-n_I, $\mathbf{B}_{12}$ is n_B-by-n_B, and $\mathbf{C}_I$ is the connectivity matrix for the interior lattice cells. For instance, if $n = 16$, $n_I = 4$, $n_B = 12$, $\varrho = 0.24$, and the sub-region configuration is

A	B	C	D
E	F	G	H
I	J	K	L
M	N	O	P

$$\varrho_{0,0} = 1 \qquad \varrho_{2,2} = 0.126290$$

$$\varrho_{0,1} = \varrho_{1,0} = 0.434110 \qquad \varrho_{0,3} = \varrho_{3,0} = 0.121540$$

$$\varrho_{0,2} = \varrho_{2,0} = 0.219580 \qquad \varrho_{1,3} = \varrho_{3,1} = 0.108020$$

$$\varrho_{1,1} = 0.294600 \qquad \varrho_{2,3} = \varrho_{3,2} = 0.083470$$

$$\varrho_{1,2} = \varrho_{2,1} = 0.179640 \qquad \varrho_{3,3} = 0.59690$$

$$(\mathbf{I} - \varrho \mathbf{C}_I) = \begin{array}{c|cccc} & F & G & J & K \\ \hline F & 1 & -0.24 & -0.24 & 0 \\ G & -0.24 & 1 & 0 & -0.24 \\ J & -0.24 & 0 & 1 & -0.24 \\ K & 0 & -0.24 & -0.24 & 1 \end{array}$$

$\mathbf{B}_1$ =

	F	G	J	K
A	0	0	0	0
B	−0.362	0	0	0
C	0	−0.362	0	0
D	0	0	0	0
E	−0.362	0	0	0
H	0	−0.362	0	0
I	0	0	−0.362	0
L	0	0	0	−0.362
M	0	0	0	0
N	0	0	−0.362	0
O	0	0	0	−0.362
P	0	0	0	0

$\mathbf{B}_{12}$ =

	A	B	C	D	E	H	I	L
A	1.243							
B	−0.407	1.393						
C	−0.017	−0.400	1.393					
D	−0.008	−0.017	−0.047	1.243				
E	−0.407	−0.002	−0.001	−0.001	1.393			
H	−0.001	−0.001	−0.002	−0.407	−0.000	1.393		
I	−0.017	−0.001	−0.000	−0.000	−0.400	−0.000	1.393	
L	−0.000	−0.000	−0.001	−0.017	−0.000	−0.400	−0.000	1.393
M	−0.008	−0.001	−0.000	−0.000	−0.017	−0.000	−0.407	−0.001
N	−0.001	−0.000	−0.000	−0.000	−0.001	−0.000	−0.002	−0.001
O	−0.000	−0.000	−0.000	−0.001	−0.000	−0.001	−0.001	−0.002
P	−0.000	−0.000	−0.001	−0.008	−0.000	−0.017	−0.001	−0.407

	M	N	O	P
M	1.243			
N	−0.407	1.393		
O	−0.017	−0.400	1.393	
P	−0.008	−0.017	−0.407	1.243

From these results one would expect the aforementioned buffer zone width to be a single lattice cell, since the 4-by-4 interior sub-region retains the simple autoregressive structure of $(\mathbf{I} - \varrho\mathbf{C})$. Furthermore, these interior cells covary only with their respective juxtaposed buffer zone lattice cells. The complicated edge effect structure is manifested in the corner lattice cells, which covary with all of their nearby closest border lattice cells. Now, in keeping with the manner in which the simultaneous model modification was written above, the variance–covariance matrix for the conditional model can be modified in such a way that

$$(\boldsymbol{\Gamma}^T)^{1/2}(\mathbf{I} - \varrho\mathbf{C})\boldsymbol{\Gamma}^{1/2} = \boldsymbol{\Omega}^{-1}. \tag{7.3}$$

The problem, then, reduces to one of specifying $\boldsymbol{\Gamma}^{1/2}$, since $\boldsymbol{\Omega}$ is obtainable from equation (5.16), matrix **C** is known, and ϱ is to be estimated. Although this first solution is based upon GLS theory, it is wanting in empirical situations because ϱ is unknown. Thus, it is useful for simulation experiments in which ϱ is a specified input, but presents estimation problems when ϱ and the edge effects must be simultaneously estimated.

The second statistical solution utilizes dummy variables and a regression framework. A buffer zone is established whose areal units also can be employed in subsequent statistical analyses so that no empirical information is lost. First, let y_i be the value of some spatial distribution variate in areal unit i, and let $X_{i1} \equiv 1$. Next, let $X_{i2} = 1$ if areal unit i is in the buffer zone, and zero otherwise. Matrix **X** equals $\langle \mathbf{X}_1, \mathbf{X}_2 \rangle$. Vector **Y** may be predicted from matrix **X** using OLS by calculating $\hat{\boldsymbol{\beta}} = (\mathbf{X}^T\mathbf{X})^{-1}\mathbf{X}^T\mathbf{Y}$. Consequently

$$\mathbf{X}^T\mathbf{X} = \begin{pmatrix} n & n_b \\ n_b & n_b \end{pmatrix}$$

where n denotes the total number of areal units, and n_b denotes the number of boundary areal units,

$$\mathbf{X}^T\mathbf{Y} = \begin{pmatrix} \sum_{i=1}^{n} & y_i \\ \sum_{b=1}^{k} & y_b \end{pmatrix}$$

where y_b denotes the value of some variate in boundary areal unit b, and

$$\hat{\boldsymbol{\beta}} = \begin{pmatrix} \bar{y}_{\not b} \\ \bar{y}_b - \bar{y}_{\not b} \end{pmatrix}$$

where $\bar{y}_b$ denotes the mean of the values taken on by boundary areal units, and $\bar{y}_{\not b}$ denotes the mean of the values taken on by the non-boundary areal units. Since $\hat{\mathbf{Y}} = \boldsymbol{X\beta}$, boundary values become a function solely of $\bar{y}_b$, whereas non-boundary values become a function solely of $\bar{y}_{\not b}$. Now an adjusted spatial autocorrelation parameter can be calculated with the following steps (Cliff and Ord, 1973):

Step 1: compute $e_i = y_i - \bar{y}_b$ for the boundary areal units, and $e_i = y_i - \bar{y}_{\not b}$ for non-boundary areal units,
Step 2: estimate the spatial autocorrelation parameter ϱ from expression (6.5) using the maximum likelihood method,
Step 3: construct the new variables $\mathbf{Y}^* = (\mathbf{I} - \hat{\varrho}\mathbf{C})\mathbf{Y}$, and then calculate $\bar{y}_b^*$ and $\bar{y}_{\not b}^*$,
Step 4: replace y_i, $\bar{y}_b$ and $\bar{y}_{\not b}$ with y_i^*, $\bar{y}_b^*$ and $\bar{y}_{\not b}^*$, respectively, and return to Step 1.

These four steps should be repeated until a stable value of $\hat{\varrho}$ is converged upon. This procedure permits a differential treatment of the buffer zone areal units while spatial autocorrelation is being assessed, but since the sequential modification of these values introduces an edge effect adjustment, their original values may be used in ensuing geographical analyses. The major drawback here has to do with a lack of understanding of the behavior of parameter estimates as lattice size changes. Also asymptotic properties may be meaningless, since they should coincide with an infinite lattice.

The third statistical solution casts the geographical boundary value problem as a missing values problem. This procedure also demonstrates why the fourth operational method is misleading. Once again an imaginary buffer zone is constructed. Values for the areal units comprising this zone are treated as missing values. Once the estimate of spatial autocorrelation has been determined for incomplete data using equations (6.3) and (6.7), the artificial buffer zone areal units should be deleted for subsequent statistical analyses. Clearly the problem with the fourth operational solution highlighted here is that it halts in this estimation procedure after completing Step 2 in the missing values iterative estimation procedure (see Chapter 6). Hence it fails to produce an unbiased estimate of ϱ.

Griffith and Amrhein (1983) and Griffith (1985) conducted a series of simulation experiments in order to evaluate the success of many of those edge effect correction techniques outlined in the previous section. These simulation experiments employed the following initial conditions:

(1) $\xi_i \sim$ iid $N(0, 1)$ such that $77.647 \leqslant \sigma^2 \leqslant 124.059$, $|\mu| < 1.645$, and |Kolmogorov–Smirnov statistic| < 0.0736 [these are the 10% level of significance critical values];
(2) the initial ξ_i were allocated to the lattice cells such that $|\hat{\varrho}| < 0.001$;
(3) $\varrho = \pm 0.1$, ± 0.5 and ± 0.9;
(4) the sub-region consisted of an 8-by-8 lattice centrally located within a 10-by-10 torus; and,
(5) the constraint $|\varrho| < 1/\lambda_{\max}$ was employed.

Estimates of ϱ for the simulation experiments were evaluated with a two-way fixed effects balanced design analysis of variance

TABLE 7.1

Two-way ANOVA Results on $\hat{\rho}$ for the Operational Solutions Based Upon 50 Simulated Surfaces

Source of Variation	Sum of Squares	Degrees of Freedom	Mean Square	F-Ratio
Including the original data				
level of spatial autocorrelation	3160	5	633	5466.45*
boundary adjustment technique	.206	3	.069	.59
interaction	4.19	15	.279	2.42*
within cross-classification groups	136.13	1176	.116	
Excluding the original data				
level of spatial autocorrelation	4160	5	832	8819.92*
boundary adjustment technique	3.280	4	.820	8.70*
interaction	15.889	20	.794	8.42*
within cross-classification groups	138.65	1470	.094	

*** Denotes a significant difference at the 0.01 level.

model. These ANOVA results produced three basic implications for the five operational solutions (see Table 7.1). First, on the average recovered values of $\hat{\varrho}$ for the finite surface differ markedly from those for the infinite surface. This finding corroborates the parallel inference drawn in Chapter 5. Second, results for the correction techniques do not differ significantly from those when boundary effects are simply ignored. Third, there is significant interaction between the performance of these correction techniques and the prevailing level of spatial autocorrelation. In general the assumptions of the ANOVA model were found to be satisfied by the artificial data that were generated, suggesting that these three implications should be quite robust.

The three statistical solutions failed to perform better (see Table 7.2). The missing values solution tended to over-correct for the edge effects bias, whereas on average the dummy variable solution had no affect whatsoever on this bias. The incomplete data solution used a simultaneous scheme, which essentially is a second-order model. As such (for the infinite case of $\mathbf{C}/4$)

$$\begin{aligned} E(X_{i,j} | \text{all other site values}) \\ &= [2\varrho/(4 + \varrho^2)]\,(x_{i-1,j} + x_{i,j-1} + x_{i+1,j} + x_{i,j+1}) \\ &- [\varrho^2/(8 + 2\varrho^2)]\,(x_{i-1,j-1} + x_{i-1,j+1} + x_{i+1,j-1} + x_{i+1,j+1}) \\ &- [\varrho^2/(16 + 4\varrho^2)]\,(x_{i-2,j} + x_{i,j-2} + x_{i+2,j} + x_{i,j+2}). \end{aligned}$$

This is a second-order model, then, because it embraces the second-order lags of (1, 1) and (2, 0). Accordingly this second-order scheme requires a buffer zone that is two areal units in width. Hence for an n_1-by-n_2 rectangular lattice, $(4n_1 + 4n_2 + 16)$ buffer zone areal units must be added. In order for the observed data to carry at least as much weight as the 'missing data', $n_1 \geqslant 9$ and $n_2 \geqslant 9$. It is not surprising, then, that $\hat{\varrho}$ tends to be over-estimated. Meanwhile, two classes of GLS solutions have been explored, since they seemed to offer the most promise (given the successful GLS solution for time series data). The first lets $\mathbf{\Gamma} = (\mathbf{I} - \varrho^{D+1}\mathbf{C}^{D+1})$ for the simultaneous model, and did furnish some encouraging results for $\varrho \to 0.25$. The second solution involved constrained forms of $\mathbf{\Gamma}$.

Again consider the simultaneous autoregressive scheme. For the conditional scheme matrix $\mathbf{\Omega}$ can be written as a finite

TABLE 7.2

T-test Results for $\hat{\rho}$ from the Statistical Correction Techniques Based Upon 50 Simulated Surfaces

Embedded ρ	Dummy Variable Technique	Generalized Least Squares Technique	Missing Data Technique
0.9	6.31***	4.09***	-9.70***
0.5	7.01***	1.90*	-4.95***
0.1	4.54***	2.45**	.98
-0.1	2.87***	1.86*	3.76***
-0.5	-1.09	1.18	11.34***
-0.9	-2.78***	-.98	14.32***

*** Denotes a significant difference at the 0.01 level.
** Denotes a significant difference at the 0.05 level.
* Denotes a significant difference at the 0.10 level.

expansion of the term (ϱC), since for an unbounded surface $\mathbf{\Omega} = (\mathbf{I} - \varrho\mathbf{C})^{-1}$. Now for the simultaneous scheme $\mathbf{\Omega}$ would need to be written as a finite expansion of the term $(\varrho\mathbf{W})$, namely

$$\sum_{d=0}^{d=\infty} (\varrho\mathbf{W})^d, \tag{7.4}$$

which, since the eigenvalues δ_i of matrix $\mathbf{W}$ are such that $|\delta_i| \leqslant 1$ for all i, will converge to $(\mathbf{I} - \varrho\mathbf{W})^{-1}$. If $(\mathbf{I} - \varrho\mathbf{W})$ is rewritten in terms of its eigenvalues and eigenvectors, the eigenvalues are $(1 - \varrho\delta_i)$, while the eigenvectors $\mathbf{E}$ are those for matrix $\mathbf{W}$. Returning to expression (7.4),

$$\sum_{d=0}^{d=\infty} (\varrho\mathbf{W})^d = \sum_{d=0}^{d=\infty} (\varrho\mathbf{E\Delta E}^{-1})^d = \mathbf{E}\left[\sum_{d=0}^{d=\infty} (\varrho\mathbf{\Delta})^d\right]\mathbf{E}^{-1},$$

and since $|\varrho\delta_i| < 1$ this matrix series must converge.

But this expansion is to infinity, and the problem under study asks what happens when d stops at $D \ll \infty$. Now

$$\sum_{d=0}^{d=D} (\varrho\mathbf{W})^d = \mathbf{E}\left[\sum_{d=0}^{d=D} (\varrho\mathbf{\Delta})^d\right]\mathbf{E}^{-1},$$

and the term in brackets reduces to

$$\sum_{d=0}^{d=D} (\varrho\delta_i)^d = [1 - (\varrho\delta_i)^{D+1}]/[1 - (\varrho\delta_i)].$$

Therefore,

$$\mathbf{E}\left[\sum_{d=0}^{d=D} (\varrho\mathbf{\Delta})^d\right]\mathbf{E}^{-1} = \mathbf{E}\left[\sum_{d=0}^{d=\infty} (\varrho\mathbf{\Delta})^d\right]\mathbf{RE}^{-1}, \tag{7.5}$$

where **R** is a diagonal matrix having entries

$$r_{ii} = 1 - (\varrho\delta_i)^{D+1}.$$

Since $\mathbf{E}^{-1}\mathbf{E} = \mathbf{I}$, equation (7.5) may be rewritten as

$$\mathbf{E}\left[\sum_{d=0}^{d=\infty} (\varrho\mathbf{\Delta})^d\right]\mathbf{E}^{-1}\mathbf{ERE}^{-1} = (\mathbf{I} - \varrho\mathbf{W})^{-1}\mathbf{ERE}^{-1}.$$

Consequently, for two-dimensional data series

$$\mathbf{\Gamma} = (\mathbf{ERE}^{-1})^{-1} = \mathbf{ER}^{-1}\mathbf{E}^{-1}.$$

Matrix $\mathbf{R}^{-1}$ can be calculated quite easily, since **R** is a diagonal matrix. Further, $\mathbf{\Gamma}$ can be calculated quite easily, since **E** is the matrix of eigenvectors from **W**. The density of $\mathbf{\Gamma}$ may be ascertained by inspecting

$$\begin{aligned}(\mathbf{ERE}^{-1})^{-1} &= [\mathbf{E}(\mathbf{I} - \varrho^{D+1}\mathbf{\Delta}^{D+1})\mathbf{E}^{-1}]^{-1} \\ &= (\mathbf{I} - \varrho^{D+1}\mathbf{W}^{D+1})^{-1}.\end{aligned}$$

Given the combinatorial basis of matrix **W**, as D increases, $\mathbf{W}^{D+1}$ as well as $(\mathbf{I} - \varrho^{D+1}\mathbf{W}^{D+1})^{-1}$ often tend to have fewer and fewer non-zero cell values, especially if D is less than the dimension of matrix **W**. Clearly as $D \to \infty$, $\varrho^{D+1} \to 0$, and hence $\mathbf{\Gamma} \to \mathbf{I}$. Meanwhile, if $D = 0$, then $\mathbf{\Gamma} = (\mathbf{I} - \varrho\mathbf{W})^{-1}$, and hence the total variance–covariance matrix becomes **I**, which would be inconsistent with a spatial autoregressive model specification.

Once more simulation experiments were conducted. The same initial conditions were utilized as before. The first 20 surfaces for each of the six embedded values of ϱ were investigated in order to determine a best value for D. Estimates of ϱ were calculated using the values $D = 1, 2, \ldots, 14$. A maximum value of 14 was chosen because the most remote areal units on each surface are 14 lags apart. Hence, in the sequence defined by expression (7.4), by the time matrix $\mathbf{W}^{14}$ is reached, all lag effects latent in the data will be captured. Inspection of these results uncovered three prominent trends. First, a unique optimal D did not exist across the sets of embedded ϱ's. Second, a unique optimal D did not exist across the set of autocorrelated surfaces for a given value of ϱ. Third, two types of trajectories were identified. These two kinds are illustrated in Figure 7.1. In Figure 7.1(a), the adjusted estimate of ϱ

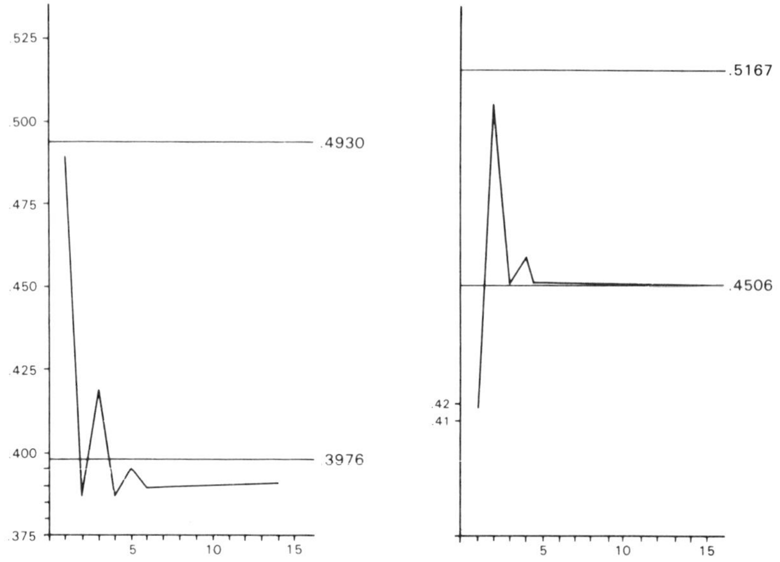

(a) Autocorrelated surface #3, $\rho = .5$, toroidal $\rho = .4930$, boundary affected $\rho = .3976$

(b) Autocorrelated surface #7, $\rho = .5$, toroidal $\hat{\rho} = .5167$, boundary affected $\hat{\rho} = .4506$

Fig. 7.1. Convergence trajectories of D for selected simulated autocorrelated surfaces.

initially is close to the unbiased estimate $\hat{\varrho}$, and then moves toward the biased estimate $\hat{\varrho}$. In Figure 7.1(b), the adjusted estimate of ϱ initially starts close to the biased estimate $\hat{\varrho}$, jumps close to the unbiased estimate $\hat{\varrho}$, and then returns to a path that moves toward the biased estimate $\hat{\varrho}$.

Given these 6 sets of 20 trajectories, the closest correspondences were found for the following couplings:

ϱ	D
0.9	6
0.5	2
0.1	1
−0.1	1 or 2
−0.5	2
−0.9	6

D was set to 1 for $\varrho = -0.1$, so that the pattern was consistent for both positive and negative values of ϱ.

TABLE 7.3

Correlations and Differences of Means t-statistics
for the Values of $\hat{\rho}$ Based Upon 500 Simulated Surfaces

	Embedded Value of ρ	Recovered Value of $\hat{\rho}$	Corrected Estimated Value of $\hat{\rho}$ for Sub-region	Uncorrected Estimated Value of $\hat{\rho}$ for Sub-region
Recovered	.9	————	.664***	.661***
Value of	.5	————	.337***	.287***
ρ	.1	————	-.011	.043
	-.1	————	-.053	.070
	-.5	————	.395***	.319***
	-.9	————	.720***	.736***
Corrected	.9	11.55***	————	.914***
Estimated	.5	3.69***	————	.868***
Value of	.1	5.34***	————	.972***
ρ for	-.1	2.28**	————	.957***
Sub-region	-.5	1.24	————	.862***
	-.9	-3.75***	————	.913***
Uncorrected	.9	36.52***	26.50***	————
Estimated	.5	19.35***	6.09***	————
Value of	.1	6.98***	2.90***	————
ρ for	-.1	.74	-5.41***	————
Sub-region	-.5	-12.89***	-16.51***	————
	-.9	-34.52***	-29.53***	————

*** Denotes a significant difference from zero at the 0.01 level.
** Denotes a significant difference from zero at the 0.05 level.

On the average, the adjusted values of $\hat{\varrho}$ were closer to the true value of ϱ than were the uncorrected values of $\hat{\varrho}$. In general this tendency was found to be more pronounced as $|\varrho| \rightarrow 1$. This finding is further supported by the difference of means for correlated samples test results appearing in Table 7.3. Moreover, the t-test values for the adjusted $\hat{\varrho}$'s with respect to the actual ϱ's tend to be substantially less than those with respect to the unadjusted $\hat{\varrho}$'s. In addition, correlations between the actual ϱ's and the adjusted $\hat{\varrho}$'s increase as $|\varrho| \rightarrow 1$. Furthermore, the adjusted $\hat{\varrho}$'s are significantly different from the unadjusted $\hat{\varrho}$'s. But, some boundary effects are still latent in the adjusted $\hat{\varrho}$'s, since they (a) tend to be significantly different from the actual ϱ's, and (b) they are very strongly correlated with the unadjusted $\hat{\varrho}$'s.

Next, let $\mathbf{\Lambda}_c$ and $\mathbf{\Lambda}_\omega$ be the diagonal matrices of eigenvalues for matrices $(\mathbf{I} - \varrho\mathbf{C})$ and $\mathbf{\Omega}^{-1}$, respectively, and let $\mathbf{E}_c$ and $\mathbf{E}_\omega$ be the corresponding matrices of right-hand eigenvectors. If matrix $\mathbf{\Gamma}$ is unconstrained, then the right hand side and the left hand side of

$(\boldsymbol{\Gamma}^T)^{1/2}(\mathbf{I} - \varrho\mathbf{C})\boldsymbol{\Gamma}^{1/2} = \boldsymbol{\Omega}^{-1}$ can be rewritten in terms of their eigenvalues and eigenvectors as

$$(\boldsymbol{\Gamma}^T)^{1/2}\mathbf{E}_c\boldsymbol{\Lambda}_c\mathbf{E}_c^T\boldsymbol{\Gamma}^{1/2} = \mathbf{E}_\omega\boldsymbol{\Lambda}_\omega\mathbf{E}_\omega^T.$$

$$(\boldsymbol{\Gamma}^T)^{1/2}\mathbf{E}_c\boldsymbol{\Lambda}_c^{1/2}\boldsymbol{\Lambda}_c^{1/2}\mathbf{E}_c^T\boldsymbol{\Gamma}^{1/2} = \mathbf{E}_\omega\boldsymbol{\Lambda}_\omega^{1/2}\boldsymbol{\Lambda}_\omega^{1/2}\mathbf{E}_\omega^T. \qquad (7.6)$$

Equation (7.6) implies that

$$\boldsymbol{\Lambda}_c^{1/2}\mathbf{E}_c^T\boldsymbol{\Gamma}^{1/2} = \boldsymbol{\Lambda}_\omega^{1/2}\mathbf{E}_\omega^T$$

$$\boldsymbol{\Gamma}^{1/2} = \mathbf{E}_c\boldsymbol{\Lambda}_c^{1/2}\boldsymbol{\Lambda}_\omega^{1/2}\mathbf{E}_\omega^T.$$

If the X_i's are iid, then $\varrho = 0$ and $\boldsymbol{\Omega} = \mathbf{I}$, $\mathbf{E}_c = \mathbf{E}_\omega$ and $\boldsymbol{\Lambda}_c = \boldsymbol{\Lambda}_\omega$, and thus $\boldsymbol{\Gamma}^{1/2} = \mathbf{I}$. In contradistinction, for reasonable distribution assumptions on X, as $n \to \infty$

$$(\mathbf{X} - \mu\mathbf{1})^T(\mathbf{I} - \varrho\mathbf{C})(\mathbf{X} - \mu\mathbf{1}) \to (\mathbf{X} - \mu\mathbf{1})^T\boldsymbol{\Omega}^{-1}(\mathbf{X} - \mu\mathbf{1}),$$

with probability one, and hence $\boldsymbol{\Gamma}^{1/2} = \mathbf{I}$. Except for these two limiting cases, $\boldsymbol{\Gamma}^{1/2}$ will be a dense matrix. This expectation seems both intuitive and reasonable, since the dependency structure involved in a first-order conditional autoregressive spatial process is multidirectional as well as two-dimensional in nature.

Recalling that earlier in this discussion the spatial variates had been dichotomized into interior and border lattice cell groups, and that the principal eigenvector of matrix $\mathbf{C}$, namely $\mathbf{E}_p$, was used to distinguish between these locations, the simplest version of $\boldsymbol{\Gamma}$ is one that both utilizes this particular set of properties and is diagonal. The question at this point, then, asks how to integrate this information into a specification of the γ_{hh} terms.

Consider the following section of a lattice sub-region:

	4	
3	0	1
	2	

Let

$$y_0|(y_j, j = 1, 2, 3, 4) \sim N\left(\sum_{j=1}^{4} \varrho y_j, \sigma_0^2\right),$$

and

$$y_4|(y_j, j = 1, 2, 3) \sim N\left(\sum_{j=1}^{3} h_j y_j, \sigma_m^2\right),$$

where the weights $h_j \neq \varrho$ since cells $j = 1, 2, 3$ are not adjacent to cell 4, and σ_m^2 is the variance terms when cell 0 is deleted from the conditional specification of y_4. Then, from algebraic manipulations of their characteristic functions,

$$y_0|(y_j, j = 1, 2, 3) \sim N\left[\sum_{j=1}^{3} \varrho(1 + h_j)y_j, (\sigma_0^2 + \varrho^2\sigma_m^2\right].$$

Thus, if site 0 is in the interior, then the coefficient of y_i is ϱ, whereas if site 0 is on the boundary and y_4 is integrated out, then the coefficient of y_j is $\varrho(1 + h_j)$. Therefore, for $\varrho > 0$, removal of a site increases the remaining coefficients and the conditional variances. Accordingly, keeping in mind the results reported in Tables 5.3 and 5.4, γ_{ii} has been defined as follows:

$$\gamma_{ii} = \begin{cases} \alpha e_{i,p}/e_{\max,p}, & \text{if lattice cell } i \text{ is a border cell, and} \\ 1, & \text{if lattice cell } i \text{ is an interior cell.} \end{cases} \tag{7.7}$$

Hence,

$$\begin{pmatrix} \mathrm{diag}\{[\alpha e_{i,p}/e_{\max,p}]^{1/2}\} & \vdots & \mathbf{0} \\ \cdots & & \cdots \\ \mathbf{0} & \vdots & \mathbf{I} \end{pmatrix}.$$

The reduced likelihood function for a first-order conditional autoregressive process may be written in the following manner. Assuming the multivariate normal model, the likelihood function becomes [after Cliff and Ord, 1981, pp. 154–155]:

$$\begin{aligned} L(\mu, \sigma^2, \varrho, \alpha) = {} & (2\pi)^{-n/2}\{\det[\sigma^2\boldsymbol{\Gamma}^{-1/2}(\mathbf{I} - \varrho\mathbf{C})^{-1}\boldsymbol{\Gamma}^{-1/2}]\}^{-1/2} \\ & \times \exp[-(\mathbf{X} - \mu\mathbf{1})^T \\ & \times \boldsymbol{\Gamma}^{1/2}(\mathbf{I} - \varrho\mathbf{C}_{00})^T(\mathbf{X} - \mu\mathbf{1})/2\sigma^2] \end{aligned}$$

$$\begin{aligned} \ln(L) = {} & \text{constant} - \ln\{\det[\sigma^2\boldsymbol{\Gamma}^{-1/2}(\mathbf{I} - \varrho\mathbf{C})^{-1}\boldsymbol{\Gamma}^{-1/2}]\}/2 \\ & - (\mathbf{X} - \mu\mathbf{1})^T\boldsymbol{\Gamma}^{1/2}(\mathbf{I} - \varrho\mathbf{C})\boldsymbol{\Gamma}^{1/2}(\mathbf{X} - \mu\mathbf{1})/2\sigma^2 \end{aligned}$$

$$\partial \ln(L)/\partial\mu = \mathbf{1}^T\mathbf{\Gamma}^{1/2}(\mathbf{I} - \varrho\mathbf{C})\mathbf{\Gamma}^{1/2}\mathbf{X} - \mu\mathbf{1}^T\mathbf{\Gamma}^{1/2} \times (\mathbf{I} - \varrho\mathbf{C})\mathbf{\Gamma}^{1/2}\mathbf{1} = 0.$$

$$\therefore \quad \hat{\mu} = \mathbf{1}^T\mathbf{\Gamma}^{1/2}(\mathbf{I} - \varrho\mathbf{C})\mathbf{\Gamma}^{1/2}\mathbf{X}/\mathbf{1}^T\mathbf{\Gamma}^{1/2}(\mathbf{I} - \varrho\mathbf{C})\mathbf{\Gamma}^{1/2}\mathbf{1}$$

$$\partial \ln(L)/\partial\sigma^2 = [-n/\sigma^2 + (\mathbf{X} - \mu\mathbf{1})^T\mathbf{\Gamma}^{1/2}(\mathbf{I} - \varrho\mathbf{C})\mathbf{\Gamma}^{1/2}(\mathbf{X} - \mu\mathbf{1})]/2 = 0$$

$$\therefore \quad \hat{\sigma}_2 = (\mathbf{X} - \mu\mathbf{1})^T\mathbf{\Gamma}^{1/2}(\mathbf{I} - \varrho\mathbf{C})\mathbf{\Gamma}^{1/2}(\mathbf{X} - \mu\mathbf{1})/n.$$

Since $\partial \ln(L)/\partial\varrho$ results in a non-linear function, nothing is gained by studying this derivative. Thus the reduced likelihood function becomes

$$\begin{aligned}
L(\mu, \sigma^2, \varrho, \alpha) &= (2\pi)^{-n/2}\{\det[\sigma^2\mathbf{\Gamma}^{-1/2}(\mathbf{I} - \varrho\mathbf{C})^{-1}\mathbf{\Gamma}^{-1/2}]\}^{-1/2} \\
&\quad \times \exp[-(\mathbf{X} - \mu\mathbf{1})^T\mathbf{\Gamma}^{1/2}(\mathbf{I} - \varrho\mathbf{C})\mathbf{\Gamma}^{1/2} \\
&\quad \times (\mathbf{X} - \mu\mathbf{1})/2\sigma^2] \\
&= (2\pi)^{-n/2}\{\sigma^{-2n}\det[\mathbf{\Gamma}^{-1/2}(\mathbf{I} - \varrho\mathbf{C})^{-1}\mathbf{\Gamma}^{-1/2}]\}^{-1/2} \\
&\quad \times \exp\{-(\mathbf{X} - \mu\mathbf{1})^T\mathbf{\Gamma}^{1/2}(\mathbf{I} - \varrho\mathbf{C})\mathbf{\Gamma}^{1/2}(\mathbf{X} - \mu\mathbf{1})/ \\
&\quad [2(\mathbf{X} - \mu\mathbf{1})^T\mathbf{\Gamma}^{1/2}(\mathbf{I} - \varrho\mathbf{C})\mathbf{\Gamma}^{1/2}(\mathbf{X} - \mu\mathbf{1})/n]\} \\
&= (2\pi)^{-n/2}(n^{-n/2})\exp(-n/2) \\
&\quad \times \{\det[\mathbf{\Gamma}^{1/2}(\mathbf{I} - \varrho\mathbf{C})\mathbf{\Gamma}^{1/2}]\}^{-1/2} \\
&\quad \times \{(\mathbf{X} - \mu\mathbf{1})^T\mathbf{\Gamma}^{1/2}(\mathbf{I} - \varrho\mathbf{C})\mathbf{\Gamma}^{1/2}(\mathbf{X} - \mu\mathbf{1})\}^{n/2}.
\end{aligned}$$

Consequently the problem reduces to that of maximizing the function

$$\{\det[\mathbf{\Gamma}^{1/2}(\mathbf{I} - \varrho\mathbf{C})\mathbf{\Gamma}^{1/2}]\}^{-1/n}(\mathbf{X} - \mu\mathbf{1})^T\dot{\mathbf{\Gamma}}^{1/2}(\mathbf{I} - \varrho\mathbf{C})\mathbf{\Gamma}^{1/2}(\mathbf{X} - \mu\mathbf{1}). \tag{7.8}$$

The Jacobian term may be evaluated in the following manner:

$$\begin{aligned}
\det[\mathbf{\Gamma}^{1/2}(\mathbf{I} - \varrho\mathbf{C})\mathbf{\Gamma}^{1/2}] &= \det(\mathbf{\Gamma}^{1/2})\det(\mathbf{I} - \varrho\mathbf{C})\det(\mathbf{\Gamma}^{1/2}) \\
&= \det(\mathbf{\Gamma})\det(\mathbf{I} - \varrho\mathbf{C}) \\
&= \left(\prod_{i=1}^{n} \alpha e_{i,p}/e_{\max,p}\right)\left[\prod_{i=1}^{n}(1 - \varrho\lambda_i)\right],
\end{aligned}$$

where λ_i are the eigenvalues of matrix $\mathbf{C}$. Recall that the MLE of μ is

$$\mathbf{1}^T\mathbf{\Gamma}^{1/2}(\mathbf{I} - \varrho\mathbf{C})\mathbf{\Gamma}^{1/2}\mathbf{X}/\mathbf{1}^T\mathbf{\Gamma}^{1/2}(\mathbf{I} - \varrho\mathbf{C})\mathbf{\Gamma}^{1/2}\mathbf{1}.$$

Making these two substitutions into equation (7.8) converts it into a function of just two parameters, and will simplify its maximization.

Another plausible specification results in the matrix $(\mathbf{D} - \varrho\mathbf{C})$, where matrix $\mathbf{D}$ is a diagonal matrix having entries defined by equation (7.7). This modified form of the inverse variance–covariance matrix preserves symmetry, and exploits the expected variance structure. But, if

$$(\mathbf{\Gamma}^T)^{1/2}(\mathbf{I} - \varrho\mathbf{C})\mathbf{\Gamma}^{1/2} = \mathbf{\Omega}^{-1},$$

then

$$\begin{aligned}(\mathbf{I} - \varrho\mathbf{C}) &= (\mathbf{\Gamma}^T)^{-1/2}\mathbf{\Omega}^{-1}\mathbf{\Gamma}^{-1/2}\\ -\varrho\mathbf{C} &= -\mathbf{I} + (\mathbf{\Gamma}^T)^{-1/2}\mathbf{\Omega}^{-1}\mathbf{\Gamma}^{-1/2}\\ \mathbf{D} - \varrho\mathbf{C} &= \mathbf{D} - \mathbf{I} + (\mathbf{\Gamma}^T)^{-1/2}\mathbf{\Omega}^{-1}\mathbf{\Gamma}^{-1/2}.\end{aligned}$$

Hence the assumption here is that $(\mathbf{D} - \varrho\mathbf{C}) = \mathbf{\Omega}^{-1}$, or

$$\mathbf{D} - \mathbf{I} + (\mathbf{\Gamma}^T)^{-1/2}\mathbf{\Omega}^{-1}\mathbf{\Gamma}^{-1/2} \doteq \mathbf{\Omega}^{-1}.$$

Similarly, a third feasible restricted version of the correction matrix $\mathbf{\Gamma}$ produces the matrix $[\mathbf{I} - \varrho(\mathbf{D}^T)^{1/2}\mathbf{C}\mathbf{D}^{1/2}]$. Here

$$\begin{aligned}\mathbf{I} - \varrho\mathbf{C} &= (\mathbf{\Gamma}^T)^{-1/2}\mathbf{\Omega}^{-1}\mathbf{\Gamma}^{-1/2}\\ (\mathbf{D}^T)^{1/2}\mathbf{D}^{1/2} - \varrho(\mathbf{D}^T)^{1/2}\mathbf{C}\mathbf{D}^{1/2} &= (\mathbf{D}^T)^{1/2}(\mathbf{\Gamma}^T)^{-1/2}\\ &\quad\times \mathbf{\Omega}^{-1}\mathbf{\Gamma}^{-1/2}\mathbf{D}^{1/2}.\end{aligned}$$

So

$$\begin{aligned}&\mathbf{I} - \varrho(\mathbf{D}^T)^{1/2}\mathbf{C}\mathbf{D}^{1/2}\\ &\quad = \mathbf{I} - (\mathbf{D}^T)^{1/2}\mathbf{D}^{1/2} + (\mathbf{D}^T)^{1/2}(\mathbf{\Gamma}^T)^{-1/2}\mathbf{\Omega}^{-1}\mathbf{\Gamma}^{-1/2}\mathbf{D}^{1/2}.\end{aligned} \tag{7.9}$$

$\mathbf{I} - \varrho(\mathbf{D}^T)^{1/2}\mathbf{C}\mathbf{D}^{1/2} = \mathbf{\Omega}^{-1}$ holds only if $(\mathbf{D}^T)^{1/2}\mathbf{D}^{1/2} = \mathbf{I}$, and $\mathbf{D}^{1/2}\mathbf{\Gamma}^{-1/2} = \mathbf{I}$, implying that $\mathbf{D}^{1/2} = \mathbf{\Gamma}^{1/2}$. Consequently, equation (7.9) becomes the desired solution when the columns of $\mathbf{\Gamma}^{1/2}$ are required to be orthonormal, and $\mathbf{D} = \mathbf{\Gamma}$. Orthonormality is the constraint evoked here, then.

Again simulation experiments were conducted in order to assess how successful these correction techniques are. In the first case, the constraint $\alpha < 4/3$ was attached to the numerical analysis, since most boundary cells were missing a single adjacent cell. Unfortunately the second and third forms consistently failed to produce estimates of ϱ that fell within the feasible parameter space. Meanwhile, the correction technique for which $\mathbf{\Gamma}$ was constrained to be a diagonal matrix that is a function of the principal eigenvalue of connectivity matrix $\mathbf{C}$, failed to remove any of the bias due to edge effects.

In this latter case, these simulation experiments employed the following initial conditions:

(1) $\xi_i \sim$ iid $N(0, 1)$;
(2) $\varrho = 0.10, 0.20$, or 0.249;
(3) the sub-region consisted of a 5-by-5 lattice centrally located within an 11-by-11 lattice; and,
(4) the two constraints $|\varrho| < 1/\lambda_{\max}$ and $0 < \alpha < 4/3$ were applied.

As a result of the constraint on α the variance, which was assumed to be unity for interior cells (without loss of generality), would tend to increase. Since 3/4 of the information tended to be available, then, an upper bound of 4/3 was assumed. Interestingly, all $\hat{\alpha}$ calculated for the simulated surfaces fell below this value. Because these experiments were exploratory in nature, and very costly to run in terms of computer resources, only 25 simulated surfaces were investigated.

Summary results from these simulation experiments are presented in Table 7.4. Let $\hat{\varrho}_u$ denote the uncorrected estimate of ϱ, and let $\hat{\varrho}_c$ denote the corrected estimate of ϱ. Comparing the results in Table 7.4 clearly demonstrates that $E(\hat{\varrho}_c) \leqslant E(\hat{\varrho}_u) < \varrho$; thus the correction technique for which $\mathbf{\Gamma}$ is constrained to be diagonal fails to remove any of the bias due to edge effects. In fact, this procedure actually may exacerbate the bias. The correlation coefficients indicate that the pairs of estimates are very similar. Consequently, the simulation experimental results are not supportive of this method.

Kuensch (1983) has taken a similar approach to defining a correction factor matrix for edge effects. His formulation is of the form

$$(\mathbf{I} - \varrho\mathbf{C} - \mathbf{\Gamma}), \qquad (7.10)$$

TABLE 7.4

Summary Statistics for Characteristics of 25 Simulated Surfaces

Sampling Distribution Statistic	original $\hat{\rho}$			corrected $\hat{\rho}$		
$\rho =$	.10	.20	.249	.10	.20	.249
Mean	.023	.125	.175	.021	.114	.173
Variance	.013	.023	.012	.013	.025	.017
Skewness	- .158	- .624	-1.342	- .125	- .542	-1.226
Kurtosis	- .067	-1.054	1.190	- .129	- .961	.759
Minimum	- .220	- .165	- .112	- .221	- .220	- .182
Maximum	.261	.285	.284	.253	.289	.289

where $\gamma_{ij} = 0$ except when both lattice cells i and j are on the boundary. He then further restricted γ_{ij} such that it equals zero unless lattice cells i and j both are on the same edge of the sub-region lattice. The foregoing results for sub-matrices $\mathbf{B}_{12}$ and $\mathbf{B}_1$ extracted from matrix $\mathbf{\Omega}^{-1}$ support these contentions to some degree. Kuensch also conducted a set of simulation experiments ($\varrho = 0.2$ and 0.249, 100 replications, a conditional Gaussian scheme) and explored 5-by-5 and 10-by-10 sub-regions. The basic conclusion he reached from this numerical exercise was that the uncorrected estimator is hardly worse in the cases of weak to moderate spatial autocorrelation (recall that the first-order correlations here are 0.2659 and 0.5898, respectively, as shown in Chapter 5). Nevertheless, Guyon (1982) has shown that the uncorrected expression yields an estimator that has the correct asymptotic behavior and is asymptotically efficient, which is not surprising since $(\mathbf{I} - \varrho\mathbf{C}) \rightarrow \mathbf{\Omega}^{-1}$ as $n \rightarrow \infty$.

7.4. CONCLUSIONS AND IMPLICATIONS

A wide variety of edge effect correction techniques have been surveyed here. The presence of edge effects has been documented here as well as elsewhere. Unfortunately findings from the research devoted to this problem imply that these edge effects are quite resistant with regard to the taking of any action, especially that involving intuitively appealing methods, to correct for them.

One should note, though, that the bias uncovered in the simulation experiments is not strictly due to edge effects. Haining (1978), for instance, has found that the MLE of ϱ is highly variable, even when the specification of $(\mathbf{I} - \varrho\mathbf{C})$ is correct. This latter source of bias arises from sampling variability. When drawing a set of iid normal random values, the sampling procedure itself will introduce error that is detected when calculating $\hat{\mu}$, $\hat{\sigma}^2$ and $\hat{\varrho}$. In various ways these errors due to sampling are magnified to some extent when a prescribed level of autocorrelation is embedded in each simulated surface.

Another source of bias arises from the non-linearity of the maximum likelihood equations. This source of bias should result in $E(\hat{\varrho}) < \varrho$, $\varrho > 0$. In this situation the best one can achieve is an unbiased estimating equation, whose form would be

$$E[\partial \ln (L)/\partial \tau|_{\varrho=\hat{\varrho}}] \quad = \quad 0.$$

A third prominent source of deviation between ϱ and $\hat{\varrho}$ has to do with lattice size. If the lattice is replicated in an appropriate way, which allows it to become larger while preserving the stationarity assumption, then $\hat{\varrho}$ needs to be made asymptotically unbiased. Thus the size of the lattice introduces variability into $\hat{\varrho}$.

In conclusion, at least for the time being spatial analysts should recognize and acknowledge the presence of edge effect biases when analysing spatially autocorrelated data. But for the most part attempts on their part to cope with these biases will be futile. Thus, for the time being they should devote their energies to more productive considerations, and temper their conclusions accordingly. They should at least acknowledge the likely inferential errors that may plague the inferences they draw, too.

7.5. REFERENCES

Bartlett, M., 1975, *The Statistical Analysis of Spatial Pattern*. London: Chapman and Hall.

Boots, B., 1985, Size effects in the spatial patterning of nonprincipal eigenvectors of planar networks, *Geographical Analysis*, Vol. 17: 74–81.

Cliff, A., and J. Ord, 1981, *Spatial Processes: Models and Applications*. London: Pion.

Cliff, A., P. Haggett, and J. Ord, 1979, Graph theory and geography, in *Applications of Graph Theory*, edited by R. Wilson and L. Beineke. New York: Academic Press, pp. 293–326.

Goldsmith, H., 1977, The exact distributions of the serial correlation coefficients and an evaluation of some approximation distributions, *Journal of Statistical Computation and Simulation*, Vol. 5: 115–134.

Griffith, D., 1983, The boundary value problem in spatial statistical analysis, *Journal of Regional Science*, Vol. 23: 377–387.

Griffith, D., 1984, Measuring the arrangement property of a system of areal units generated by partitioning a planar surface, in *Recent Developments in Spatial Data Analysis*, edited by G. Bahrenberg, M. Fischer, and P. Nijkamp. Hants, England: Gower, pp. 191–199.

Griffith, D., 1985, An evaluation of correction techniques for boundary effects in spatial statistical analysis: contemporary methods, *Geographical Analysis*, Vol. 17: 81–88.

Griffith, D., and C. Amrhein, 1983, An evaluation of correction techniques for boundary effects in spatial statistical analysis: traditional methods, *Geographical Analysis* Vol. 15: 352–360.

Guyon, X., 1982, Parameter estimation for a stationary process on a *d*-dimensional lattice, *Biometrika*, Vol. 69: 95–105.

Haggett, P., 1980, Boundary problems in quantitative geography, in *Die Bedeutung von Grenzen in der Geographie*, edited by H. Kishimoto. Zurich: Kummerley and Frey, pp. 59–67.

Haggett, P., 1981, The edges of space, in *European Progress in Spatial Analysis*, edited by R. Bennett. London: Pion, pp. 51–70.

Haining, R., 1978, *Specification and Estimation Problems in Models of Spatial Dependence*. Evanston, Ill.: Northwestern University.

Haining, R., 1978, Estimating spatial interaction models. *Environment and Planning* A, Vol. 10: 305–320.

Johnston, J., 1972, *Econometric Methods* (2nd ed.). New York: McGraw-Hill.

Krumbein, W., 1966, *A Comparison of Polynomial and Fourier Models in Map Analysis*, ONR Task No. 338–078. Washington, D.C.: Office of Naval Research.

Kuensch, H., 1983, Approximations to the maximum likelihood equations for some Gaussian random fields, *Scandinavian Journal of Statistics*, Vol. 10: 239–246.

Ripley, B., 1981, *Spatial Statistics*. New York: Wiley.

Wonnacott, R., and T. Wonnacott, 1977, *Econometrics*. New York: Wiley.

CHAPTER 8

Multivariate Models of Spatial Dependence

Chapter 4 presents several extensions from a univariate context to a multivariate context. These extensions include the replacement of $\mu\mathbf{1}$ with $\mathbf{X}\boldsymbol{\beta}$ (a trend surface model with autocorrelated errors), a fixed effects multiple linear regression model with autocorrelated errors (for ANOVA), the two-groups discriminant function model with autocorrelated errors, and the bivariate regression model with autocorrelated random variables (a variable effects regression model). This chapter will be concerned with the relationships between spatial autocorrelation and multivariate data sets. One of the first attempts at exploring this issue was reported by Lebart (1969), who assumed the ϱ_j, $j = 1, 2, \ldots, p$, autocorrelation parameters for a multivariate data set were the same, and in fact equal to unity. Later, Griffith (1986) outlined a taxonomy of autocorrelation problems for the linear regression model. His discussion alluded to the preoccupation of statisticians either with the case of no autocorrelation in any of these regression model terms, or a concern only with the problem of a spatially autocorrelated error term. In contrast, spatial econometricians have focused much of their attention on an autocorrelated dependent variable. Both groups seemed to ignore problems associated with autocorrelated independent variables. Earlier both Griffith (1980) and Bivand (1980) observed the effects of spatial autocorrelation on a correlation coefficient when autocorrelation is present in one or more of the variables, pointing out that the product moment correlation coefficient r will tend to be inflated when both X and Y are highly positively spatially autocorrelated. Meanwhile, Griffith (1979) demonstrated that MANOVA and discriminant function analysis results can be altered quite dramatically once data are cleaned of spatial autocorrelation effects. More recently Wartenberg (1985) described what he calls multivariate spatial autocorrelation. In his discussion he generalizes the Moran Coefficient from its matrix form of $(\mathbf{X} - \bar{x}\mathbf{1})^T\mathbf{W}(\mathbf{X} - \bar{x}\mathbf{1})$, where vector $\mathbf{X}$ is n-by-1, to its z-score multivarite form of $\mathbf{Z}^T\mathbf{W}\mathbf{Z}$, where matrix $\mathbf{Z}$ is n-by-p. But the resulting matrix here may not be symmetric, and will yield negative eigenvalues. The more traditional counterpart of Wartenberg's formulation is $n\mathbf{S}^{-1/2}\mathbf{Z}^T\mathbf{C}\mathbf{Z}\mathbf{S}^{-1/2}/[(n - 1)\mathbf{1}^T\mathbf{C}\mathbf{1}]$, where matrix $\mathbf{Z}$ is a centered version of the original X variables (i.e., only their respective means have been subtracted from them), and $\mathbf{S}^{1/2}$ is a diagonal matrix of sample standard deviations. This latter form at least is symmetric. Finally, Streitberg (1979) has discussed the modification of selected multivariate techniques using a spatial moving average scheme, which because of invertibility is connected to the conditional and simultaneous models.

8.1. A MULTIVARIATE NORMAL PROBABILITY DENSITY FUNCTION WITH SPATIAL AUTOCORRELATION

The multivariate normal density function was selected for study here for two reasons. First, one of the two models used extensively in spatial autocorrelation analysis derives from this particular probability density function. Second, this density forms the basis for much of the inferential and pseudo-inferential multivariate statistical theory.

In its classical form, for n independent observations and p variates, the multivariate normal probability density function is written as

$$L_1 = \prod_{i=1}^{i=n} (2\pi)^{-p/2}|{}_p\mathbf{\Sigma}|^{-1/2} \exp\,[-({}_p\mathbf{X}_i - {}_p\boldsymbol{\mu})^T\,{}_p\mathbf{\Sigma}^{-1}({}_p\mathbf{X}_i - {}_p\boldsymbol{\mu})/2], \tag{8.1}$$

where ${}_p\mathbf{\Sigma}$ denotes a p-by-p variance–covariance matrix, ${}_p\boldsymbol{\mu}$ denotes the p-by-1 vector of variable means, and ${}_p\mathbf{X}_i$ denotes the p-by-1 vector of variable values observed on observation i. Equation (8.1) also may be written as

$$L_1 = (2\pi)^{-np/2} |{}_p\boldsymbol{\Sigma}|^{-n/2} \times \exp\left[-(1/2) \sum_{i=1}^{i=n} ({}_p\mathbf{X}_i - {}_p\boldsymbol{\mu})^T {}_p\boldsymbol{\Sigma}^{-1} ({}_p\mathbf{X}_i - {}_p\boldsymbol{\mu}) \right],$$

or

$$L_1 = (2\pi)^{-np/2} |{}_p\boldsymbol{\Sigma}|^{-n/2} \exp\left[-(1/2) \langle {}_p\mathbf{X}_i - {}_p\boldsymbol{\mu} \rangle^T \times ({}_n\mathbf{I} \otimes {}_p\boldsymbol{\Sigma}^{-1}) \langle {}_p\mathbf{X}_i - {}_p\boldsymbol{\mu} \rangle \right],$$

where ${}_n\mathbf{I}$ is an n-by-n identity matrix, $\otimes$ denotes the Kronecker product (see Appendix 8A), and

$$\langle {}_p\mathbf{X}_i - {}_p\boldsymbol{\mu} \rangle = \begin{pmatrix} {}_p\mathbf{X}_1 - {}_p\boldsymbol{\mu} \\ \hline \vdots \\ \hline {}_p\mathbf{X}_n - {}_p\boldsymbol{\mu} \end{pmatrix},$$ which is a partitional np-by-1 vector.

This third version of equation (8.1) employs a vector that is partitioned into n sub-vectors, each of which is p-by-1. In contrast, one common form of equation (8.1) used to model spatially autocorrelated data is expressed as equation (4.7). In other words, ${}_p\boldsymbol{\Sigma}$ of equation (8.1) has been equated with $[({}_n\mathbf{I} - \varrho\ {}_n\mathbf{W})^T ({}_n\mathbf{I} - \varrho\ {}_n\mathbf{W})]^{-1}$ of equation (4.7), and hence this latter equation actually constitutes a single multivariate observation. The difference between equations (8.1) and (4.7) is akin to that between Q-mode and R-mode factor analysis.

Next, consider a truly multivariate data set, where p measures are made on n areal units. Let ${}_p\boldsymbol{\Sigma} = {}_p\mathbf{L}_p\mathbf{L}^T$, a decomposition commonly referred to in multivariate analysis. One should note that this decomposition is not a unique one. For instance, ${}_p\mathbf{L}$ could be a relatively dense matrix, constructed from the eigenvalues and eigenvectors of matrix ${}_p\boldsymbol{\Sigma}$. Or, ${}_p\mathbf{L}$ could be lower-triangular, a form that results from the Cholesky decomposition of the variance–covariance matrix. Clearly, since the eigenvector solution could be orthogonally rotated, other solutions are possible, too. The non-uniqueness of this decomposition will not alter those results reported here. Now equations (8.1) and (4.7) may be accommodated by writing the multivariate normal density function as

$$L_2 = (2\pi)^{-np/2} |[({}_p\mathbf{L}^{-1} \otimes {}_n\mathbf{I}) \langle({}_n\mathbf{I} - \varrho_j\, {}_n\mathbf{W})\rangle_{\text{diag}}]^T [({}_p\mathbf{L}^{-1} \otimes {}_n\mathbf{I})$$
$$\times \langle {}_n\mathbf{I} - \varrho_j\, {}_n\mathbf{W})\rangle_{\text{diag}}]|^{1/2} \exp \{(-1/2) \langle {}_n\mathbf{X}_j - \mu_j\, {}_n\mathbf{1}^T\rangle$$
$$\times [\langle({}_n\mathbf{I} - \varrho_j\, {}_n\mathbf{W})\rangle^T_{\text{diag}} ({}_p\mathbf{\Sigma}^{-1} \otimes {}_n\mathbf{I}) \langle({}_n\mathbf{I} - \varrho_j\, {}_n\mathbf{W})\rangle_{\text{diag}}]$$
$$\times \langle {}_n\mathbf{X}_j - \mu_j\, {}_n\mathbf{1}\rangle\}$$
$$= (2\pi)^{-np/2} \left\{\prod_{j=1}^{j=p} \left[\prod_{i=1}^{i=n} (1 - \varrho_j \lambda_i)\right]\right\} |{}_p\mathbf{\Sigma}|^{-n/2}$$
$$\times \exp [(-1/2) \langle {}_n\mathbf{X}_j - \mu_j\, {}_n\mathbf{1}\rangle^T [\langle({}_n\mathbf{I} - \varrho_j\, {}_n\mathbf{W})\rangle^T_{\text{diag}}$$
$$\times ({}_p\mathbf{\Sigma}^{-1} \otimes {}_n\mathbf{I}) \langle({}_n\mathbf{I} - \varrho_j\, {}_n\mathbf{W})\rangle_{\text{diag}}] \langle {}_n\mathbf{X}_j - \mu_j\, {}_n\mathbf{I}\rangle\}, \quad (8.2)$$

where

$$\langle {}_n\mathbf{X}_j - \mu_j\, {}_n\mathbf{1}\rangle = \begin{pmatrix} {}_n\mathbf{X}_1 - \mu_1\, {}_n\mathbf{1} \\ \hline \vdots \\ \hline {}_n\mathbf{X}_p - \mu_p\, {}_n\mathbf{1} \end{pmatrix},$$ which is a partitioned np-by-1 vector,

$\langle({}_n\mathbf{I} - \varrho_j\, {}_n\mathbf{W})\rangle_{\text{diag}}$ is an np-by-np block-diagonal matrix, having p diagonal blocks, each being n-by-n, and λ_i are the n eigenvalues of matrix ${}_n\mathbf{W}$.

The sums-of-squares-and-cross-products term of equation (8.2) also can be expressed as

$$\langle {}_p\mathbf{X}_i - \boldsymbol{\mu}_p\rangle^T [{}_n\mathbf{I} \otimes {}_p\mathbf{\Sigma}^{-1} - {}_n\mathbf{W} \otimes ({}_p\mathbf{\Sigma}^{-1}\, {}_p\langle\varrho_j\rangle_{\text{diag}}) - {}_n\mathbf{W}^T$$
$$\otimes ({}_p\langle\varrho_j\rangle_{\text{diag}}\, {}_p\mathbf{\Sigma}^{-1}) + {}_n\mathbf{W}^T\, {}_n\mathbf{W} \otimes ({}_p\langle\varrho_j\rangle_{\text{diag}}\, {}_p\mathbf{\Sigma}^{-1}\, {}_p\langle\varrho_j\rangle_{\text{diag}})]$$
$$\times \langle {}_p\mathbf{X}_i - \boldsymbol{\mu}_p\rangle.$$

If $\varrho_j = \varrho$ for all j, then ${}_p\langle\varrho_j\rangle_{\text{diag}} = \varrho\, {}_p\mathbf{I}$, and hence the inverse variance–covariance term of this expression becomes

$${}_n\mathbf{I} \otimes {}_p\mathbf{\Sigma}^{-1} - \varrho\, {}_n\mathbf{W} \otimes {}_p\mathbf{\Sigma}^{-1} - \varrho\, {}_n\mathbf{W}^T \otimes {}_p\mathbf{\Sigma}^{-1} + \varrho^2\, {}_n\mathbf{W}^T\, {}_n\mathbf{W} \otimes {}_p\mathbf{\Sigma}^{-1}$$
$$= ({}_n\mathbf{I} - \varrho\, {}_n\mathbf{W})^T ({}_n\mathbf{I} - \varrho\, {}_n\mathbf{W}) \otimes {}_p\mathbf{\Sigma}^{-1},$$

and the Jacobian reduces to $|{}_n\mathbf{I} - \varrho\, {}_n\mathbf{W}|^p$. This, in essence, is the type of model Lebart was considering. But he assumed all of the ϱ_j's were the same, and, more specifically, that $\varrho = 1$. Accordingly,

$$|_n\mathbf{I} - \varrho\ _n\mathbf{W}|^p = \left|\prod_{i=1}^{i=n} (1 - \varrho\lambda_i)\right|^p .$$

Since $\lambda_{\max} = 1$ by construction, this determinant is zero, and hence $L_2 = 0$. Therefore, it is not surprising that Lebart was discouraged by the results he obtained. Furthermore, equation (8.2) states that there are $p(p + 5)/2$ parameters to be estimated. Clearly n will need to be very large in order for a proper number of degrees of freedom to be maintained.

If $\varrho = 0$, then equation (8.2) reduces to equation (8.1). Similarly, if $p = 1$, then equation (8.2) reduces to equation (4.7). Another attribute of equation (8.2) worth mentioning is that the sums-of-squares-and-cross-products term could have been written as

$$\langle_n\mathbf{X}_j - \mu_j\ _n\mathbf{1}\rangle^T[(_p\mathbf{L}^{-1} \otimes\ _n\mathbf{I})\ \langle_n\mathbf{I} - \varrho\ _n\mathbf{W}\rangle_{\text{diag}}]^T[(_p\mathbf{L}^{-1} \otimes\ _n\mathbf{I})$$
$$\times\ \langle_n\mathbf{I} - \varrho\ _n\mathbf{W}\rangle_{\text{diag}}]\ \langle_n\mathbf{X}_j - \mu_j\ _n\mathbf{1}\rangle.$$

A suitable permutation matrix can be applied to the partitioned vectors in order to have either n partitions, each of size p, or p partitions, each of size n.

While equation (8.2) was the type of model utilized by Lebart, there is no reason to constrain the ϱ_j's so that they all are the same value. In fact, this constraint seems a bit silly. Consider, for example, the classical multivariate problem corresponding to equation (8.1). As was mentioned earlier, Lebart's restriction would be equivalent to assuming that $\sigma_j^2 = \sigma^2$, for all j. Clearly applied statisticians would find such an assumption quite unpalatable in most situations, even though such an assumption would help to simplify equation (8.1), as well as considerably reduce the number of parameters that would need to be estimated.

8.2. DISCERNING LATENT STRUCTURE IN MULTIVARIATE SPATIAL DATA

Much work has been completed on the multivariate analysis of time series. This work will serve here as a point of departure.

Basically, multiple time series relations are studied by first removing the latent autocorrelation structure each embraces (Vandaele, 1983). This process is referred to as "prewhitening" the data, and amounts to first obtaining an appropriate univariate autoregressive model (as well as an appropriate moving average model) for each data series $\mathbf{X}_j$ under study, and then calculating the cross correlations of the underlying random variates ξ_j. Cross correlations allow the transformed variables to be inspected to see what type of relationships exists amongst them. The primary rationale for this approach stems from the fact that highly autocorrelated data series render cross correlation coefficients that are difficult to interpret, and even misleading at times. Moreover, it is quite possible that two variates ξ_j and ξ_k (white noise) which are not related will have corresponding $\mathbf{X}_j$ and $\mathbf{X}_k$ variables (autocorrelated) that exhibit high spurious correlation when each of these latter two variables is markedly autocorrelated. Thus in properly measuring the degree of association it is important to prewhiten the data.

The formulation of equation (8.2) is in keeping with this maxim. The term

$$\begin{pmatrix} {}_n\mathbf{X}_1 - \mu_1\,{}_n\mathbf{1} \\ \hline \vdots \\ \hline {}_n\mathbf{X}_p - \mu_p\,{}_n\mathbf{1} \end{pmatrix}^T \langle({}_n\mathbf{I} - \varrho_j\,{}_n\mathbf{W})\rangle^T_{\text{diag}}$$

actually is a prewhitening transformation, since it yields the partitioned np-by-n matrix

$$\langle({}_n\mathbf{X}_1 - \mu_1\,{}_n\mathbf{1})^T({}_n\mathbf{I} - \varrho_1\,{}_n\mathbf{W}) \vdots \cdots \vdots ({}_n\mathbf{X}_p - \mu_p\,{}_n\mathbf{1})^T({}_n\mathbf{I} - \varrho_p\,{}_n\mathbf{W})\rangle.$$

Since for this data set the inverse variance–covariance matrix is ${}_p\mathbf{\Sigma}^{-1} \otimes {}_n\mathbf{I}$, one can see why the product moment correlation coefficient becomes equation (4.28). As an aside, viewing regression coefficients in terms of correlation coefficients (if the variables are converted to z-scores, then the regression coefficients are identical to the product moment correlation coefficients) reinforces the intuitive appeal of equation (4.28).

But equation (8.2) implies that a prewhitening procedure that parallels the one followed in time series analysis is inappropriate for spatial data series, since the Jacobian term

$$\prod_{j=1}^{j=p}\left[\prod_{i=1}^{i=n}(1-\varrho_j\lambda_i)\right]$$

does not vanish. In addition, inspection of the MLE's reveals strongly articulated patterns of dependencies. For instance,

$$\begin{aligned}{}_p\hat{\boldsymbol{\mu}} &= [({}_p\mathbf{I}\otimes{}_n\mathbf{1})^T\langle{}_n\mathbf{I}-\varrho_j\,{}_n\mathbf{W}\rangle^T_{\text{diag}}\,({}_p\boldsymbol{\Sigma}^{-1}\otimes{}_n\mathbf{I})\,\langle{}_n\mathbf{I}-\varrho_j\,{}_n\mathbf{W}\rangle_{\text{diag}}\\ &\quad\times({}_p\mathbf{I}\otimes{}_n\mathbf{1})]^{-1}\,({}_p\mathbf{I}\otimes{}_n\mathbf{1})^T\langle{}_n\mathbf{I}-\varrho_j\,{}_n\mathbf{W}\rangle^T_{\text{diag}}({}_p\boldsymbol{\Sigma}^{-1}\otimes{}_n\mathbf{I})\\ &\quad\times\langle{}_n\mathbf{I}-\varrho_j\,{}_n\mathbf{W}\rangle_{\text{diag}}\,\langle{}_n\mathbf{X}_j\rangle\end{aligned}$$

where $\langle{}_n\mathbf{X}_j\rangle$ is an np-by-1 vector consisting of p partitions, each containing an n-by-1 vector ${}_n\mathbf{X}_j$. This last estimate simplifies to a vector whose individual elements are given by

$$\hat{\mu}_j = {}_n\mathbf{1}^T({}_n\mathbf{I}-\varrho_j\,{}_n\mathbf{W})^T({}_n\mathbf{I}-\varrho_j\,{}_n\mathbf{W})_n\mathbf{X}_j/[n(1-\varrho_j)^2]$$

only if ${}_p\boldsymbol{\Sigma} = {}_p\langle\sigma_j^2\rangle_{\text{diag}}$. Further, it simplifies to the classical statistical results only if $\varrho_j = 0$ for all j. And, it simplifies to equation (4.3) only if $p = 1$.

As was illustrated with the bivariate regression model in Chapter 4, estimation of ${}_p\boldsymbol{\mu}$ is far more difficult, in general, because matrix ${}_p\boldsymbol{\Sigma}$ is embedded in the variance–covariance matrix, and hence does not factor out. Moreover,

$$\begin{aligned}\partial L({}_p\boldsymbol{\mu},{}_p\boldsymbol{\Sigma})/\partial_p\boldsymbol{\mu} &\Rightarrow (\langle{}_n\mathbf{X}_j\rangle-{}_p\boldsymbol{\mu}\otimes{}_n\mathbf{1})^T\langle({}_n\mathbf{I}-\varrho_j\,{}_n\mathbf{W})\rangle^T_{\text{diag}}\\ &\quad\times({}_p\boldsymbol{\Sigma}^{-1}\otimes{}_n\mathbf{I})\,\langle({}_n\mathbf{I}-\varrho_j\,{}_n\mathbf{W})\rangle_{\text{diag}}({}_p\mathbf{I}\otimes{}_n\mathbf{1}) = \mathbf{0}. \qquad (8.3)\end{aligned}$$

Because of its algebraic location in equation (8.3), matrix ${}_p\boldsymbol{\Sigma}^{-1}$ cannot be factored out. If $p = 1$, then ${}_p\boldsymbol{\Sigma}^{-1} = \sigma^{-2}$, a scalar, and it does factor out. If $\varrho_j = \varrho$ for all j, then as was mentioned earlier, the inverse variance–covariance matrix becomes ${}_p\boldsymbol{\Sigma}^{-1}\otimes({}_n\mathbf{I}-\varrho\,{}_n\mathbf{W})^T({}_n\mathbf{I}-\varrho\,{}_n\mathbf{W})$, and so matrix ${}_p\boldsymbol{\Sigma}^{-1}$ can be removed. Obviously, as a special case, if $\varrho = 0$, the case of no spatial autocorrelation, matrix ${}_p\boldsymbol{\Sigma}^{-1}$ can be factored out. Finally, if matrix ${}_p\boldsymbol{\Sigma}^{-1}$ is diagonal, then

$$\begin{aligned}&\langle({}_n\mathbf{I}-\varrho_j\,{}_n\mathbf{W})\rangle^T_{\text{diag}}({}_p\boldsymbol{\Sigma}^{-1}\otimes{}_n\mathbf{I})\,\langle({}_n\mathbf{I}-\varrho_j\,{}_n\mathbf{W})\rangle_{\text{diag}}\\ &\quad = ({}_p\boldsymbol{\Sigma}^{-1}\otimes{}_n\mathbf{I})\,\langle({}_n\mathbf{I}-\varrho_j\,{}_n\mathbf{W})\rangle^T_{\text{diag}}\langle({}_n\mathbf{I}-\varrho_j\,{}_n\mathbf{W})\rangle_{\text{diag}},\end{aligned}$$

and thus again matrix ${}_p\boldsymbol{\Sigma}^{-1}$ can be factored out.

The important feature of this modelling approach is that the researcher is able to correctly identify the q components that underly a data structure. In other words, one is trying to determine some matrix $\mathbf{P}$ that will permit statistically independent linear combinations of variables to be constructed. The presence of non-zero spatial autocorrelation complicates the estimation of matrix $\mathbf{P}$. Once the matrix of raw data $\mathbf{X}$ is transformed by prewhitening it to the matrix of values ξ, then matrix $\mathbf{P}$ can be estimated from the correlation matrix based upon these latter transformed variables. Then, synthetic variates can be constructed with either $\boldsymbol{\xi}\mathbf{P}$ or $\mathbf{XP}$, depending upon whether or not a researcher wants spatial autocorrelation to be present in the final results. In most cases one should want to preserve the original structure of the raw data.

8.3. ESTIMATION PROBLEMS

Clearly the estimation problems alluded to in the previous section are both formidable and lead to numerically intensive work. Because of the non-linear nature of the likelihood function to be evaluated, a functional form established with equation (4.5), iterative techniques will need to be employed.

On general estimation procedure would proceed with the following six steps:

Step 1: estimate μ_j and σ_j assuming $\varrho_j = 0$,

Step 2: estimate ϱ_j assuming $\langle\sigma_j^2\rangle_{\text{diag}}$ and using ${}_p\hat{\boldsymbol{\mu}}$ and $\hat{\sigma}_j^2$ from Step 1,

Step 3: estimate ${}_p\boldsymbol{\Sigma}$ using the formula

$$\hat{\sigma}_{jk} = ({}_n\mathbf{X}_j - \hat{\mu}_j\,{}_n\mathbf{1})^T({}_n\mathbf{I} - \hat{\varrho}_j\,{}_n\mathbf{W})^T({}_n\mathbf{I} - \hat{\varrho}_k\,{}_n\mathbf{W}) \times ({}_n\mathbf{X}_k - \hat{\mu}_k\,{}_n\mathbf{1}),$$

Step 4: estimate ${}_p\boldsymbol{\mu}$ using equation (8.4),

Step 5: maximize equation (8.3) for $\{\varrho_j\}$, substituting ${}_p\hat{\boldsymbol{\Sigma}}^{-1}$ from Step 3 and ${}_p\hat{\boldsymbol{\mu}}$ from Step 4, and

Step 6: repeat Steps 3 thru 5 until convergence occurs for the set $\{\hat{\varrho}_j\}$.

Steps 1 and 2 provide reasonable, rather than arbitrary, starting values for the iterative process. Step 3 furnishes updated estimates of the elements of ${}_p\boldsymbol{\Sigma}$, so that the inverse and determinant of this matrix can be calculated at each iteration. Step 4 supplies an estimate of ${}_p\boldsymbol{\mu}$ so that each variable X_j can be centered at each iteration. Given ${}_p\boldsymbol{\Sigma}^{(\tau-1)}$ and ${}_p\boldsymbol{\mu}^{(\tau-1)}$, for iteration $(\tau - 1)$, then the function to be maximized in Step 5, for iteration τ, becomes

$$\sum_{j=1}^{j=p}\sum_{i=1}^{i=n} \ln(1 - \varrho_j\lambda_i) - (n/2)\ln|{}_p\boldsymbol{\Sigma}^{(\tau-1)}|$$
$$- \sum_{j=1}^{j=p}\sum_{k=1}^{k=p} s_{jk}^{(-1)(\tau-1)}({}_n\mathbf{X}_j - \mu_j^{(\tau-1)}\,{}_n\mathbf{1})^T$$
$$\times ({}_n\mathbf{I} - \varrho_k\,{}_n\mathbf{W})({}_n\mathbf{X}_k - \mu_k^{(\tau-1)}\,{}_n\mathbf{1})/2.$$

This last expression has p unknowns (one for each ϱ_j), and so will have a greater chance of achieving a local rather than a global optimum. In addition, the p constraints $|\varrho_j| < 1$ must be attached here.

As was pointed out for equation (8.3), this procedure can be simplified if evidence exists supporting the hypothesis of $\varrho_j = \varrho$ for all j. Then

$$L_3 = (2\pi)^{-np/2}|{}_n\mathbf{I} - \varrho\,{}_n\mathbf{W}|^p|{}_p\boldsymbol{\Sigma}|^{-n/2}$$
$$\times \exp\{(-1/2)\langle{}_p\mathbf{X}_i - {}_p\boldsymbol{\mu}\rangle^T[({}_n\mathbf{I} - \varrho\,{}_n\mathbf{W})^T$$
$$\times ({}_n\mathbf{I} - \varrho\,{}_n\mathbf{W}) \otimes {}_p\boldsymbol{\Sigma}^{-1}]\langle{}_p\mathbf{X}_i - {}_p\boldsymbol{\mu}\rangle\},$$

which is a simpler function to maximize. For instance, the MLE's of ${}_p\boldsymbol{\mu}$ and ${}_p\boldsymbol{\Sigma}$ are

$$\frac{\partial \ln(L_3)}{\partial_p\boldsymbol{\mu}} = 0 \Rightarrow \hat{\mu}_j = {}_n\mathbf{1}^T({}_n\mathbf{I} - \varrho\,{}_n\mathbf{W})^T$$
$$\times ({}_n\mathbf{I} - \varrho\,{}_n\mathbf{W})\,{}_n\mathbf{X}_j/[n(1 - \varrho)^2],$$

$$\frac{\partial \ln(L_3)}{\partial_p\boldsymbol{\Sigma}^{-1}} = 0 \Rightarrow {}_p\hat{\boldsymbol{\Sigma}} = \langle{}_p\mathbf{X}_i - {}_p\mu\rangle^T[({}_n\mathbf{I} - \varrho\,{}_n\mathbf{W})^T$$
$$\times ({}_n\mathbf{I} - \varrho\,{}_n\mathbf{W}) \otimes {}_p\mathbf{I}]\langle{}_p\mathbf{X}_i - {}_p\boldsymbol{\mu}\rangle/n.$$

Hence, the tradeoffs between a simple solution, on the one hand, and permitting ϱ to vary over the p variables, on the other hand, simply may be too costly. However, rather than pursuing Lebart's suggestion of setting $\varrho = 1$, the single parameter should be estimated together with $_p\boldsymbol{\mu}$ and $_p\boldsymbol{\Sigma}$. Further, even if p different values of ϱ_j are obtained, if they are almost identical in magnitude, then a much simpler model could be worked with by reestimating a single value for them. Since they are nearly identical, the specification error that would be introduced should be negligible.

The parameter estimates determined for equation (8.2) document the intimate relationship that exists between correlated variables and correlated observations. Obviously, neither of these sources of data structure should be ignored. Furthermore, their parameter estimates need to be calculated in a simultaneous fashion. Returning to the taxonomy discussed by Griffith (1986), non-zero spatial autocorrelation potentially could have serious consequences for multiple regression analysis. For instance, it will alter the observed multicollinearity amongst the predictor variables. Or, it could be the source of observed correlation between the X's and the model's error term. Consequently, there is a need to fully explore and contrast all eight classes of problems outlined in the taxonomy proposed by Griffith.

8.4. SELECTED EMPIRICAL EXAMPLES

Two empirical examples will be presented in this section in an attempt to illustrate the major concepts emphasized in this chapter. The first example deals with 1981 crime data for the City of Buffalo. Geographic spillovers are expected because crime displacement is related to prevailing levels of law enforcement. Moreover, if there is an increase in law enforcement in a specific geographic location, criminals tend to shift their crimes to other geographic areas. This phenomenon has been documented elsewhere to some extent by Fabrikant (1979, 1980), Gabor (1981), Hakim et al. (1979). Mehay (1977), and Rengert (1980).

The second example has to do with the 1969 spatial organization of agricultural production in Puerto Rico. Geographic spillovers are expected here because of the spatial distribution of common climate, topography and soil features, and because of Von Thuenen types of economic influences emanating from the island's major urban areas. This first source, a missing variables type of factor, has been documented to some degree for the high plains of the United States by Haining (1978).

8.4.1. *An Empirical Example: 1981 Buffalo Crime Data*

The selected 1981 crime data were collected for the police precincts of the City of Buffalo. These data appear in Table 8.1, while the accompanying map appears in Figure 8.1. In order to partial out size effects, manpower, calls for police service, and each of the crime variables was divided by precinct population. These per capita rates then were subjected to extensive analysis.

The first step in this analysis was to construct the matrix of Moran Coefficients, as described by Wartenberg. These statistics appear in Table 8.2. The eigenstructure of this matrix also appears in Table 8.2. Clearly the interpretation of these results is not straight forward. A number of the eigenvalues are negative. Matrix **C** rather than matrix **W** had to be used because this latter matrix was not symmetric. Further, the first eigenvector indicates that all of the variables display a markedly similar pattern of cross spatial autocorrelation indices. In addition, the second and third eigenvectors uncover the dimensions of murder and arson.

Returning to a traditional principal components analysis of these data, whose results appear in Table 8.3, three dimensions have been extracted. Apparently these results are consistent with the multivariate Moran Coefficient matrix results. The second component is arson, the third is murder, and the first captures all other variables. Together these three components account for 98.6% of the variation displayed in the Buffalo crime data (partly due to a very small sample size of 14 versus 11 variables). In contrast, Lebart's first spatial differences data yielded those results reported in Table 8.4. Again only three prominent dimensions are present, although evidence exists that possibly just two prominent dimensions are present (i.e., only two eigenvalues exceed unity for

TABLE 8.1

City of Buffalo Crime Data for 1981

Precinct	Manpower	Population	Calls for police service	Murder	Rape	Robbery	Serious assault
3	53	7695	32342	0	20	347	68
4	41	17354	15673	5	19	166	87
5	47	37558	26876	2	7	83	52
6	48	39950	31008	4	36	273	103
7	41	9846	12827	0	16	83	22
8	47	26539	22429	4	27	159	63
9	39	28822	14992	1	0	38	16
10	42	28077	24404	3	27	153	75
11	40	29897	14789	2	4	51	26
12	61	36005	32911	8	64	282	157
13	46	29502	19461	1	5	49	28
15	40	33998	13967	3	4	29	28
16	62	60671	40503	4	35	313	99
17	48	42863	23547	0	11	92	23

TABLE 8.1 (continued)

Precinct	Burglary	Larceny-theft	Auto theft	Arson	Prosti-tution	Precinct 3	4	5	6	7	8	9	10	11	12	13	15	16	17
3	734	1373	492	13	220	0	1	0	1	1	0	0	1	0	0	0	0	0	0
4	405	354	200	23	74	1	0	0	1	1	1	0	0	0	1	0	0	0	0
5	660	713	395	48	1	0	0	0	1	0	0	0	1	0	0	1	0	0	1
6	877	799	482	33	15	1	1	1	0	0	0	0	1	0	1	0	0	1	1
7	375	314	143	72	0	1	1	0	0	0	1	1	0	0	0	0	1	0	0
8	702	412	259	46	1	0	1	0	0	1	0	1	0	1	1	0	0	0	0
9	355	382	228	44	0	0	0	0	0	1	1	0	0	1	0	0	1	0	0
10	744	558	281	53	7	1	0	1	1	0	0	0	0	0	0	0	0	0	0
11	318	318	190	23	0	0	0	0	0	0	1	1	0	0	1	0	0	1	0
12	1327	607	407	80	10	0	1	0	1	0	1	0	0	1	0	0	0	1	0
13	524	444	229	30	0	0	0	1	0	0	0	0	0	0	0	0	0	0	1
15	412	396	233	33	0	0	0	0	0	1	0	1	0	0	0	0	0	0	0
16	1335	1004	604	65	3	0	0	0	1	0	0	0	0	1	1	0	0	0	1
17	816	772	314	24	0	0	0	1	1	0	0	0	0	0	0	1	0	1	0

the initial solution in this case), and these three dimensions have a combined statistical explanatory power of 98.5%. Exactly the same dimension labels pertain to this set of components as was used before. Comparing the results between Table 8.3 and 8.4 suggests that little has changed in the correlation patterns of the

Precinct Population and Manpower for the City of Buffalo

PRECINCT	POPULATION	MANPOWER
3	7,695	54
4	17,354	45
5	37,558	47
6	39,950	61
7	9,846	42
8	26,539	48
9	24,822	47
10	28,077	45
11	29,897	43
12	36,005	59
13	29,502	47
15	33,998	43
16	60,671	67
17	42,863	48

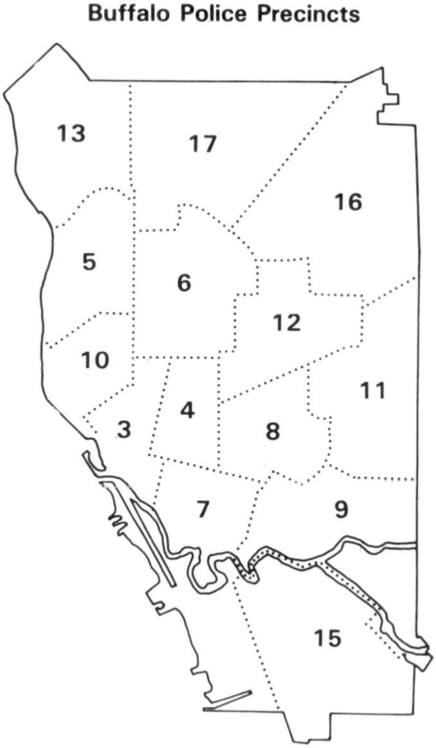

Fig. 8.1.

rotated components structure matrix. About the only noticeable alteration is that the relative importance of the individual rotated components has shifted slightly.

Next equation (4.13) was used to estimate each of the eight ϱ_j values. These results, together with the accompanying principal components results, appear in Table 8.5. But the ϱ_j's were found to be approximately zero, as was expected given the magnitudes of the corresponding conventional Moran Coefficients (i.e., the diagonal entries appearing in Table 8.2), resulting in a rotated components structure matrix that is almost identical to that appearing in Table 8.3.

Finally, in order to demonstrate the impacts that spatial autoregressive modelling can have upon multivariate analysis, the ϱ_j's

TABLE 8.2

Multivariate Moran Coefficients for the Standardized Buffalo Crime Data

Variable	Manpower	Calls for police service	Murder	Rape	Robbery	Serious assault	Burglary
Manpower	.1478						
Calls for police service	.0866	.0394					
Murder	.1456	.1352	.1134				
Rape	.1521	.1279	.2476	.2064			
Robbery	.0980	.0516	.1531	.1628	.0665		
Serious assault	.1563	.1222	.1903	.2534	.1458	.2328	
Burglary	.0804	.0404	.1669	.1209	.0601	.1373	.0342
Larceny-theft	.0620	.0142	.1043	.1026	.0215	.0851	.0173
Auto theft	.0694	.0214	.1220	.1181	.0314	.1013	.0263
Arson	.1612	.1405	.1159	.0591	.1572	.1562	.0995
Prostitution	.0755	.0208	.0998	.1354	.0262	.0911	.0334

TABLE 8.2 (continued)

Variable	Larceny-theft	Auto theft	Arson	Prostitution	Selected Eigenvectors $\underline{E}_1$	$\underline{E}_2$	$\underline{E}_3$
Manpower					.3185	-.0669	.0416
Calls for police service					.2246	-.2219	-.1404
Murder					.3974	.1498	.7159
Rape					.4337	.2348	-.4002
Robbery					.2709	-.2326	-.0650
Serious assault					.4356	-.0062	.2210
Burglary					.2302	-.1418	-.4107
Larceny-theft	-.0102				.1604	-.3238	-.0977
Auto theft	-.0044	.0017			.1875	-.3075	-.1301
Arson	.1493	.1490	-.0473		.2973	.6565	-.2044
Prostitution	-.0102	-.0022	.1936	-.0181	.1914	-.4050	.1260
Eigenvalue	____	____	____	____	1.2208	-.4052	-.1459

TABLE 8.3

Principal Components Structure Matrices
for the Raw Standardized Buffalo Crime Data

Variable	Principal Components Initial Loadings Matrix PC I	PC II	PC III	Varimax Rotated Loadings Matrix PC I	PC II	PC III
Manpower	.9503	-.1325	.2067	.8931	.3890	-.1202
Calls for police service	.9940	-.0778	-.0638	.9856	.1258	-.1045
Murder	-.1523	.9785	.0766	-.1210	-.0831	.9823
Rape	.8563	.3002	.3610	.7923	.4649	.3314
Robbery	.9922	.0314	-.1081	.9962	.0677	-.0025
Serious assault	.9070	.4140	-.0046	.9109	.1023	.3922
Burglary	.9880	.0003	.0501	.9632	.2253	-.0112
Larceny-theft	.9725	-.1438	-.1797	.9818	.0178	-.1856
Auto theft	.9815	-.0921	-.1576	.9890	.0341	-.1315
Arson	.2315	-.2169	.9445	.0536	.9910	-.0878
Prostitution	.9556	-.0328	-.2698	.9857	-.0878	-.0879
Eigenvalue	8.3065	1.3208	1.2192	8.0779	1.4473	1.3213
Per cent of variance accounted	75.5	12.0	11.1	73.4	13.2	12.0
Cumulative per cent of variance accounted	75.5	87.5	98.6	73.4	86.6	98.6

were respectively equated with the set of values $\{1, -1, -1, 1, 1, 1, 1, 1, -1, -1, -1\}$ (see Table 8.6). The resulting rotated principal components structure matrix signifies that the variable dependencies in question have begun to register noticeable changes. First of all, nearly 10% less of the overall variation in the data set is being accounted for. Second, the relative importance of the three components has shifted considerably, with the first component accounting for approximately 11% less of the total variation. Third, although murder remains a single, distinct component, calls for police service, auto theft and prostitution have become split between the composite component and the arson component. Furthermore, the remaining high loadings onto the composite component tend to be slightly lower. In contrast with those results reported in Table 8.4, where $\varrho_j = 1$ for all j, different

TABLE 8.4

Principal Components Structure Matrices
for the First Spatial Differences Standardized Buffalo Crime Data

	Principal Components					
	Initial Loadings Matrix			Varimax Rotated Loadings Matrix		
Variable	PC I	PC II	PC III	PC I	PC II	PC III
Manpower	.9405	.2648	-.0016	.8540	.3281	-.3429
Calls for police service	.9975	-.0223	-.0499	.9571	.0571	-.2803
Murder	-.5100	-.5297	.6707	-.2678	-.2206	.9328
Rape	.7747	.3922	.4499	.7493	.6272	.0406
Robbery	.9897	-.1317	-.0225	.9761	-.0286	-.2091
Serious assault	.8916	-.2595	.3590	.9746	.0191	.2023
Burglary	.9809	.1014	.0887	.9413	.2275	-.2061
Larceny-theft	.9826	-.1249	-.1349	.9479	-.0737	-.3086
Auto theft	.9850	-.1222	-.1017	.9556	-.0562	-.2812
Arson	-.0088	.9687	.2022	-.1640	.9503	-.2222
Prostitution	.9422	-.3047	-.1121	.9486	-.2271	-.2041
Eigenvalue	8.2997	1.6617	.8740	7.8059	1.5691	1.4604
Per cent of variance accounted	75.5	15.1	8.0	71.0	14.3	13.3
Cumulative per cent of variance accounted	75.5	90.6	98.5	71.0	85.2	98.5

results are yielded when the ϱ_j vary. This same finding is alluded to in Griffith (1979).

8.4.2. *An Empirical Example: 1969 Agricultural Production in Puerto Rico*

The spatial distributions of eight agricultural variables have been analyzed using several multivariate techniques (Griffith, 1979). Size effects were adjusted for by converting each measure to a density. Selected results for this data analysis appear in Table 8.7. The raw data consisted of 72 areal units, and had two significant dimensions that facilitated discriminating amongst the five agricultural administrative regions of Puerto Rico (see Figure 8.2).

TABLE 8.5

Principal Components Structure Matrices
for the MLE Spatial Autoregressive Model Residuals
Based Upon the Standardized Buffalo Crime Data

Variable	MLE $\hat{\rho}$	Principal Components Initial Loadings Matrix PC I	PC II	PC III	Varimax Rotated Loadings Matrix PC I	PC II	PC III
Manpower	.0542	.9514	-.1566	.1897	.8901	.3927	-.1386
Calls for police service	.0487	.9939	-.0672	-.0754	.9837	.1262	-.1202
Murder	.0615	-.1744	.9590	.1884	-.1256	-.0832	.9813
Rape	.0701	.8539	.2484	.3947	.7925	.4728	.3082
Robbery	.0671	.9914	.0424	-.1113	.9961	.0651	-.0232
Serious assault	.0806	.9075	.4125	.0339	.9192	.0991	.3744
Burglary	.0341	.9879	-.0018	.0494	.9621	.2283	-.0254
Larceny-theft	.0058	.9742	-.1135	-.1900	.9799	.0245	-.1931
Auto theft	.0058	.9828	-.0651	-.1620	.9875	.0411	-.1396
Arson	-.0758	.2381	-.3219	.9110	.0547	.9895	-.0912
Prostitution	-.0035	.9571	.0080	-.2629	.9852	-.0773	-.0933
Eigenvalue		8.3227	1.3032	1.2104	8.0756	1.4540	1.3067
Per cent of variance accounted		75.7	11.9	11.0	73.4	13.2	11.9
Cumulative per cent of variance accounted		75.7	87.5	98.5	73.4	86.6	98.5

The first of these two dimensions might be labelled rural propinquity (areal units weighted the highest having a lot of renting farm families and coffee and sugarcane production, and little tobacco and banana/plantain production, and vice versa). The second of these dimensions might be labelled a rural-urban agricultural linkage (areal units weighted the highest having a lot of farms and milk and coffee production, and little banana/plantain production, and vice versa).

The spatial autocorrelation indices (MC and GR) suggest that each of these eight density measures displays a moderate degree of positive autocorrelation. In other words, there is a noticeable

TABLE 8.6

Principal Components Structure Matrices
for the Contrived Spatial Autoregressive Model Residuals
Based Upon the Standardized Buffalo Crime Data

	Principal Components					
	Initial Loadings Matrix			Varimax Rotated Loadings Matrix		
Variable	PC I	PC II	PC III	PC I	PC II	PC III
Manpower	.9212	-.1865	-.1152	.9011	.2151	-.1960
Calls for police service	.8911	.3115	-.1303	.7637	.5507	.1469
Murder	-.0165	.6995	.6822	-.0356	-.0038	.9766
Rape	.8262	.1180	.0671	.7798	.2672	.1469
Robbery	.9374	-.2694	.1154	.9778	.0061	-.0926
Serious assault	.9043	-.0070	.3392	.9326	.0191	.2507
Burglary	.9426	-.2248	.0494	.9608	.0827	-.1071
Larceny-theft	.9093	-.3520	.0488	.9558	-.0116	-.1986
Auto theft	.8940	.2552	-.1123	.7818	.5015	.1196
Arson	.2714	.5583	-.6474	.0195	.8951	-.0528
Prostitution	.8908	.2464	-.0581	.7908	.4575	.1514
Eigenvalue	7.4038	1.3195	1.0684	6.9038	1.6903	1.1977
Per cent of variance accounted	67.3	12.0	9.7	62.8	15.4	10.9
Cumulative per cent of variance accounted	67.3	79.3	89.0	62.8	78.1	89.0

tendency for similar values to cluster on the map of Puerto Rico for each of these agricultural variates. Unfortunately the information provided by these two indices is somewhat inconsistent ($r = -0.053$); they agree on the nature, but not the degree, of the prevailing relationship. Because a randomization sampling perspective was used in this particular analysis, MLE's for ϱ could not be calculated. Rather, the GR index was used to estimate this spatial autocorrelation parameter. To begin with, equation (1.2) was rewritten in matrix form as

$$(n - 1)\,(\mathbf{X} - \bar{x}\mathbf{1})^T[(\mathbf{C1} + \mathbf{C}^T\mathbf{1})_{\text{diag}} - 2\mathbf{C}]\,(\mathbf{X} - \bar{x}\mathbf{1})/$$
$$[2\mathbf{1}^T\mathbf{C1}(\mathbf{X} - \bar{x}\mathbf{1})^T(\mathbf{X} - \bar{x}\mathbf{1})].$$

TABLE 8.7

Numerical Results for the Puerto Rican Data

Variable	GR	MC	$\hat{\rho}$	Normalized Pattern of Discriminant Function Loadings: Autocorrelation DF1	Autocorrelation DF2	No autocorrelation DF1
Number of farms	.7302	.3499	.5618	.031	-.241**	.006
Amount of farm land	.7751	.1085	.4446	.010	.022	.056
Milk production	.7129	.1975	.6184	-.011	-.013*	-.014
Coffee production	.6948	.5181	.5770	.242	-.084*	.115
Sugarcane production	.5324	.3157	1.3046	.056	.009	-.050
Tobacco production	.7802	.4361	.4185	-.001	-.026	-.010
Bananas/plantains production	.8982	.3538	.2226	-.732	.892**	-.682*
Number of renting farm families	.6465	.3664	.4538	.633	-.371**	.719
Eigenvalue	____	____	____	2.045	.635	.491
Percent of variance accounted	____	____	____	68.0	21.0	48.0

* Denotes a significant difference at the 5% level.
** Denotes a significant difference at the 1% level.

Then this expression was modified to

$$(n - 1)\mathbf{X}^T(\mathbf{I} - \varrho\mathbf{W})^T(\mathbf{I} - \mathbf{1}\mathbf{1}^T/n)\,[(\mathbf{C}\mathbf{1} + \mathbf{C}^T\mathbf{1})_{\text{diag}} - 2\mathbf{C}]$$
$$\times\,(\mathbf{I} - \mathbf{1}\mathbf{1}^T/n)\,(\mathbf{I} - \varrho\mathbf{W})\mathbf{X}\,\{2\mathbf{1}^T\mathbf{C}\mathbf{1}[(\mathbf{I} - \varrho\mathbf{W})\mathbf{X}$$
$$- \mathbf{1}^T(\mathbf{I} - \varrho\mathbf{W})\mathbf{X}/n]^T[(\mathbf{I} - \varrho\mathbf{W})\mathbf{X} - \mathbf{1}^T(\mathbf{I} - \varrho\mathbf{W})\mathbf{X}/n]\}^{-1}.$$

This second expression was equated with unity (the value for a random spatial distribution), and then solved for ϱ. Except for sugarcane production, the estimates of ϱ obtained from this estimation procedure are quite reasonable.

Next each X_j was transformed to a set of spatially random values using the estimates of the ϱ_j's obtained here coupled with equation (1.5). The resulting ξ_j's were subjected to a discriminant function analysis, and produced a single significant function accounting for roughly 49% of the variation displayed by these estimated residual variates. Hence prewhitening of these

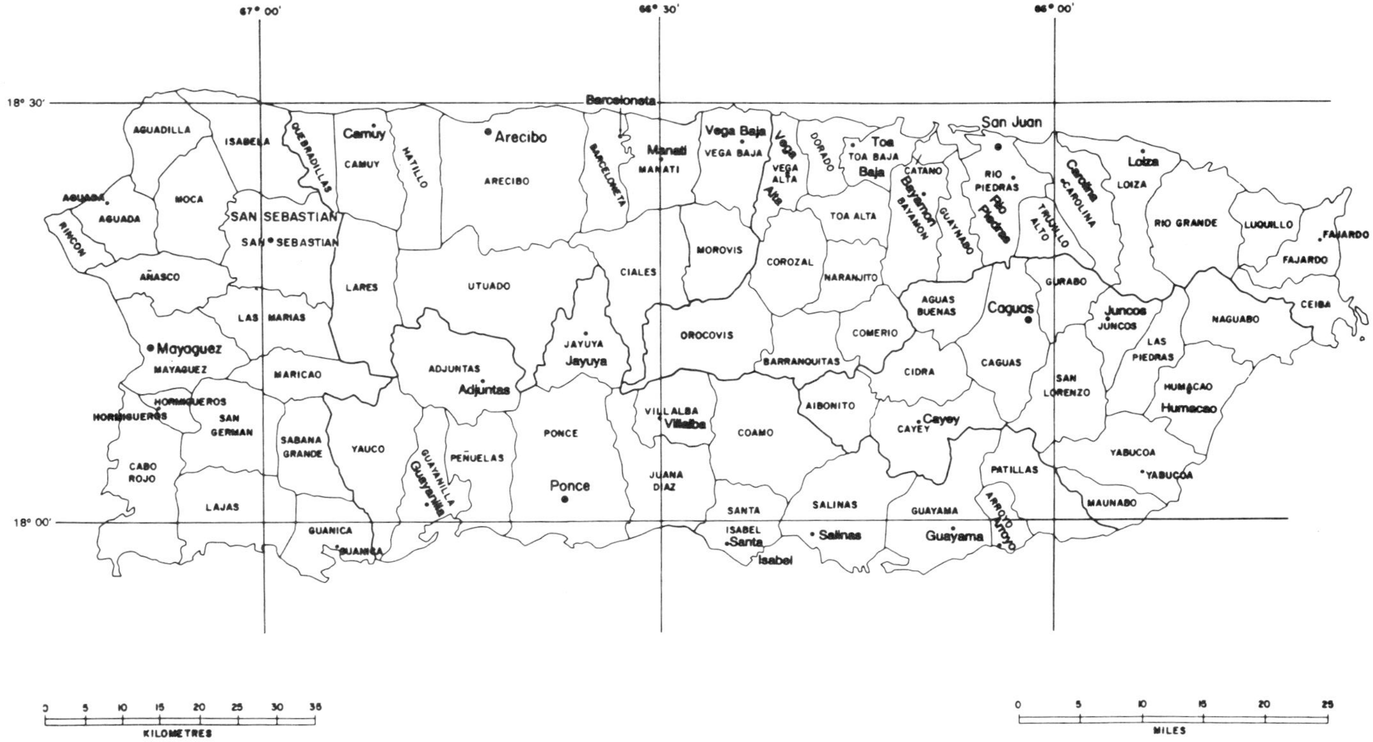

Fig. 8.2. Agricultural administrative regions and municipios of Puerto Rico.

agricultural data extract a sizeable amount of structure that can be thought of as being spurious. This lone dimension can be labelled urban-rural contrast (areal units weighted the highest having little banana/plantain production, and vice versa). Clearly the latent multivariate relations for these data are strongly linked with the governing spatial autoregressive processes.

8.5. EXTENSIONS TO MULTIVARIATE MODELS IN GENERAL

The key to multivariate spatial data analysis here is to estimate the inverse variance–covariance matrix

$$\langle({}_n\mathbf{I} - \varrho_j\, {}_n\mathbf{W})\rangle^T({}_p\mathbf{\Sigma}^{-1} \otimes {}_n\mathbf{I})\,\langle({}_n\mathbf{I} - \varrho_j\, {}_n\mathbf{W})\rangle.$$

Once this matrix has been estimated, which really means once the set $\{\varrho_j\}$ has been estimated, then a number of multivariate models that assume normality can be evaluated. On the one hand, structure latent in the prewhitened data matrix ξ is being sought. Accordingly estimation of the above matrix will yield ${}_p\hat{\mathbf{\Sigma}}$, which then can be subjected to, for instance, a canonical correlation analysis, a principal components analysis, or a multivariate regression analysis. For regionalization problems, which involve fixed effects specifications, it permits an adjusted version of the within regions and between regions covariation matrices to be constructed.

On the other hand, because spatial dependency structures are being incorporated directly into the multivariate model, specification errors could be present. In other words, one should not necessarily have a strong commitment to a particular spatial dependency structure. By reestimating a multivariate model with different dependency structure assumptions, the effects of this commitment can and should be studied. Certainly, given the overall tone of the spatial statistics literature, a conditional, a simultaneous, and a first-order moving average scheme should be entertained and explored.

Since alternative specifications are possible, then criteria for choosing the most appropriate model need to be established. Following the published findings on time series analysis, one important empirical criterion would be whether or not the matrix ξ conforms to a multivariate normal distribution. Using asymptotic theory, a second criterion would involve testing the significance of the set of estimates $\{\hat{\varrho}_j\}$, determining which are likely to differ from zero in the population. Unfortunately although tests for additional spatial autocorrelation structure in the residuals would be useful, such tests are not yet available (Cliff and Ord, 1981). In fact, a suitable test statistic for classical multivariate techniques is available only for fixed effects or unautocorrelated stochastic predictor regression residuals. A generalization of the likelihood ratio test that permits one to choose between the three alternative spatial dependency structures mentioned here would be valuable. But, of course, the multivariate problem is further complicated by the likely possibility that no single spatial dependency structure characterizes all of the variables under study.

As Streitberg (1979) notes, the techniques considered in this chapter mostly deal with the analysis of interrelationships among variables. Although some insights were gained for discriminant function analysis (in the case of a fixed effects type of statistical model specification), both this technique and cluster analysis are oriented toward the analysis of interrelationships among observations rather than among variables. For discriminant analysis the problem becomes one of finding a sensible way to classify new areal units when these observations are known to be spatially autocorrelated, but this effect cannot be partialed out (possibly without recalibrating the discriminant functions). In other words, research should not just focus on matrix ξ, but it must also address issues concerning the original matrix **X**. Just as in time series analysis, where the serial correlation structure is retained in order to make sound and accurate forecasts, spatial autocorrelation structure must be incorporated into data analyses in order to better understand spatial processes. For example, if a principal components analysis is undertaken, then the product moment correlation coefficient version of matrix ${}_p\mathbf{\Sigma}$ can be estimated. From this matrix would be obtained a set of coefficients denoted by matrix **P** that would allow selected linear combinations to be constructed from the matrix of residuals ξ. These same coefficients would be the appropriate ones to use for extracting counterpart linear combinations from matrix **X**. For cluster analysis the problem becomes one of finding a sensible (dis)similarity measure when observations are spatially autocorrelated, so that the sorts of spurious relations mentioned earlier in this chapter are avoided.

In the two empirical examples presented in this chapter, two different estimation procedures were employed in order to estimate the set $\{\varrho_j\}$. The first was based upon the multivariate normal probability model, whereas the second was based upon the randomization sampling perspective. In this second case exchangeability becomes a crucial issue. Far more work needs to be done with this particular estimation procedure, though. Several univariate alternatives to it have been proposed recently by Ancot, Paelinck and Prins (1986). In support of the first approach, however, results presented in Chapter 2 strongly suggest that matrix ξ often will not differ significantly from its multivariate normal counterpart.

Therefore, a multivariate normal probability model approach seems quite reasonable to pursue. The crux of the problem is to estimate both ${}_p\boldsymbol{\Sigma}$ and the set of p values $\{\varrho_j\}$. Once these estimates have been obtained, most multivariate questions can be addressed.

8.6. CONCLUDING COMMENTS

An attempt was made here to suggest considerations that need to be taken into account in order to spell out a fully specified multivariate probability density function for spatially autocorrelated data. The main finding is that there is an urgent need to reappraise multivariate analyses of geographical data. Such a reappraisal must transcend the somewhat simplistic, but novel and imaginative, approach put forth by Wartenberg (1985). The aim of such an endeavor in part should be to better understand statistical properties of parameter estimates when complications that lie dormant in data sets are aroused by the presence of spatial autocorrelation. Clearly the relationships between spatial autocorrelation and multivariate statistical theory merit a more careful and a more comprehensive articulation.

8.7. REFERENCES

Ancot, J., J. Paelinck, and P. Prins, 1986, Some new estimators in spatial econometrics, *Working Paper* No. 1986/1, Foundations of Empirical Economic Research, Netherlands Economic Institute, Rotterdam.

Bivand, R., 1980, A Monte Carlo study of correlation coefficient estimation with spatially autocorrelated observations, *Quaestiones Geographicae*, 6: 5–10.

Buffalo Police Department, 1981, *Annual Report*. Buffalo, N. Y.

Cliff, A., and J. Ord, 1973, *Spatial Autocorrelation*. London: Pion.

Cliff, A., and J. Ord, 1981, *Spatial Processes*. London: Pion.

Fabrikant, R., 1979, The distribution of criminal offenses in an urban environment: a spatial analysis of criminal spillovers and of juvenile offenders, *American Journal of Ecomonics and Sociology*, 38: 32–47.

Fabrikant, R., 1980, Inter-jurisdictional spillovers of urban police services: comment, *Southern Economic Journal*, 46: 955–961.

Gabor, T., 1981, The crime displacement hypothesis: an empirical examination, *Crime and Deliquency*, 27: 390–404.

Griffith, D., 1979, Urban dominance, spatial structure and spatial dynamics: some theoretical conjectures and empirical implications, *Economic Geography*, 55: 95–113.

Griffith, D., 1980, Towards a theory of spatial statistics, *Geographical Analysis*, 12: 325–339.

Griffith, D., 1986, An overview of the spatial statistics literature, *Working Paper* No. 1986/6, Foundations of Empirical Economic Research, Netherlands Economic Institute, Rotterdam.

Haining, R., 1978, A spatial model for high plains agriculture, *Annals*, Association of American Geographers, 68: 493–504.

Hakim, S., A. Ovadia, E. Sagi, and J. Weinblat, 1979, Interjurisdictional spillover of crime and police expenditure, *Land Economics*, 55: 200–212.

Johnson, R., and D. Wichern, 1982, *Applied Multivariate Statistical Analysis*. Englewood Cliffs, N.J.: Prentice-Hall.

Lebart, L., 1969, Analyse statistique de la contiquite, *Publications de l'Istitut Statistique de l'Universite de Paris*, 18: 81–112.

Mehay, S., 1977, Interjurisdictional spillovers of urban police services, *Southern Economic Journal*, 43: 1352–1359.

Regert, G., 1980, Spatial aspects of criminal behavior, in *Crime: A Spatial Perspective*, edited by D. Georges-Abeyie and K. Harries. New York: Columbia University Press, pp. 47–57.

Streitberg, B., 1979, Multivariate models of dependent spatial data, in *Exploratory and Explanatory Statistical Analysis of Spatial Data*, edited by C. Bartels and R. Ketallapper. The Hague: Martinus Nijhoff, pp. 139–177.

Vandaele, W., 1983, *Applied Time Series and Box Jenkins Models*. New York: Academic Press.

Wartenberg, D., 1985, Multivariate spatial correlation: a method for exploratory geographical analysis, *Geographical Analysis*, 17: 263–283.

APPENDIX 8A

Rules for Kronecker Products

1. Multiplication:

$$\mathbf{A} \otimes \mathbf{B} = \begin{pmatrix} a_{11}\mathbf{B} & a_{12}\mathbf{B} & \cdots & a_{1n}\mathbf{B} \\ a_{21}\mathbf{B} & a_{22}\mathbf{B} & \cdots & a_{2n}\mathbf{B} \\ \cdot & \cdot & \cdot & \cdot \\ \cdot & \cdot & \cdot & \cdot \\ \cdot & \cdot & \cdot & \cdot \\ a_{n1}\mathbf{B} & a_{n2}\mathbf{B} & \cdots & a_{nn}\mathbf{B} \end{pmatrix}$$

2. Matrix product of two Kronecker products:

$$(\mathbf{A} \otimes \mathbf{B})(\mathbf{C} \otimes \mathbf{D}) = \mathbf{AC} \otimes \mathbf{BD}$$

3. Inverse of Kronecker product:

$$(\mathbf{A} \otimes \mathbf{B})^{-1} = \mathbf{A}^{-1} \otimes \mathbf{B}^{-1}$$

4. Transpose of Kronecker product:

$$(\mathbf{A} \otimes \mathbf{B})^T = \mathbf{A}^T \otimes \mathbf{B}^T$$

5. Determinant of Kronecker product:

$$|_m\mathbf{A} \otimes {}_n\mathbf{B}| = |_m\mathbf{A}|^n |_n\mathbf{B}|^m$$

6. Kronecker product obeys the distributive and associative laws.

CHAPTER 9

Simulation Experimentation in Spatial Analysis

Quite often sampling distributions for spatial statistics are unknown, are exceedingly complex in form, or do not have closed form solutions for their probability density functions or parameter values. Further, although asymptotic variances can be obtained (see Chapter 4), spatial statistics may behave quite differently when small, finite lattices are studied (possibly due to boundary effects; see Chapter 7), or when irregular lattices are studied. In these various cases, where analytical solutions are elusive for test statistic problems, the identification of estimation properties, or the like, researchers have resorted to simulation experiments in order to gain insights into spatial statistical analysis.

Simulating two-dimensional spatial autocorrelated surfaces can serve two primary purposes. First, it can furnish necessary information for testing a null hypothesis, especially about the presence of zero spatial autocorrelation. Second, it can illuminate tendencies that exist over a range of possible levels of spatial autocorrelation. In either case the simulation experimentation seeks to supply a basis for drawing proper inferences.

9.1. TESTING A NULL HYPOTHESIS OF ZERO SPATIAL AUTOCORRELATION

In Chapter 4 the asymptotic variance of $\hat{\varrho}$ was established, and can and should be used for very large n. But most geographic problems fall into the realm of small sample theory, meaning that the asymptotic variance becomes inappropriate. Hope (1968) has proposed a simplified Monte Carlo significance test procedure to handle this situation. This procedure consists of the comparison of the statistic for an observed geographic distribution with those for random samples generated in accordance with the null hypothesis. A level of significance is chosen, and the outcome of the statistical test is determined by whether or not the test statistic for the geographic distribution under study is one of the extreme values when combined with the reference set of statistics calculated for each of the random geographic distributions.

For example, a 5% level of significance could be selected, and $\hat{\varrho}_0$ calculated for the conditional spatial autoregressive model from the observed map pattern. Then 199 geographic distributions could be generated, each being a set of n independent drawings from a normal random variable $X \sim (\mu, \sigma^2)$. This reference set conforms with a null hypothesis of $\varrho = 0.0$, whereas this sampling perspective is that of multivariate normality. Next $\hat{\varrho}$ could be calculated for each of the 199 geographic distributions, and the total set of 200 estimates could be ranked to determine whether or not $\hat{\varrho}_0$ is one of the ten extreme values (five largest or five smallest).

If evidence exists that supports the contention of non-normality, then the randomization sampling perspective should be employed. In this instance total enumeration of the sample space could involve as many as $n!$ maps. Thus, parallelling the previous example, 199 random permutations of the map values under study could be generated, and then $\hat{\varrho}$ calculated for each of these geographic distributions. Again, the total set of 200 estimates would be ranked to see where $\hat{\varrho}_0$ falls relative to the other estimates.

Traditionally the level of significance (α) is set at 10%, 5% or 1%. The alternative hypothesis being a one- or two-tailed situation depends upon the research problem being investigated. What remains to be determined here, then, is the necessary number of simulated surfaces. Hope (1968) has shown that if the level of significance is $1/r$, where r is the number of maps generated under the null hypothesis, the power of this Monte Carlo test procedure is a monotonically increasing function of the size of the reference set, $(r - 1)$, and the limit is the power of the corresponding uniformly most powerful test (a test is most powerful if it has the smallest Type II error for a given alternative hypothesis value among all tests whose Type I error is no larger than α, and the test is uniformly most powerful if it is most powerful against every possible given alternative hypothesis valuc). The Type I error is exactly $1/r$. For a fixed level of significance, α, increasing r would increase the power of this Monte Carlo test procedure. Haining (1978) suggests that r should be about 400–500, in order to achieve an acceptable level of power. One should keep in mind, though, that restricting the reference set to a small number of random samples generated, in accordance with the null hypothesis being tested, facilitates a comparison of these samples with the observed data, which is not as cumbersome as it would be if the size of the reference set were large.

9.2. GENERATING AUTOCORRELATED PSEUDO-RANDOM NUMBERS FOR TWO-DIMENSIONAL SURFACES

Frequently the null hypothesis under study is that $\varrho = a \neq 0.0$, or the research problem is one of exploring small lattice properties associated with a spatial autoregressive scheme. In this context the simulation experimentation undertaken must be modified. The construction of spatially autocorrelated surfaces differs markedly from that for time series [see, for example, Lakhan (1981)], because, as was mentioned in earlier chapters of this monograph, two-dimensional planar surface dependencies are multi-directional, include feedback loops, may not be directionally symmetric, and have more extensive boundary effects than do one-dimensional time series.

In Chapter 5 the variance-covariance matrix for the infinite, regular lattice surface was presented, for any value of ϱ, and was denoted as $\mathbf{\Omega}$. Without loss of generality, simulation experiments involving the normality assumption can be simplified by assuming that, for areal unit i, $\xi_i \sim N(0, 1)$. The problem then is one of determining x_i such that

$$\mathbf{X}^T\mathbf{\Omega}^{-1}\mathbf{X} = \xi^T\xi. \tag{9.1}$$

If $\mathbf{D}^T\mathbf{D} = \mathbf{\Omega}$, then equation (9.1) implies that

$$(\mathbf{D}^T)^{-1}\mathbf{X} = \xi, \quad \text{or} \quad \mathbf{X} = \mathbf{D}^T\xi. \tag{9.2}$$

Now the problem becomes one of properly specifying matrix $\mathbf{D}$.

Besag (1974) argues that matrix **D** should be upper triangular, and hence matrix $\mathbf{\Omega}^{-1}$ should be subjected to a Cholesky decomposition. This decomposition can be obtained as follows [see Quandt (1983)]:

$$d_{11} = 1,$$

$$d_{1j} = \omega_{1j}, \quad j = 2, 3, \ldots, n,$$

$$d_{ii} = \left[1 - \sum_{p=1}^{i-1} d_{pi}^2\right]^{1/2}, \quad i = 2, \ldots, n,$$

and

$$d_{ij} = \left[\omega_{ij} - \sum_{p=1}^{i-1} d_{pi} d_{pj}\right] / d_{ii},$$

$$j = i + 1, \ldots, n \quad \text{and} \quad i = 2, \ldots, n,$$

where ω_{ij} is the (i, j) cell entry of matrix $\mathbf{\Omega}$. This formulation casts $x_1 = \xi_1$, $x_2 = f(\xi_1, \xi_2)$, $x_3 = f(\xi_1, \xi_2, \xi_3)$, and so forth. It is equivalent to rewriting the multidirectional model in unidirectional form, as was done with the univariate, bidirectional model in Chapter 5. This also is the approach used by the IMSL subroutine package to generate multivariate normal random variates (GGNSM).

An alternative method for specifying matrix **D** is to decompose matrix $\mathbf{\Omega}^{-1}$ into its eigenvalues and eigenvectors. Hence,

$$\mathbf{\Omega} = \mathbf{E\Lambda E}^T, \quad \text{or} \quad \mathbf{\Omega}^{-1} = \mathbf{E\Lambda}^{-1}\mathbf{E}^T.$$

Now $\mathbf{D}^T = \mathbf{E\Lambda}^{1/2}$, so that

$$\mathbf{X} = \mathbf{E\Lambda}^{1/2}\boldsymbol{\xi}. \tag{9.3}$$

Equation (9.3) represents a decomposition of matrix $\mathbf{\Omega}$ into orthogonal components, rather than unidirectional dependencies. In the limit, the entries of matrix $\mathbf{\Lambda}$ are equivalent to $(1 - \varrho\lambda_i)^{1/2}$, where λ_i are the eigenvalues of matrix **C**. Also, since matrix $\mathbf{\Omega}$ is symmetric, $\mathbf{E}^T\mathbf{E} = \mathbf{I}$.

9.3. BACKGROUND

Presently there are two papers that outline methods for generating two-dimensional autocorrelated surfaces. The first of these two to have appeared is by Goodchild (1980), and presents a procedure that begins by generating n pseudo-random numbers from the Gaussian distribution (in fact, any distribution could be selected to sample from). Next, these numbers are permuted over a lattice until a target level of spatial autocorrelation is achieved. The permutations in question result from a random swapping of lattice values. If a given swap moves the observed level of spatial autocorrelation towards the targeted value, this swap is retained. Otherwise, the values are returned to their respective preceding lattice positions, and a new pair of values is randomly selected for swapping.

Goodchild's method is wanting on theoretical grounds. Foremost, if a set of independent and identically distributed pseudo-random numbers is generated, and then arranged haphazardly on a lattice, then the spectral representation of this two-dimensional surface realization is given by

$$f(\lambda_1, \lambda_2) = \sigma^2/(4\pi^2), \tag{9.4}$$

where λ_1 and λ_2 represent the frequency domains for the orthogonal pairs of axes involved. If, however, these values are permuted in such a way that some first-order autoregressive pattern is detected in the resulting spatial distribution, then the spectral representation would be expected to equal

$$f(\lambda_1, \lambda_2) = \sigma^2/\{4\pi^2[1 - 2\cos(\lambda_1) - 2\cos(\lambda_2)]^2\}. \tag{9.5}$$

With Goodchild's method, though, since the pseudo-random numbers are independent, equation (9.5) is incorrect, and any attempts to empirically estimate the spectrum for a surface constructed with his procedure will tend to show that this spectrum is not well behaved. Abnormal behavior of the spectrum would arise primarily because this permutation approach ignores indirect effects associated with the feedback loops of multi-directional dependencies, which are manifested in the covariation of pseudo-random numbers over an autocorrelated two-dimensional surface, and also because the realized surface is not necessarily the outcome of a homogeneous process operating over the planar surface in question (stationarity). Moreover, for equation (9.2), matrix $\mathbf{D}^T$ is not changed, while the elements of vector ξ_0 are permuted in such a way that $\mathbf{D}^T\xi_0$ is similar to $\mathbf{D}^{*T}\xi_p$, where $\mathbf{D}^{*T}$ is the target matrix and ξ_p is the permuted vector.

Haining, Griffith and Bennett (1983) devised a method for simulating two-dimensional autocorrelated surfaces that tries to correct Goodchild's error. Their procedure uses equation (9.3). First r sets of n pseudo-random numbers are generated. Since pseudo-random number generators are autonomous sequential machines, and hence selection of these r sets of n numbers is not left to chance, one pragmatic criterion that can be imposed for determining selection is that these sets of numbers pass a series of statistical tests (Ziegler, 1976). Furthermore, as many empirical tests as are practical should be performed on these sets of pseudo-random numbers (Law and Kelton, 1982). Haining, Griffith and Bennett suggest that the r sets of values be tested for, in the case of desiring Gaussian variates, (1) multivariate normality, and (2) $\boldsymbol{\mu} = \mu\mathbf{1}$ (where $\mathbf{1}$ is an n-by-1 vector all of whose elements are unity) and $\boldsymbol{\Sigma} = \sigma^2\mathbf{I}$. They maintain that only a sample that passes these two tests should be utilized. Tests for multivariate normality appear in Malkovich and Afifi (1973), while tests for H_0: $\boldsymbol{\mu} = \mathbf{0}$ and $\boldsymbol{\Sigma} = \mathbf{I}$ appear in Anderson (1958). Goodchild has no controls on the quality of his pseudo-random numbers.

One of the most troublesome initial conditions to deal with in the simulation of two-dimensional autocorrelated surfaces has to do with edge effects (see Griffith, 1983). If a near infinite lattice were utilized, then a buffer zone along the periphery, which embraces the observations responsible for transient behavior (i.e., behavior of characteristics of the output sequence during a phase in which these characteristics oscillate about those for the steady state), could be deleted. However, since this approach is infeasible for most practical purposes (the size of the lattices would have to be enormous), Haining, Griffith and Bennett suggest that the lattice be mapped onto a torus. This technique is similar to that used in time series simulations (see Goldsmith, 1977). But as was shown in Chapter 5, this torus mapping of two-dimensional lattices yields incorrect correlograms. Moreover, the spatial autocorrelations fail to converge on zero in the limit as the number of lags increases. However, the covariance matrix based upon equation (5.16) can serve this end.

In conducting a simulation experiment, the quality of the pseudo-random number generator is of importance. If a generator is to yield a normal random variate with mean μ and variance σ^2, then the sampling distribution should behave within the bounds found analytically for the null hypothesis. A Wilk–Shapiro, Kolmogorov–Smirnov, or chi-square test can be conducted on the resulting frequency distribution, whereas the sample mean $\bar{x}$ and the sample variance s^2 for each random number generator sequence can be tested against the population parameter values of μ and σ^2. If the results are acceptable, then the pseudo-random number generator should be employed. Many commercial packages, such as IMSL, furnish such test information about their generators.

There are two additional cautions to be heeded when working with pseudo-random number generators. Both again are concerned with pilot study tests for a simulation experiment. First of all, Law and Kelton (1982) have emphasized the importance of pairwise independence of the sequence of pseudo-random numbers in order to obtain correct simulation results. This property of the generated pseudo-random numbers can be ensured in a prudent way by using Goodchild's (1980) random swapping algorithm, with a target spatial autocorrelation value of zero ($\pm\varepsilon$). With this approach, equation (9.4) would indeed characterize both the expected and observed spectra. If, as Goodchild notes, a set of pseudo-random numbers cannot be permuted to the target level ($\pm\varepsilon$), then this set should be discarded, and a new one should be selected. This procedure is somewhat of a variant on the acceptance–rejection method (see Kronmal and Peterson, 1981).

The empirical tests referred to above are local in nature, and refer only to the segment of the generator cycle that yielded a particular set of pseudo-random numbers. One should not misconstrue this criterion as one that refers to truncated distributions. Moreover, a representative sample is being sought under the null hypothesis (see Lehman, 1977). And, a simulation is being conducted in order to control for deviations from this assumption. The issue is that of exactness. Law and Kelton (1982) argue that one should use an algorithm that results in random variables with exactly the desired distribution. Robustness can always be assessed within the simulation experiment itself.

A closely affiliated consideration is concerned with the use of joint as opposed to marginal probability distributions. Haining, Griffith and Bennett (1983) suggest that multivariate normality should be considered, hence recommending the use of a joint probability distribution. Because this property of the resulting r vectors of ξ is for their joint distribution, it does not ensure that each vector ξ is Gaussian. Since these ξ vectors constitute replications of a simulation, then one could argue that each vector should be normally distributed in order to eliminate error due to poor random numbers (see Schafer, 1974). Clearly if each ξ vector conforms to a normal distribution, then the accompanying joint distribution will be multivariate normal, given the way equation (9.2) is constructed. Again this property refers to a testing of the random number generator, rather than the use of truncated distributions theory. Rather than a strict test criterion being applied to each vector ξ, however, a simultaneous inference method such as the Bonferroni one should be used. Since the simulation sample drawings are independent, the joint confidence level associated with r generated two-dimensional surface sets of n numbers will be exactly $(1 - \alpha)$. Hence the size of the critical region for a two-tailed situation will be $\alpha/(2r)$, and for a one-tailed situation will be α/r. Clearly the criterion here seeks to eliminate any extremely deviant samples. In addition, Law and Kelton (1982) contend that the resulting random variables should have exactly the desired distribution. On the other hand, Law and Carson (1979) have found that non-normality does not appear to present too severe a problem to the drawing of inferences from simulation experiment results, at least for queing theory problems.

In either of these two cases, independence of replications is accomplished by using a different sequence of pseudo-random numbers for each simulation run, together with the same initial conditions (Sargent, 1976; Law and Kelton, 1982). Griffith and Amrhein (1983) tested for this condition with principal components analysis, and found that it was satisfied. The restriction encountered here is that one cannot test more replications than there are pseudo-random numbers in a given two-dimensional simulated autocorrelation surface.

9.5. VARIANCE REDUCTION TECHNIQUES

One of the fundamental advantages for conducting simulation experiments is to exercise control over many exogenous and endogenous sources of variation. The goal is to increase statistical accuracy over that attainable from the same sample size using random sampling. One prominent way of achieving this control is through variance reduction techniques. Moreover, if the variance of the output from a simulation experiment can somehow be reduced, without disturbing expected values, greater precision can be attained for the same values of r and n. But one must keep in mind that the misuse of variance reduction techniques can lead to poorer results than random sampling offers. One of the clearest ways of achieving a variance reduction is to restrict the simulation experiment to lattice configurations of areal units. This restriction will help minimize edge effects, too.

As mentioned earlier, the same random numbers could be used in different simulation experiments in order to increase simulation ouput precision. The basic idea here is to increase the confidence that any observed differences are due to differences in experimental design, rather than to fluctuations of the random numbers themselves. One variant of this common pseudo-random numbers approach for the case of two-dimensional autocorrelated surfaces, which will permit variance reduction over replications of simulation runs, is to generate a single set of pseudo-random numbers, and then select r of the $n!$ possible geographic distributions, which result from permuting these n values over the lattice, in such a way that these r sets are mutually orthogonal. Now each of the marginal distributions is identically the same, with identical means and variances. If these permutations are selected in such a way that the measured spurious spatial autocorrelation is essentially zero, then further variance reduction can be achieved. Law and Kelton (1982) note, though, that to date no evidence exists to demonstrate that the use of common random numbers guarantees variance reduction. Furthermore, for the cases where it does seem to reduce variance, there is no way of ascertaining a priori the magnitude of this reduction. Meanwhile, Roach and Wright (1977) have asserted that common random numbers are the most popular sampling technique used in simulation work.

Now the random swapping approach endorsed by Goodchild (1980), combined with the deletion of any non-normal ξ vectors whose sample mean and variance deviate significantly from μ and σ^2, will help reduce variance in the pilot simulation experiment, too. But in the final simulation experiment these controls will result in a deterioration of the inferential basis of the experiment. Moreover, inferences can be drawn only about the restricted types of surfaces generated, rather than surfaces in general. Therefore, in the final experiment these variance reduction constraints should serve as blocking criterion for partitioning the total variance of the simulation experiment.

9.6. SELECTING THE NUMBER OF REPLICATIONS r

One problematic consideration alluded to earlier in this chapter has to do with the appropriate number of elements for a given sequence (i.e., the lattice size), as well as the approximate number of replications to be utilized in an experiment (i.e., the number of surfaces to be generated). These two points are intimately linked to initial conditions, steady state characteristics, and their interrelationships. Heidelberger and Welch (1981) point out that running a simulation experiment for a length that is shorter than the optimal length would not provide the necessary information for drawing inferences, whereas running this experiment for a length that is beyond the optimal length would be a waste of computer resources. The two aspects of this issue are the size of an individual simulation run, and the number of run replications to be made. Determination of run size and replication size requires the formulation of stopping rules in order to control length. The simplest rule is to make r independent replications of a terminating simulation, where the length of each replication is predetermined and set to n (Law and Kelton, 1982). This stopping rule is useful where only rough but valid inferences are to be drawn. Presumably the poorest criterion for establishing r and n is a budget constraint. Law (1980) set r to 500, whereas Haining, Griffith and Bennett have suggested that $300 \leqslant r \leqslant 500$ for spatially autocorrelated surfaces. A value for both n and r is determined, in part, by the number of transient elements involved.

Hence setting a value of n for two-dimensional autocorrelated surfaces is simplified substantially when a torus is employed. Griffith (1981) and Haining, Griffith and Bennett (1983) have suggested that $n \geqslant 100$ for autocorrelated surfaces. If transient runs are present, Thomas (1974) points out that $r > 10$.

One approach that could be used here is to choose an n whose asymptotic variances of parameter estimates do not differ by more than some prespecified amount from those asymptotic results presented in Chapter 5. Again this value could be determined during the pilot studies that are used to set up the simulation experiment. Griffith (1981) and Haining, Griffith and Bennett (1983) have suggested that $n \geqslant 100$ for autocorrelated surfaces.

In terms of the number of replications (i.e., surfaces), if precise valid inferences are to be drawn, then r should result from a sequential stopping rule. The goal of such a rule is to determine the number of samples that are necessary to achieve an acceptable degree of representativeness. Gafarian, Ancker and Morisaku (1978) have found that none of the five common rules used to determine minimum run length in a single replication proved to be satisfactory. Two of these popular stopping decision rules are (1) r should be increased until the ratio of the confidence interval half-length to the point estimate is less than a prespecified constant (Law and Carson, 1979; Heidelberger and Welch, 1981), and (2) r should be increased until a sufficiently small mean-squared-error term is achieved (Donnelly and Shannon, 1981). This first stopping rule is difficult to implement, though. Since the distributional theory for estimating a confidence interval usually is based upon asymptotic results, r must be sufficiently large in order to ensure that a valid confidence interval has been calculated. In addition, the width of the estimated confidence interval must be sufficiently small to enable useful conclusions to be drawn from the simulation experiment. Thus, a value for r can be established for the appropriate confidence interval by employing a suitable stopping rule, leading to a circular argument (Lavenberg and Sauer, 1977). Therefore, many researchers prefer the mean-squared-error decision tool (Donnelly and Shannon, 1981). Heidelberger and Welch (1981) suggest that the confidence interval approach can be salvaged by studying a single pilot simulation run, and using the results from this pilot run to determine subsequent run lengths. Because of the consecutive incrementation of r, then, the use of sequential stopping rules is facilitated by the examination of marginal rather than joint frequency distributions (see Thomas, 1974; Law and Kelton, 1982).

The steady state results of interest here are those from the set of r replications, rather than from the run size n. Because these replications are independent, batch means can be used to assess some of the steady state characteristics. For instance, if $r > n$, then the r replications of vector ξ could be partitioned into b batches for evaluation (see Sargent, 1976). Meanwhile, the simulation experiment results can be analyzed by sequentially constructing confidence intervals on the statistic θ under consideration. These replications have to do with the generation of vector ξ, while the experiments require vector $\mathbf{X}$ of equation (9.2) to be calculated for different levels of spatial autocorrelation (e.g., ± 0.075, ± 0.150, ± 0.225). The initial conditions for these replications have to do with small sample size, and hence when r becomes reasonably large (presumably greater than or equal to 100, according to classical statistical small sample theory) then $\theta(r)$ will tend to stabilize. Convergence on a stable value is slow, though, since precision is a function of $r^{1/2}$ (Fishman, 1973). On the other hand, this type of transient behavior does not necessitate the deletion of any of the initial replications. As $\theta(r)$ stabilizes with increasing r, $t_{r-1,1-\alpha/2}[s_{\theta}^2(r)/r]^{1/2}$ will tend to decrease, ultimately permitting an optimal stopping criterion to be invoked. Consequently, the steady state characteristics being sought can be obtained with the three-step procedure outlined in Law and Kelton (1982), when the additional restriction of $r > 10$ is added.

9.7. ANALYSIS OF THE SIMULATION RESULTS FOR CHAPTER 6

The simulation experiments conducted and whose results are reported in Chapter 6 will be evaluated in this section. As has been mentioned in this chapter already, while many criteria exist for assessing the compatibility of simulation schemes with underlying statistical assumptions, no single criterion appears to be superior or most informative. Hence, it is sensible to examine simulated data in as many different ways as are available, since each statistical test is tuned to subtly different nuances in these data. Accordingly a selected battery of tests have been performed on the simulated data used in Chapter 6.

The first set of tests employed the batch approach, and analyzed the initial set of random numbers to determine whether or not they conformed to the assumptions of independence, normality, $\mu = 0$ and $\sigma^2 = 1$. The batches were considered to be of size 81 (the map lattice size), and so there were 400 such batches. For each batch (map), the Wilk–Shapiro correlation coefficient, sample mean, sample variance, spatial autoregressive parameter, and GLS mean were calculated. Time series plots of these five variables reveal no unusual sequential features (see Figures 9.1–9.5). The plot for the Wilk–Shapiro statistic, using an approximate critical value of 0.97 (1% level), indicates that only simulation map # 320 deviates markedly from normality. The plot for the sample mean, using approximate critical values of ± 0.286 (1% level), indicates that only simulation maps # 188 and # 326 deviate markedly from zero. The plot for the sample standard deviation, using an approximate 99% confidence interval of (0.7998, 1.2057), indicates that only simulation maps # 5, # 72, # 142, # 188, # 232 and # 366 deviate markedly from unity. The plot for the spatial autoregressive parameter, using asymptotic critical values of ± 0.215 (1% level), indicates that only simulation map # 156 deviates markedly from zero. Finally, the plot for the GLS mean, using asymptotic critical values of ± 0.286 (1% level), indicates that none of the simulation maps deviate markedly from zero.

The next set of tests addressed the notion of independence of map results. Figures 9.6 and 9.7 display the autocorrelation and partial autocorrelation correlograms for the five simulation variables. Clearly no conspicuous sequential trends are apparent in these two figures. This contention is reinforced by the Box–Jenkins estimation results presented in Table 9.1. In general a first-order autoregressive (AR) and a first-order moving average (MA) term were fitted to the given sequences of simulated maps. In none of these five cases is either parameter significant (i.e., all t-statistics are nearly equal to zero). Further, the map sample mean, and the GLS map mean were found not to be significantly different from zero. Meanwhile, although the estimate $\hat{\rho}$ was

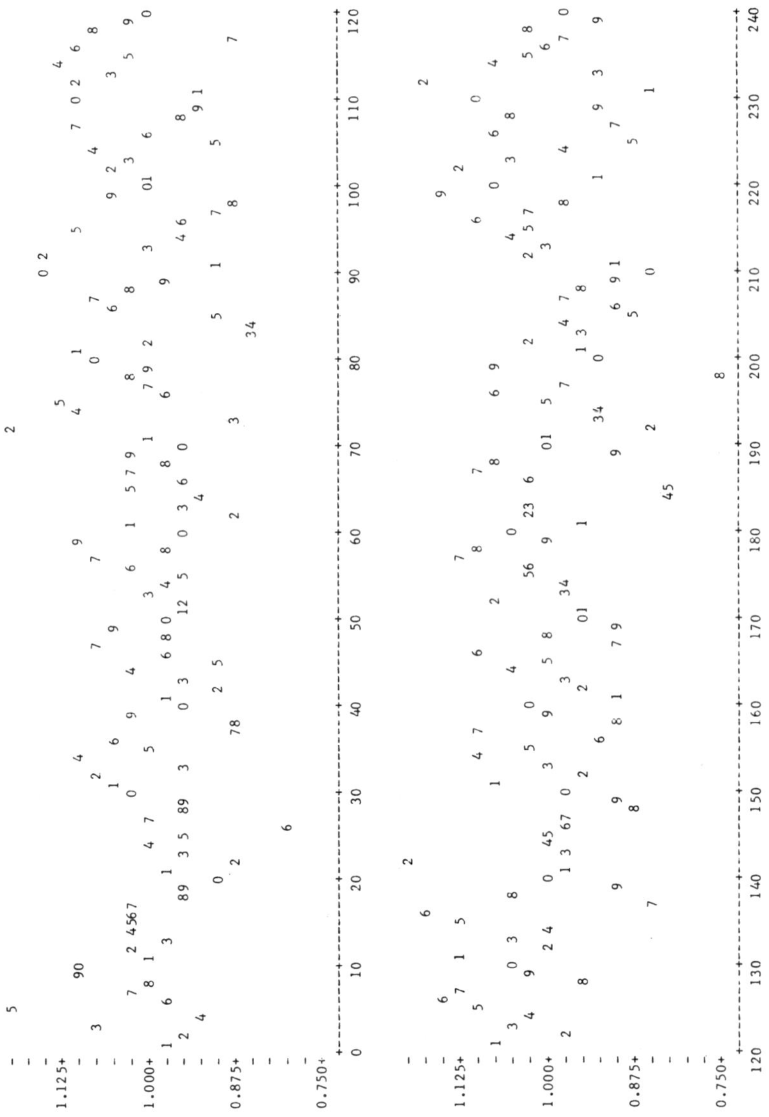
1.125
1.000
0.875
0.750
0 10 20 30 40 50 60 70 80 90 100 110 120
1.125
1.000
0.875
0.750
120 130 140 150 160 170 180 190 200 210 220 230 240

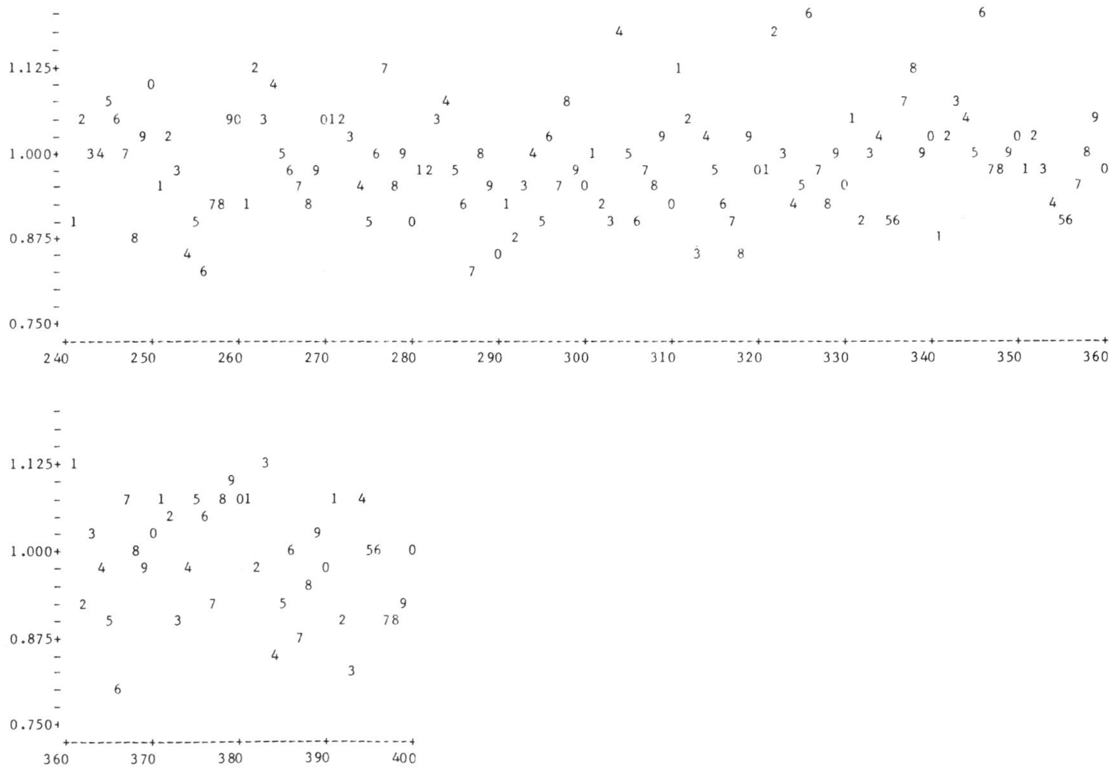

Fig. 9.1. Time series plot of Wilk–Shapiro statistic.

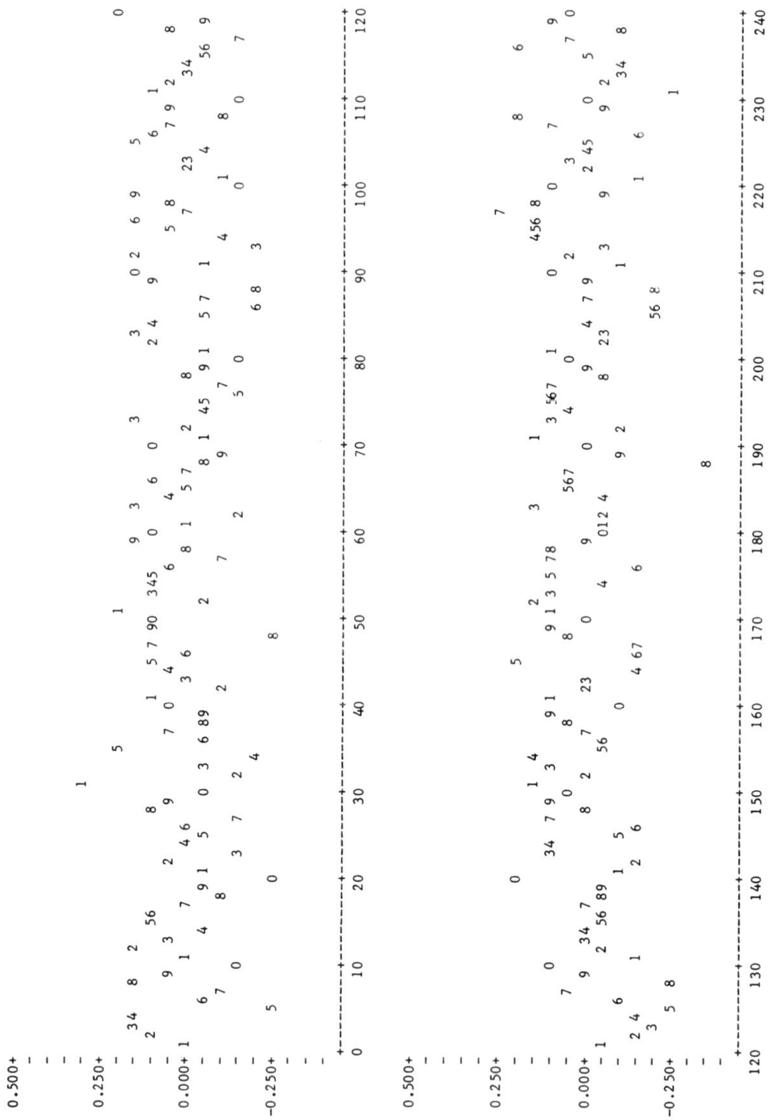
0.500+
0.250+
0.000+
-0.250+
0
10
20
30
40
50
60
70
80
90
100
110
120
0.500+
0.250+
0.000+
-0.250+
120
130
140
150
160
170
180
190
200
210
220
230
240

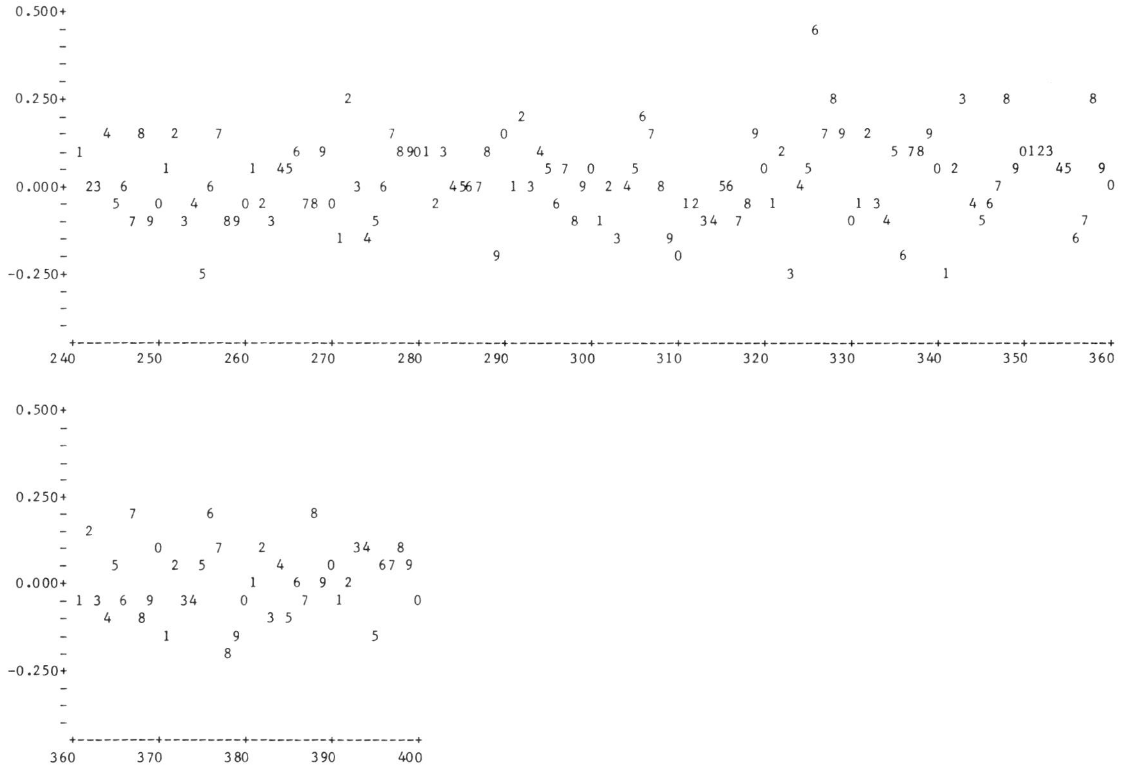

Fig. 9.2. Time series plot of map sample mean.

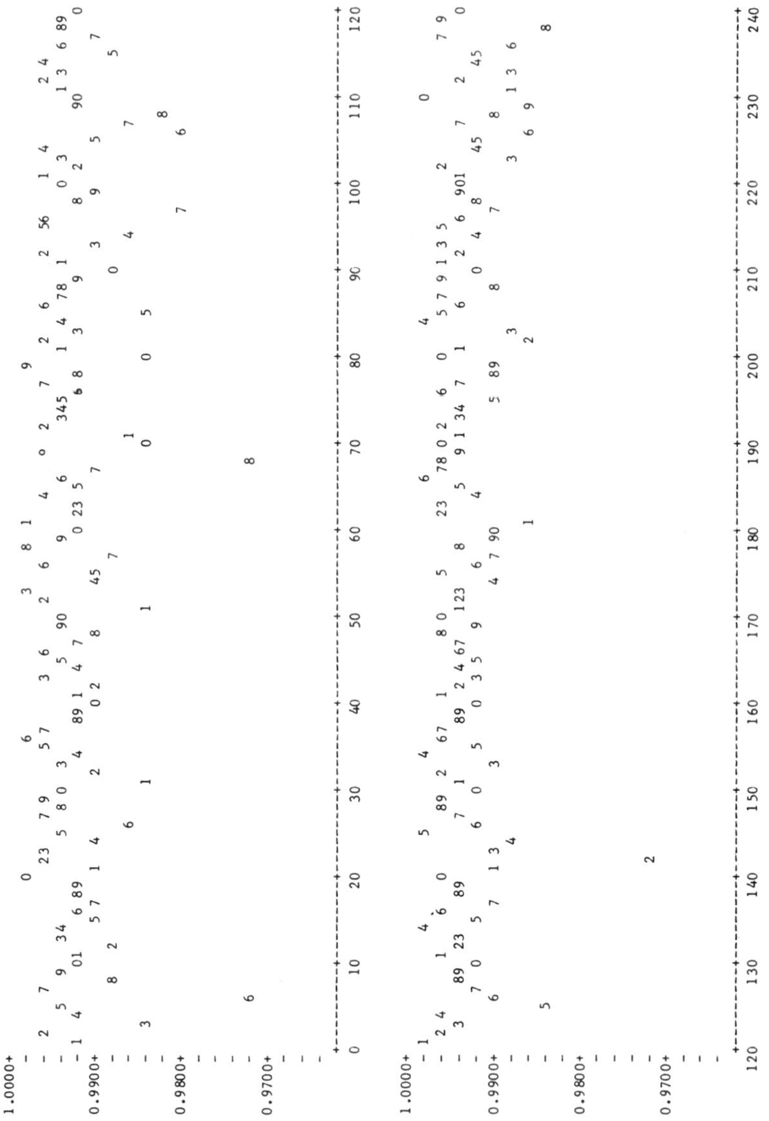
1.0000
0.9900
0.9800
0.9700
0 10 20 30 40 50 60 70 80 90 100 110 120
1.0000
0.9900
0.9800
0.9700
120 130 140 150 160 170 180 190 200 210 220 230 240

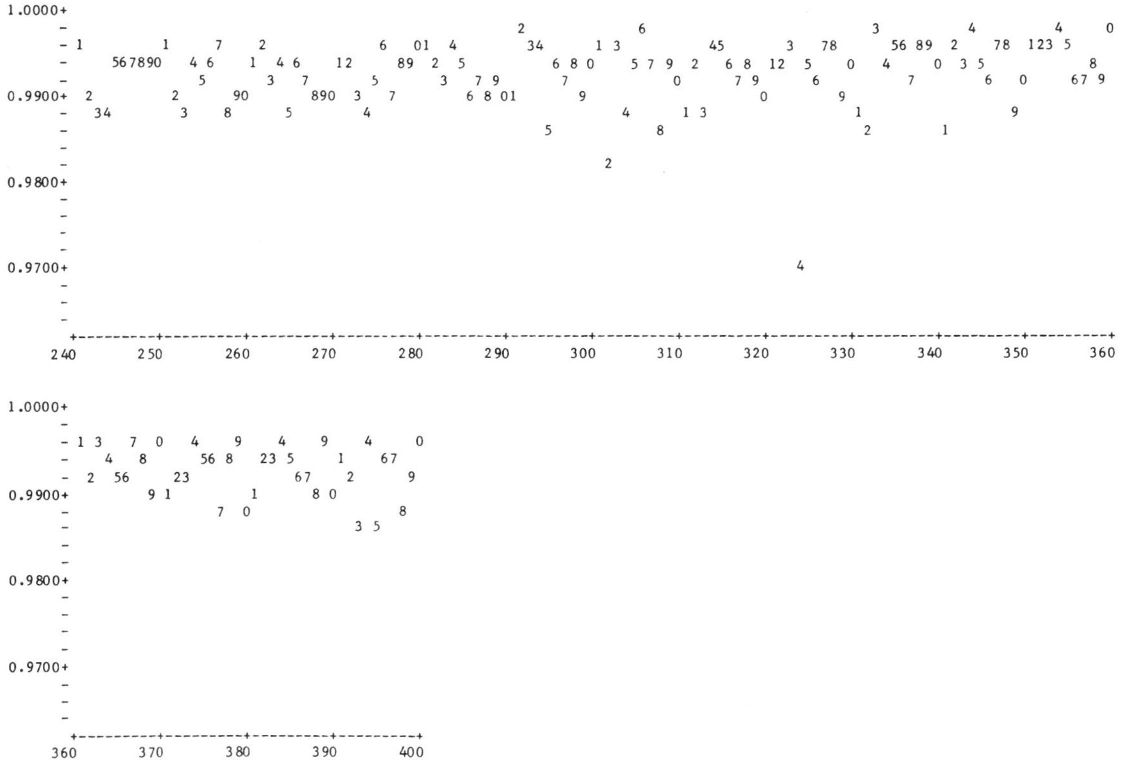

Fig. 9.3. Time series plot of map sample variance.

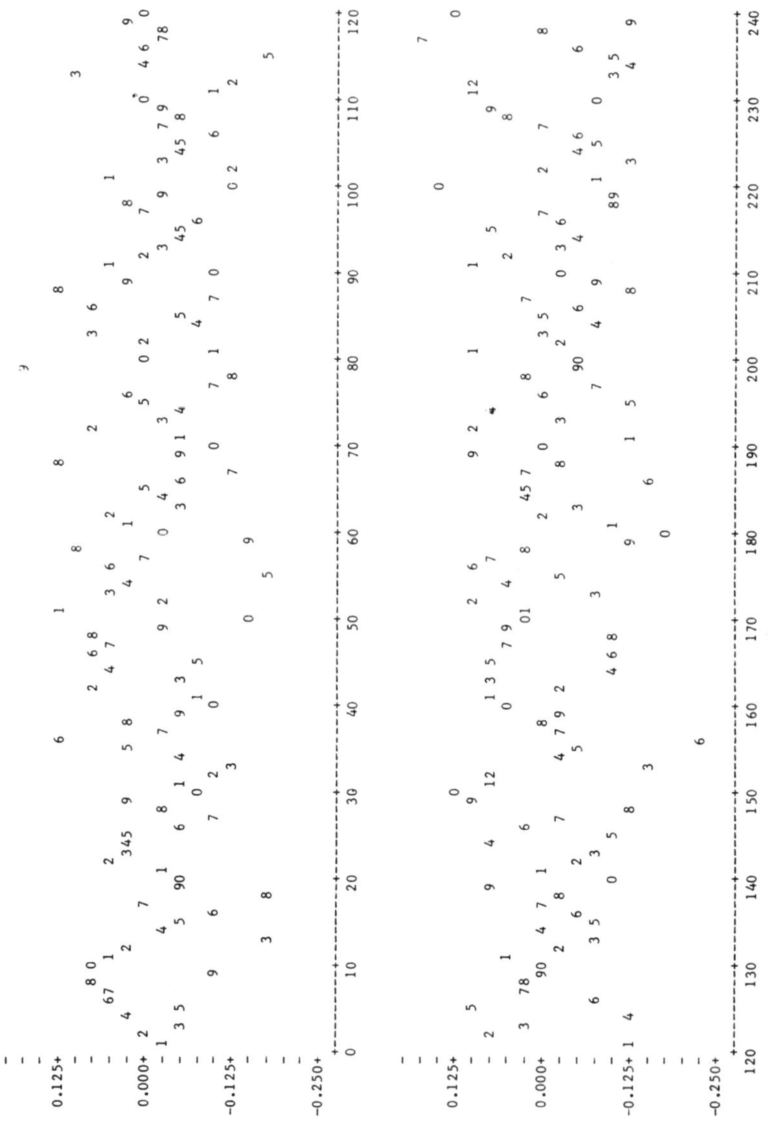
0.125+
0.000+
-0.125+
-0.250+
0 10 20 30 40 50 60 70 80 90 100 110 120
0.125+
0.000+
-0.125+
-0.250+
120 130 140 150 160 170 180 190 200 210 220 230 240

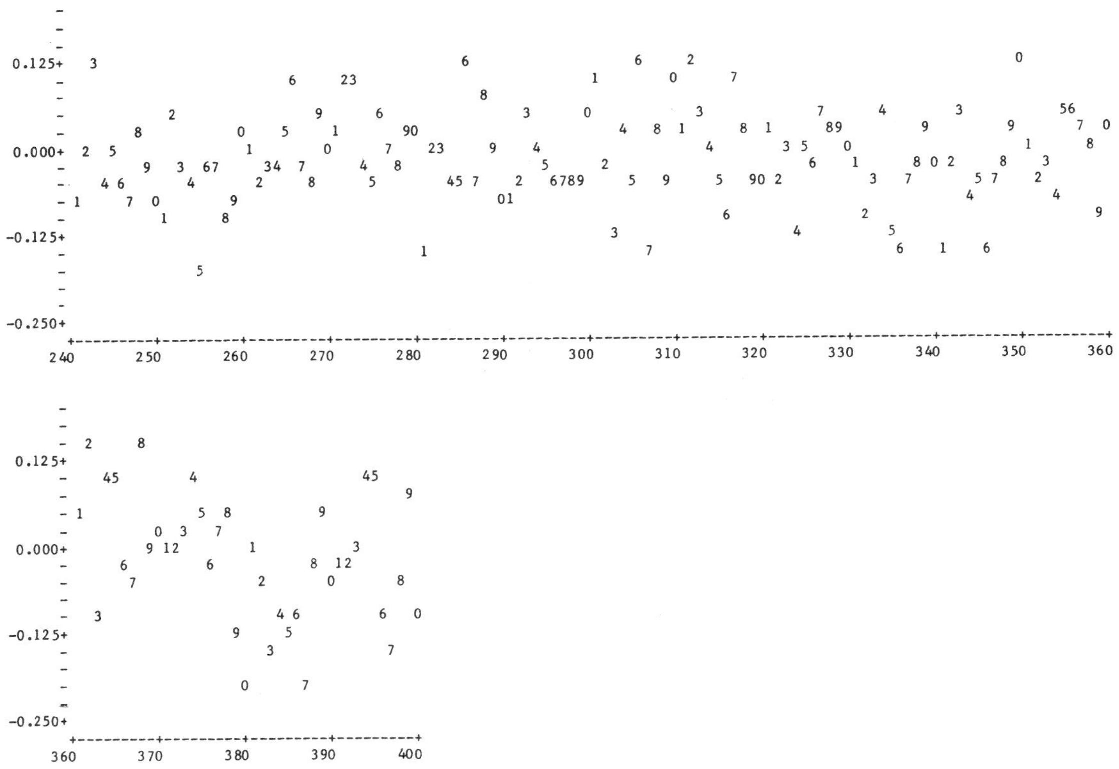

Fig. 9.4. Time series plot of the spatial autoregressive parameter.

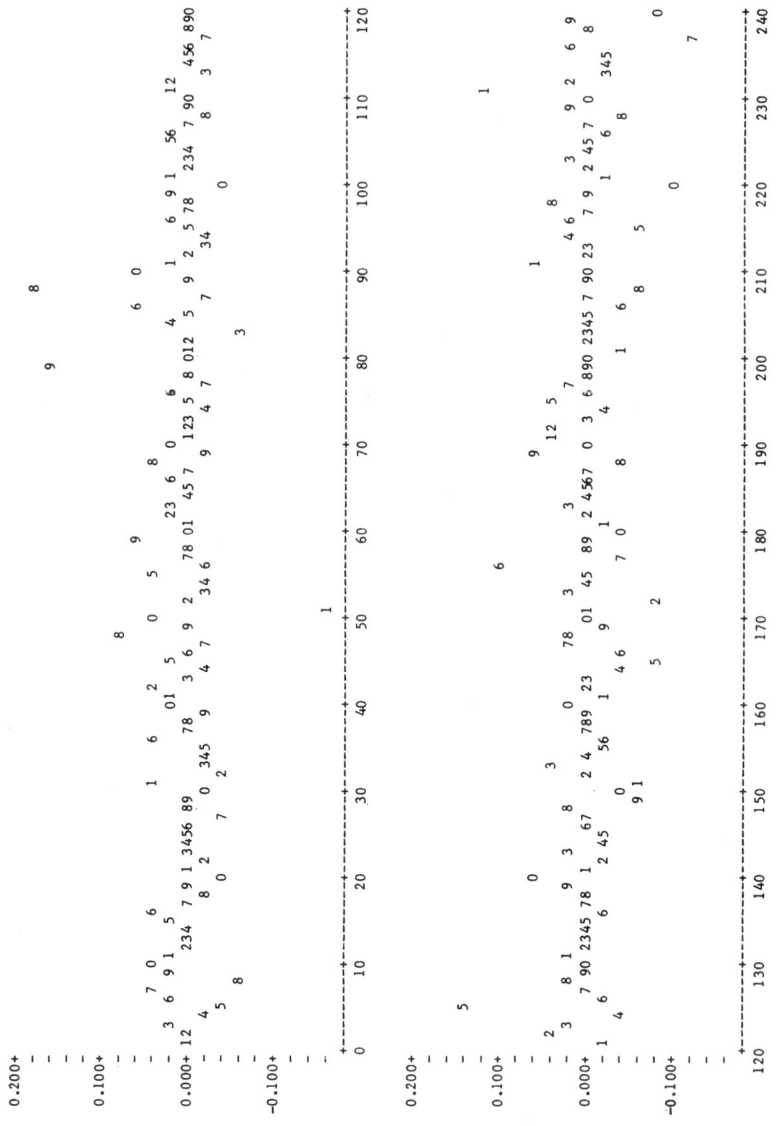
0.200+
0.100+
0.000+
-0.100+
0 10 20 30 40 50 60 70 80 90 100 110 120
0.200+
0.100+
0.000+
-0.100+
120 130 140 150 160 170 180 190 200 210 220 230 240

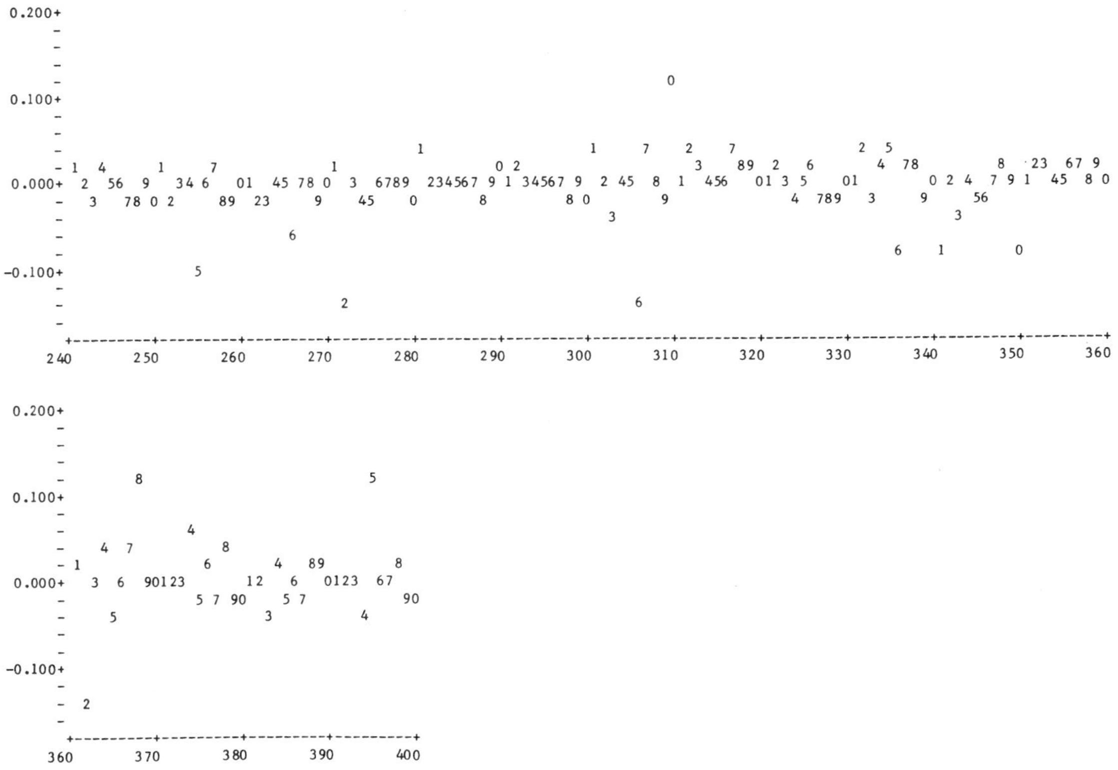

Fig. 9.5. Time series plot of the GLS mean.

```
          -1.0 -0.8 -0.6 -0.4 -0.2  0.0  0.2  0.4  0.6  0.8  1.0
            +----+----+----+----+----+----+----+----+----+----+
 1  -0.042                          XX
 2   0.050                           XX
 3   0.051                           XX
 4  -0.095                         XXX
 5   0.044                           XX
 6  -0.070                         XXX
 7   0.005                           X
 8   0.019                           X
 9   0.102                           XXXX
10  -0.036                          XX
11  -0.003                           X
12   0.009                           X
13   0.012                           X
14  -0.019                           X
15  -0.031                          XX
16   0.016                           X
17   0.112                           XXXX
18   0.017                           X
19  -0.052                          XX
20   0.010                           X
21   0.019                           X
22   0.028                           XX
23  -0.000                           X
24  -0.068                         XXX
25   0.067                           XXX
26   0.031                           XX
27   0.010                           X
28  -0.026                          XX
29   0.018                           X
30   0.023                           XX
```

(a) Wilk-Shapiro statistic

```
          -1.0 -0.8 -0.6 -0.4 -0.2  0.0  0.2  0.4  0.6  0.8  1.0
            +----+----+----+----+----+----+----+----+----+----+
 1   0.049                           XX
 2  -0.018                           X
 3  -0.041                          XX
 4   0.022                           XX
 5  -0.011                           X
 6  -0.006                           X
 7   0.035                           XX
 8  -0.020                           X
 9  -0.012                           X
10   0.011                           X
11  -0.079                         XXX
12  -0.015                           X
13   0.010                           X
14   0.079                           XXX
15  -0.021                          XX
16  -0.002                           X
17  -0.133                        XXXX
18  -0.019                           X
19  -0.055                          XX
20   0.024                           XX
21   0.037                           XX
22   0.072                           XXX
23   0.017                           X
24   0.099                           XXX
25  -0.027                          XX
26  -0.063                         XXX
27  -0.050                          XX
28   0.017                           X
29   0.023                           XX
30  -0.123                        XXXX
```

(b) sample mean

Fig. 9.6. Correlograms for the simulation statistics.

```
            -1.0 -0.8 -0.6 -0.4 -0.2  0.0  0.2  0.4  0.6  0.8  1.0
            +----+----+----+----+----+----+----+----+----+----+
 1   0.029                           XX
 2   0.091                           XXX
 3   0.042                           XX
 4   0.045                           XX
 5  -0.028                          XX
 6   0.007                           X
 7  -0.001                           X
 8   0.020                           XX
 9  -0.027                          XX
10  -0.050                          XX
11   0.062                           XXX
12   0.071                           XXX
13   0.016                           X
14  -0.046                          XX
15   0.113                           XXXX
16  -0.074                         XXX
17  -0.014                           X
18   0.007                           X
19  -0.035                          XX
20   0.054                           XX
21   0.011                           X
22   0.022                           XX
23  -0.013                           X
24   0.045                           XX
25  -0.050                          XX
26   0.021                           XX
27  -0.004                           X
28  -0.028                          XX
29  -0.057                          XX
30  -0.026                          XX
```

(c) sample variance

```
            -1.0 -0.8 -0.6 -0.4 -0.2  0.0  0.2  0.4  0.6  0.8  1.0
            +----+----+----+----+----+----+----+----+----+----+
 1  -0.025                          XX
 2  -0.049                          XX
 3   0.012                           X
 4   0.027                           XX
 5   0.036                           XX
 6   0.028                           XX
 7  -0.016                           X
 8  -0.053                          XX
 9   0.007                           X
10  -0.056                          XX
11   0.038                           XX
12   0.024                           XX
13   0.030                           XX
14  -0.081                         XXX
15  -0.032                          XX
16  -0.036                          XX
17   0.060                           XX
18  -0.054                          XX
19  -0.119                        XXXX
20  -0.034                          XX
21  -0.020                          XX
22   0.023                           XX
23  -0.105                        XXXX
24  -0.025                          XX
25  -0.007                           X
26   0.011                           X
27  -0.018                           X
28   0.032                           XX
29  -0.098                         XXX
30  -0.015                           X
```

(d) $\hat{\rho}$

Fig. 9.6. Continued

```
            -1.0 -0.8 -0.6 -0.4 -0.2  0.0  0.2  0.4  0.6  0.8  1.0
              +----+----+----+----+----+----+----+----+----+----+
 1  -0.076                           XXX
 2   0.028                             XX
 3   0.067                             XXX
 4  -0.100                          XXXX
 5  -0.003                             X
 6  -0.004                             X
 7   0.081                             XXX
 8   0.031                             XX
 9   0.031                             XX
10  -0.034                            XX
11  -0.048                            XX
12  -0.021                            XX
13   0.067                             XXX
14   0.037                             XX
15   0.026                             XX
16   0.002                             X
17   0.038                             XX
18   0.035                             XX
19   0.007                             X
20   0.083                             XXX
21   0.074                             XXX
22   0.078                             XXX
23  -0.056                            XX
24  -0.044                            XX
25  -0.034                            XX
26  -0.039                            XX
27   0.026                             XX
28  -0.080                           XXX
29   0.056                             XX
30  -0.008                             X
```

(e) GLS map mean

Fig. 9.6. Continued

TABLE 9.1

Test Statistics Based Upon the Sequence of 400 Simulated 9-by-9 Lattice Maps

Simulation	ARIMA	Parameter	Estimates	Wilk-Shapiro	Maximum
Variable	AR(1)	MA(1)	constant	statistic	Batch Variance F-ratio
Wilk-Shapiro statistic	-.2609	-.2144	1.2516	.922	2.2177
	(-.26)	(-.21)	(7885.38)		
map sample mean	-.1066	-.1572	.0057	.998	1.2925
	(-.11)	(-.16)	(.88)		
map sample variance	.0295	——	.9634	.999	1.1608
	(.59)		(241.52)		
	——	-.0250	.9926		
		(-.50)	(245.08)		
spatial auto-regressive parameter	.3551	.3900	-.0103	.998	1.2323
	(.31)	(.34)	(-4.61)		
GLS map mean	-.1630	-.0870	.0000	.923	1.6471
	(-.25)	(-.13)	(0.00)		

NOTE: t-statistics are in parentheses beneath the parameter estimates they correspond to.

```
             -1.0 -0.8 -0.6 -0.4 -0.2  0.0  0.2  0.4  0.6  0.8  1.0
              +----+----+----+----+----+----+----+----+----+----+
 1  -0.042                            XX
 2   0.048                             XX
 3   0.056                             XX
 4  -0.094                           XXX
 5   0.032                             XX
 6  -0.061                           XXX
 7   0.007                             X
 8   0.014                             X
 9   0.118                             XXXX
10  -0.046                            XX
11  -0.013                             X
12  -0.001                             X
13   0.039                             XX
14  -0.032                            XX
15  -0.019                             X
16   0.008                             X
17   0.122                             XXXX
18   0.009                             X
19  -0.057                            XX
20  -0.009                             X
21   0.043                             XX
22   0.028                             XX
23   0.010                             X
24  -0.071                           XXX
25   0.052                             XX
26   0.023                             XX
27   0.028                             XX
28  -0.036                            XX
29   0.021                             XX
30   0.004                             X
```

(a) Wilk-Shapiro statistic

```
             -1.0 -0.8 -0.6 -0.4 -0.2  0.0  0.2  0.4  0.6  0.8  1.0
              +----+----+----+----+----+----+----+----+----+----+
 1   0.049                             XX
 2  -0.020                            XX
 3  -0.039                            XX
 4   0.026                             XX
 5  -0.015                             X
 6  -0.006                             X
 7   0.037                             XX
 8  -0.025                            XX
 9  -0.008                             X
10   0.015                             X
11  -0.085                           XXX
12  -0.005                             X
13   0.010                             X
14   0.068                             XXX
15  -0.024                            XX
16   0.002                             X
17  -0.134                          XXXX
18  -0.004                             X
19  -0.060                            XX
20   0.018                             X
21   0.037                             XX
22   0.066                             XXX
23   0.011                             X
24   0.114                             XXXX
25  -0.032                            XX
26  -0.062                           XXX
27  -0.043                            XX
28  -0.018                             X
29   0.018                             X
30  -0.132                          XXXX
```

(b) sample mean

Fig. 9.7. Correlograms of partial autocorrelations for the simulation statistics.

```
            -1.0 -0.8 -0.6 -0.4 -0.2  0.0  0.2  0.4  0.6  0.8  1.0
                +----+----+----+----+----+----+----+----+----+----+
 1    0.029                              XX
 2    0.090                              XXX
 3    0.037                              XX
 4    0.035                              XX
 5   -0.037                             XX
 6   -0.001                              X
 7    0.001                              X
 8    0.021                              XX
 9   -0.026                             XX
10   -0.054                             XX
11    0.069                              XXX
12    0.079                              XXX
13    0.009                              X
14   -0.065                            XXX
15    0.101                              XXXX
16   -0.073                            XXX
17   -0.021                             XX
18    0.019                              X
19   -0.046                             XX
20    0.069                              XXX
21    0.021                              XX
22    0.019                              X
23   -0.036                             XX
24    0.034                              XX
25   -0.036                             XX
26    0.002                              X
27   -0.001                              X
28   -0.030                             XX
29   -0.037                             XX
30   -0.032                             XX
```

(c) sample variance

```
            -1.0 -0.8 -0.6 -0.4 -0.2  0.0  0.2  0.4  0.6  0.8  1.0
                +----+----+----+----+----+----+----+----+----+----+
 1   -0.025                             XX
 2   -0.050                             XX
 3    0.010                              X
 4    0.026                              XX
 5    0.039                              XX
 6    0.033                              XX
 7   -0.011                              X
 8   -0.053                             XX
 9   -0.001                              X
10   -0.064                            XXX
11    0.035                              XX
12    0.024                              XX
13    0.042                              XX
14   -0.072                            XXX
15   -0.033                             XX
16   -0.052                             XX
17    0.051                              XX
18   -0.060                             XX
19   -0.106                           XXXX
20   -0.041                             XX
21   -0.024                             XX
22    0.016                              X
23   -0.103                           XXXX
24   -0.033                             XX
25   -0.007                              X
26    0.004                              X
27   -0.012                              X
28    0.022                              XX
29   -0.113                           XXXX
30   -0.025                             XX
```

(d) $\hat{\rho}$

Fig. 9.7. Continued

```
          -1.0 -0.8 -0.6 -0.4 -0.2  0.0  0.2  0.4  0.6  0.8  1.0
            +----+----+----+----+----+----+----+----+----+----+
 1  -0.076                         XXX
 2   0.023                            XX
 3   0.071                            XXX
 4  -0.092                         XXX
 5  -0.021                          XX
 6  -0.005                            X
 7   0.096                            XXX
 8   0.037                            XX
 9   0.028                            XX
10  -0.047                          XX
11  -0.045                          XX
12  -0.021                          XX
13   0.083                            XXX
14   0.044                            XX
15   0.014                            X
16  -0.023                          XX
17   0.049                            XX
18   0.062                            XXX
19   0.031                            XX
20   0.068                            XXX
21   0.074                            XXX
22   0.083                            XXX
23  -0.054                          XX
24  -0.055                          XX
25  -0.037                          XX
26  -0.027                          XX
27   0.005                            X
28  -0.103                        XXXX
29   0.018                            X
30  -0.006                            X
```

(e) GLS map mean

Fig. 9.7. Continued

found to be significantly different from zero (probably due to the bias that is introduced into its estimation by using nonlinear techniques), the observed difference is negligible.

These five summary statistics were tested for both normality and variance homogeneity, too. In this first case, with a sample size of 400 one would speculate that both the Wilk–Shapiro statistic and the GLS map mean deviated markedly, but not dramatically, from normality. In contrast, the map sample mean and variance, and the spatial autoregressive parameter did have frequency distributions that conformed to the normal distribution. In the second case of variance homogeneity, using an approximate (interpolated) critical value of 1.64 (1% level), only the Wilk–Shapiro statistic suffers from inhomogeneity over the sequence of 400 simulations (the GLS map mean is too close to say for certain).

Another set of statistical evaluations has to do with the trajectories followed by sample statistics as the number of simulations increases. In part evaluation of these types of trajectories helps to determine whether or not $r = 400$ and $n = 81$ are reasonable numbers of replications and lattice sizes for the simulation experiment. Summary results for these assessments appear in Table 9.2, and the trajectories themselves are displayed in Figure 9.8. For all

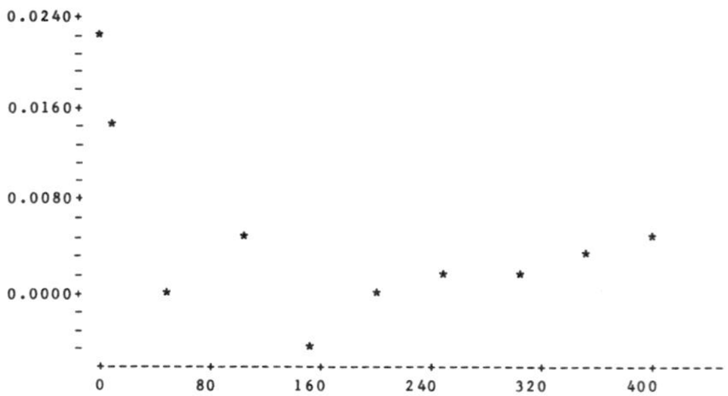

(a) sample mean

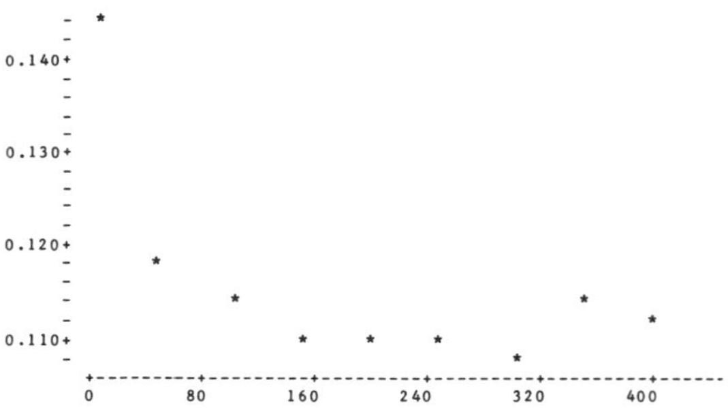

(b) variance of the sample mean

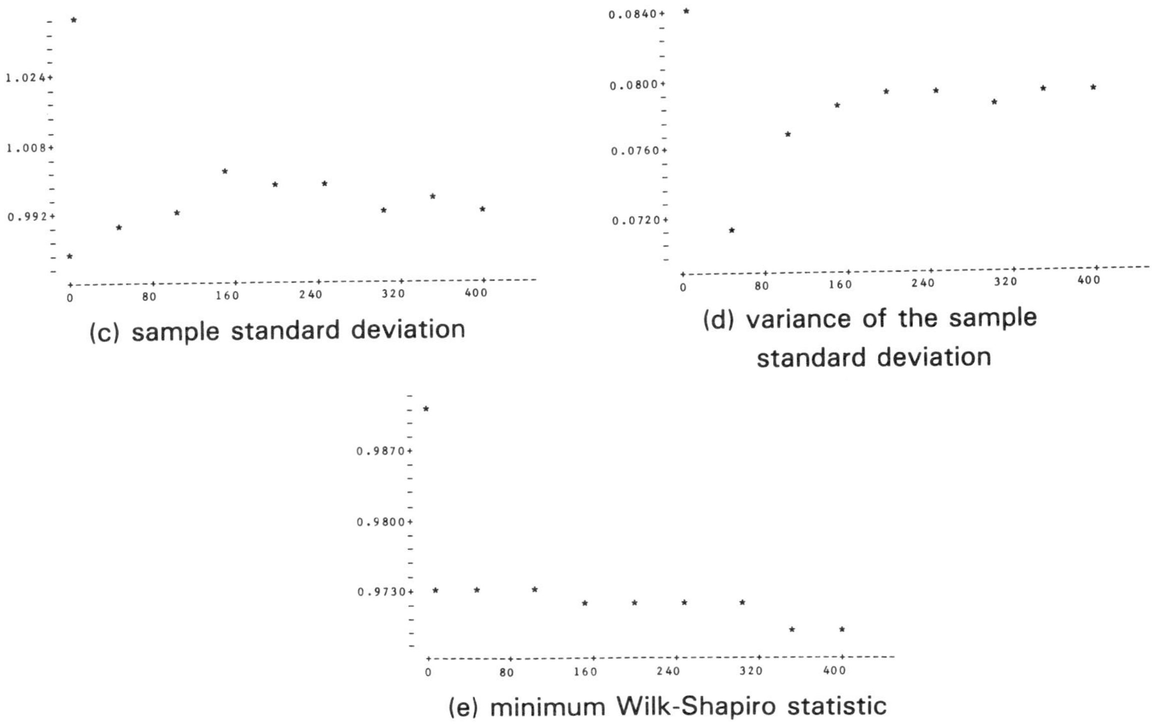

Fig. 9.8. Cumulative average simulation trajectories.

TABLE 9.2

Summary Statistics Trajectories
for the 400 Simulated 9-by-9 Lattice Map Patterns

Sequence Size	Minimum Wilk-Shapiro	$\bar{x}_{\bar{x}}$	$s_{\bar{x}}$	MSE($\hat{\mu}$)	$\bar{x}_s$	s_s	MSE($\hat{\sigma}$)
r							
1	.99140	.02255	______	.00051	.98147	______	.00034
10	.97242	.01430	.14390	.01883	1.03590	.08400	.00763
50	.97242	.00020	.11740	.01350	.98860	.07110	.00508
100	.97242	.00510	.11310	.01269	.99319	.07655	.00585
150	.97152	-.00459	.11071	.01220	1.00250	.07870	.00616
200	.97152	.00077	.10951	.01193	.99850	.07945	.00628
250	.97152	.00098	.11020	.01210	.99730	.07921	.00626
300	.97152	.00225	.10879	.01180	.99343	.07864	.00621
350	.96930	.00394	.11314	.01278	.99373	.07925	.00630
400	.96930	.00512	.11185	.01251	.99262	.07893	.00627
n							
1	.99772	-.04546	______	.00207	.96345	______	.00134
10	.99756	.00220	.04690	.00199	.99270	.03550	.00119
20	.99688	-.00260	.05410	.00279	.99260	.02951	.00088
30	.99688	-.00155	.05272	.00269	.99642	.03833	.00143
40	.99624	-.00001	.04830	.00227	.99433	.03766	.00142
50	.99624	.00138	.04876	.00233	.99315	.03627	.00134
60	.99624	.00305	.04833	.00231	.99490	.03572	.00128
70	.99472	.00245	.04905	.00238	.99498	.03423	.00118
80	.99307	.00477	.04843	.00234	.99551	.03323	.00111
81	.99307	.00512	.04823	.00232	.99541	.03303	.00110

five statistics (i.e., the minimum Wilk–Shapiro statistic, the mean and standard deviation of the map mean, and the mean and standard deviation of the map standard deviation) radical transient behavior is observed at the beginning of a trajectory (usually for about the first 100 map simulations). The map mean and its accompanying standard deviation do not yet appear to be converging by the time r reaches 400. These trends are reinforced by the mean squared error terms appearing in Table 9.2. Unfortunately the evidence presented here suggests that $r \gg 400$, and so $r = 400$ is an insufficiently low number of simulated maps.

Also presented in Table 9.2 are summary statistics for the size of n. Convergence is more apparent in these series, but still has not been achieved for $n = 81$.

Table 9.3 reports the summary results for the maps for which $\varrho = 0.075, 0.150$ and 0.225. Common random numbers were used

TABLE 9.3

Summary Statistics Trajectories
for the Different Values of ρ

ρ	r	$\overline{x}_{\hat{\rho}}$	$s_{\hat{\rho}}$	GLS $\overline{x}_{\overline{x}}$	$s_{\overline{x}}$
0.000	1	-.02679	———	-.00005	———
	10	.00530	.05800	.00249	.02976
	50	-.02082	.06877	.00212	.02523
	100	-.01780	.07238	.00523	.03674
	150	-.01921	.07123	.00456	.03422
	200	-.01757	.07316	.00320	.03341
	250	-.01658	.07396	.00123	.03454
	300	-.01625	.07194	-.00030	.03365
	350	-.01616	.07201	-.00044	.03443
	400	-.01604	.07325	-.00000	.03480
0.075	1	.04078	———	-.00068	———
	10	.06540	.05150	.00750	.07040
	50	.04638	.06763	.00409	.05681
	100	.04821	.07044	.00952	.09146
	150	.04644	.07107	.01193	.08673
	200	.04834	.07405	.00837	.08654
	250	.04952	.07384	.00330	.08755
	300	.04997	.07207	-.00035	.08452
	350	.05000	.07188	-.00193	.08657
	400	.04991	.07291	-.00143	.08628
0.150	1	.11266	———	-.01091	———
	10	.12890	.04230	.03060	.20910
	50	.11829	.06153	.00690	.17300
	100	.11976	.06165	.02250	.26890
	150	.11722	.06419	.03250	.25510
	200	.11850	.06834	.03370	.31540
	250	.11978	.06744	.01810	.30570
	300	.12004	.06607	.00750	.29010
	350	.12016	.06555	.00020	.29010
	400	.11990	.06599	.00070	.28290
0.225	1	.18814	———	-.16106	———
	10	.20236	.03008	.29500	1.08700
	50	.20130	.04617	.01100	.93600
	100	.20449	.04079	.03300	1.12400
	150	.20166	.04386	.11280	1.15180
	200	.20053	.04849	.10590	1.16860
	250	.20130	.04731	.06520	1.17080
	300	.20092	.04692	.02520	1.13460
	350	.20123	.04618	-.01190	1.19950
	400	.20125	.04569	-.00150	1.20140

here as a variance reduction technique, and are those evaluated above. Again, and for all three values of ϱ, convergence is not attained. Convergence is very close at hand for the GLS mean, however.

Consequently, presumably $r = 1000$ and $n = 121$ would be much better, although far more costly.

9.8. CONCLUDING COMMENTS

Simulation experiments in quantitative spatial analysis tend to be very costly, because of the need for moderately large lattices, and the intensive numerical work that must go into parameter estimation. Two useful variance reduction techniques that help control this problem are the use of common random numbers, and the use of regular lattices. But the findings reported in this chapter suggest that guidelines found in the literature for setting r and n are inappropriate. Hence, either a number of the other variance reduction techniques should be shown to be useful, or simulation experiments need to be enlarged. This former suggestion constitutes a fruitful area for subsequent research, especially considering the costs involved with experiment enlargement.

9.9. REFERENCES

Anderson, T., 1958, *An Introduction to Multivariate Statistical Analysis*. New York: Wiley.

Besag, J., 1974, Spatial interaction and the statistical analysis of lattice systems, *Journal of the Royal Statistical Society*, Vol. 36B: 192–236.

Donnelly, J., and R. Shannon, 1981, Minimum mean-squared-error estimators for simulation experiments, *Communications of the ACM*, Vol. 24: 253–259.

Fishman, G., 1971, Estimating sample size in computing simulation experiments, *Management Science*, Vol. 18: 21–38.

Fishman, G., 1972, Variance reduction in simulation studies, *Journal of Statistical Computation and Simulation*, Vol. 1: 173–182.

Fishman, G., 1973, Variance reduction for normal variates in Monte Carlo studies, *Journal of Statistical Computation and Simulation*, Vol. 2: 365–374.

Gafarian, A., C. Ancker, and T. Morisaku, 1978, Evaluation of commonly used rules for detecting 'steady state' in computer simulation, *Naval Research Logistics Quarterly*, Vol. 25: 511–529.

Goldsmith, H., 1977, The exact distribution of the serial correlation coefficients and an evaluation of some approximate distributions, *Journal of Statistical Computation and Simulation*, Vol. 5: 115–134.

Goodchild, M., 1980, Algorithm 9: simulation of autocorrelation for aggregate data, *Environment and Planning* A, Vol. 12: 1073–1081.

Griffith, D., 1981, Interdependence in space and time: numerical and interpretative considerations, in *Dynamic Spatial Models*, edited by D. Griffith and R. MacKinnon. Alphen aan den Rijn: Sijhoff and Noordhoff, pp. 258–287.

Griffith, D., 1983, The boundary value problem in spatial statistical analysis, *Journal of Regional Science*, Vol. 23: 377–387.

Griffith, D., and C. Amrhein, 1983, An evaluation of correction techniques for boundary effects in spatial statistical analysis: traditional methods, *Geographical Analysis*, Vol. 15: 352–360.

Haining, R., 1978, A spatial model for high plains agriculture, *Annals*, Association of American Geographers, Vol. 68: 493–504.

Haining, R., D. Griffith, and R. Bennett, 1983, Simulating two-dimensional autocorrelated surfaces, *Geographical Analysis*, Vol. 15: 247–255.

Heidelberger, P., and P. Welch, 1981, A spectral method for confidence interval generation and run length control in simulations, *Communications of the ACM*, Vol. 24: 233–245.

Hope, A., 1968, A simplified Monte Carlo significance test procedure, *Journal of the Royal Statistical Society*, Series B, Vol. 30: 582–598.

Kronmal, R., and A. Peterson, 1981, A variant of the acceptance–rejection method for computer generation of random variables, *Journal of the American Statistical Association*, Vol. 76: 446–451.

Lakhan, V., 1981, Generating autocorrelated pseudo-random numbers with specific distributions, *Journal of Statistical Computation and Simulation*, Vol. 12: 303–309.

Lavenberg, S., and C. Sauer, 1977, Sequential stopping rules for the regenerative method of simulation, *IBM Journal of Research and Development*, Vol. 21: 545–558.

Law, A., 1980, Statistical analysis of the output data from terminating simulation, *Naval Research Logistics Quarterly*, Vol. 27: 131–143.

Law, A., and J. Carson, 1979, A sequential procedure for determining the length of a steady-state simulation, *Operations Research*, Vol. 27: 1011–1025.

Law, A., and W. Kelton, 1982, *Simulation Modeling and Analysis*. New York: McGraw-Hill.

Lehman, R., 1977, *Computer Simulation and Modeling: An Introduction*. New York: Wiley.

Malkovich, J., and A. Affifi, 1973, On tests for multivariate normality, *Journal of the American Statistical Association*, Vol. 68: 176–179.

Marriott, F., 1979, Barnard's Monte Carlo tests: how many simulations? *Applied Statistics*, Vol. 28: 75–77.

Quandt, R., 1983, Computational problems and methods, in *Handbook of Econometrics*, Vol. 1, edited by Z. Griliches and M. Intriligator. Amsterdam: North-Holland, pp. 699–764.

Roach, W., and R. Wright, 1977, Optimal antithetic sampling plans. *Journal of Statistical Computation and Simulation*, Vol. 5: 99–114.

Sargent, R., 1976, Statistical analysis of simulation output data, *Simuletter* (proceedings of the symposium on the simulation of computer systems IV), Vol. 7(4): 39–50.

Schafer, R., 1974, On assessing the precision of simulations, *Journal of Statistical Computation and Simulation*, Vol. 3: 67–69.

Thomas, M., 1974, A simple sequential procedure for sampling termination in simulation investigations. *Journal of Statistical Computation and Simulation*, Vol. 3: 161–164.

Vasilopoulos, A., 1983, Generating correlated random variables for quality control applications, in *Simulation Symposium*, 16th annual record of proceedings, edited by L. Holdbrook. Silver Spring, Maryland: IEEE Computer Society Press, pp. 105–119.

Ziegler, B., 1976, *Theory of Modelling and Simulation*. New York: Wiley.

CHAPTER 10

Summary and Conclusions

This book has presented special topics in spatial statistics concerning spatial autoregression that go beyond indexing the nature and degree of spatial autocorrelation latent in a data set, simple trend surface models, description of the spectral counterparts to selected spatial autoregressive models, inclusion of spatially lagged variables, or generalized least squares solutions for linear spatial models. Attention has been directed to the properties of sufficiency, consistency, efficiency and biasedness for statistics calculated on spatial data sets. The issues of configurational information, edge effects, missing data, multivariate linear statistical models, and simulation experimentation have been explored. In other words, selected themes of an advanced nature have been treated here.

10.1. SUMMARY

A number of nontrivial complications are encountered when a researcher analyzes spatial data. Foremost, the classical independence of observations assumption is violated. This violation erodes an inferential basis for statistical analyses, and uncovers complications that tend to lie dormant in classical statistical analysis. Complex locational information is latent in these sorts of data. The complexity arises because spatial interdependencies are multidirectional, usually asymmetric, and two-dimensional in nature. Therefore the likelihood equation ends up including a Jacobian term that does not reduce to unity. Classical statistics fail to capture this locational information, resulting in their being insufficient for spatial data series. Clearly, then, the formulation of a theory of spatial statistics in part requires a number of obstructing problems to be solved. Prominent amongst these problems is that of small sample size, which characterizes an overwhelming number of spatial statistics situations. Methods for circumventing this problem, such as use of the jackknife procedure, merit considerable attention. Differences in distributional properties for finite versus infinite, and regular versus irregular, lattice surfaces need to be identified.

Another complication is associated with the use of data transformations. Because most geographical distributions contain some degree of positive spatial autocorrelation, the relation between violation of the independence of observations assumption and assumptions about variance stability, linearity, and normality needs to be understood. Evidence exists suggesting that diagnostics performed on raw data may well be misleading. Presumably the approaches taken in time series analysis should be followed here.

An in-depth study of areal unit configuration and locational information reveals that the configuration of areal units provides the underlying ordering that leads to a violation of the assumption of independent observations. Hence since at least some information contained in a given geographic fact is a function of the arrangement of areal units, a thorough understanding of its properties is needed. This understanding is attained in part by quantifying configurational structures, and noting that the limiting values of spatial autocorrelation indices are strictly a function of areal unit configuration. This quantification is achieved with use of the principal eigenvalue, and its associated eigenvector, of the connectivity matrix under study. The principal eigenvalue appears to be a quadratic function of the number of areal units. Under the assumption of random partitioning the principal eigenvalue appears to be normally distributed. And, the principal eigenvector has a natural interpretation in terms of combinatorics.

Findings for a generalized canonical correlation analysis of 1971 Toronto data indicate that the configurational information referred to here correlates very strongly with geographic land use patterns, journey-to-work spatial interaction patterns, and with the factorial ecological structure of the city. Moreover, the implication is that results of geographic data analyses are due, at least in part, to the underlying arrangement of areal units.

The problematic points raised here can be reduced to a single issue, namely that inferential problems arise when traditional statistical analyses are carried out on spatial data series having dependent observations. Thus a primary aim of spatial autocorrelation analysis should be to posit properly specified models for spatial data. These models should contain explicit autoregressive terms, and exploit the spatial structure that is of importance. Mostly this reformulation of classical linear statistical models has been along the lines of generalized least squares solutions to autocorrelated errors models, either for the pure autoregressive model or the autoregressive trend surface model. In each case both conditional and simultaneous specifications have been employed. The univariate model has been treated almost exclusively. Insights concerning the bivariate model are presented here. Letting X be a random variable, with a non-zero correlation between X and Y, introduces considerable complexity into autoregressive analysis. This complexity stems from the Jacobian term as well as the convolution of classical correlation and spatial autocorrelation in the covariance matrix. Finally, the fixed effects ANOVA model and its special case of the two-groups t-test can have Type I and Type II errors dramatically altered in the presence of spatial autocorrelation.

Often, then, traditional hypothesis testing statistics are inappropriate for drawing proper inferences about parameters when these modified models are required. In many cases parameter estimates remain unbiased. But standard errors tend to change. A handle can be obtained for this problem by inspecting asymptotic variances. These results already have been outlined for the simultaneous model, while those for the conditional model are presented here. Because most geographical problems fall into the realm of small sample theory, though, small sample estimates may be more appropriate, and can be obtained by conducting Monte Carlo simulation experiments.

Returning to the statistical properties of efficiency and consistency, comparisons between the traditional and the modified statistics prove to be interesting. First, these statistics are equally efficient if zero spatial autocorrelation is present, whereas the modified statistic is always more efficient if non-zero spatial autocorrelation is present. This latter efficiency improves as the degree of autocorrelation increases, and decreases as sample size increases. With regard to consistency, for the autocorrelated errors model both the constant and the variable spatial mean parameter estimates are at least weakly consistent. The spatial autocorrelation parameter also has been shown to be consistent.

Spectral representations of spatial autoregressive models are worth studying because they provide useful information about both finite and infinite regular lattice situations. In this first case, spectral representations suggest characteristics have to do with edge effects. In this second case, they supply a benchmark against which irregular lattice results can be compared. Spectral analysis also has been employed in spatial statistics to measure two-dimensional shapes.

Besides the complexities mentioned thus far, the field of spatial statistics also embraces several prominent difficult, bothersome, and unresolved problems. One concern has to do with incomplete data sets, or empirical situations involving missing data values. Another comes from the difficulties arising from dividing geographical landscapes into a finite number of areal units, and from dealing with finite surfaces. A third refers to the relationships between spatial autocorrelation and multivariate data sets (other than the spatially autocorrelated errors model). A fourth problem has to do with the ignorance that exists presently pertaining to sampling distributions for spatial statistics.

Geographical data sets sometimes contain missing observations that need to be estimated. Such estimates allow statistical parameters to be estimated, or can serve as input to mathematical optimization models. The reasons for missing data vary. Certainly one for census data is suppression for reasons of confidentiality. For others, such as satellite imagery or weather stations, monitoring equipment malfunctioning could lead to missing data. An exact maximum likelihood solution is discussed here, for multivariate normal spatial data sets satisfying the first-order spatial Markov property with a constant mean. Results can be extended quite easily to the simultaneous model, the variable spatial mean model, and to a case of constrained maximum likelihood in which certain exogenous information is available for the missing data.

Edge effects have been alluded to repeatedly. For instance, the finite nature of a two-dimensional surface has been found to influence the decision regarding whether or not to use data transformations. A relation seems to exist between the shape of a region and the estimate of spatial autocorrelation latent in data distributed over it. The spatial linear operator fails to yield the spectral density estimates of the covariance matrix, for finite regular lattices. Moreover, edge effects are concerned with inferential errors.

To date a variety of edge effects correction techniques have been proposed. Unfortunately findings from the research devoted to this problem imply that these edge effects are quite resistent with regard to the taking of any action to successfully correct for them. Neither additive nor multiplicative forms have been fruitful. At least for the time being, then, spatial analysts should recognize and acknowledge the presence of edge effect bias, even though they are unable to correct for these effects.

Multivariate models beyond the fixed effects type deserve treatment now, too. As has been seen for the regression parameter estimates in the presence of spatially autocorrelated errors, taking into account spatial autocorrelation for multivariate data leads to an alteration in factor structures, for instance. An empirical example has shown that multivariate analysis of variance and discriminant function analysis results can be altered quite dramatically, too.

Finally, the problems of small sample size and lack of closed forms for sampling distribution functions for most spatial statistics (in some cases due to the existence of an irregular lattice) mean simulation experimentation may be an invaluable tool for understanding spatial statistics. For example, although asymptotic variances can be obtained for particular autoregressive structures, spatial statistics may behave quite differently for small, finite lattices, or for irregular lattices. Overall, then, simulation may be a very productive route to pursue in the development of spatial statistical knowledge.

10.2. CONCLUSIONS

In conclusion, developments in spatial statistics have been concerned with the drawing of proper inferences, and should encompass a fuller explication of results already derived. In this first instance, consider the estimation of a trend surface model describing the geographic distribution by census tract of income in Houston in 1980. The general model that has been explored is of the form

$$(\mathbf{Y} - \mathbf{X}\boldsymbol{\beta})^T(\mathbf{I} - \varrho\mathbf{C})\,(\mathbf{Y} - \mathbf{X}\boldsymbol{\beta}). \tag{10.1}$$

Rewriting expression (10.1) in its expanded form yields

$$[\mathbf{Y}^T\mathbf{Y} - 2\boldsymbol{\beta}^T\mathbf{X}^T\mathbf{Y} + \boldsymbol{\beta}^T\mathbf{X}^T\mathbf{X}\boldsymbol{\beta}] - \varrho[\mathbf{Y}^T\mathbf{C}\mathbf{Y} - 2\boldsymbol{\beta}^T\mathbf{X}^T\mathbf{C}\mathbf{Y} + \boldsymbol{\beta}^T\mathbf{X}^T\mathbf{C}\mathbf{X}\boldsymbol{\beta}]. \tag{10.2}$$

Letting $\varrho = 0$ reduces this model in such a way that OLS procedures are applicable, and hence only the first bracketed term of expression (10.2) is of concern. Using the extra sum of squares principle, this first component of expression (10.2) conditionally decomposes the variance as follows:

Source	Extra sum of squares	Degrees of freedom
Simple autoregressive	0.0	0
Mean	246388.6	1
Linear terms	13907.3	2
Quadratic terms	3977.2	3
Cubic terms	6692.4	4
Quartic terms	2677.9	5
Quintic terms	5760.9	6
Sextic terms	2254.3	7
Error	50932.4	332
Total	332591.0	360

If the sextic equation is considered to be the full model here, then $s_\xi^2 = 153.4$, and the respective F-ratios are 1606.1, 45.3, 8.6, 10.9, 3.5, 6.3 and 2.1. The first six of these statistics are significant. In other words, a fifth-order trend surface model should be employed (Unwin, 1975). The regression equation for this sixth-order of trend surface model is, with the accompanying parameter estimate t-statistics in parentheses,

$$\begin{aligned}\hat{Y} = {} & \underset{(12.5)}{25.105} + [\underset{(-8.0)}{-7.200X_1} - \underset{(-0.8)}{0.909X_2}] + [\underset{(3.2)}{1.045X_1^2} - \underset{(-1.8)}{0.931X_1X_2} \\ & - \underset{(-1.7)}{1.006X_2^2}] + [\underset{(4.1)}{0.221X_1^3} + \underset{(1.8)}{0.208X_1^2X_2} + \underset{(3.5)}{0.486X_1X_2^2}\end{aligned}$$

$$
\begin{aligned}
&- \underset{(0.0)}{0.002X_2^3}] + [\underset{(-2.4)}{-0.029X_1^4} - \underset{(-0.1)}{0.001X_1^3X_2} - \underset{(-0.9)}{0.040X_1^2X_2^2} \\
&+ \underset{(2.0)}{0.095X_1X_2^3} + \underset{(1.2)}{0.046X_2^4}] + [\underset{(-2.9)}{-0.002X_1^5} - \underset{(-1.0)}{0.003X_1^4X_2} \\
&- \underset{(-1.0)}{0.005X_1^3X_2^2} - \underset{(-2.2)}{0.011X_1^2X_2^3} - \underset{(-2.3)}{0.009X_1X_2^4} \\
&+ \underset{(0.6)}{0.001X_2^5}] + [\underset{(1.5)}{0.000X_1^6} + \underset{(-0.1)}{0.000X_1^5X_2} + \underset{(0.9)}{0.001X_1^4X_2^2} \\
&+ \underset{(0.8)}{0.001X_1^3X_2^3} + \underset{(0.2)}{0.000X_1^2X_2^4} - \underset{(-1.9)}{0.002X_1X_2^5} - \underset{(-0.7)}{0.001X_2^6}]
\end{aligned}
\tag{10.3}
$$

Besides the constant, the only significant terms here appear to be those containing X_1 axis measures.

Estimating ϱ, rather than setting it equal to zero, has yielded the following results:

Source	$\hat{\varrho}$	Extra sum of squares	Degrees of freedom
Simple autoregressive	0.1750	246389.0	1
Mean	0.1750	49701.1	1
Linear terms	0.1747	752.1	2
Quadratic terms	0.1747	273.2	3
Error		35475.6	353
Total		332591.0	360

If the quadratic equation is assumed to be the full model here, and the unweighted sum of squares terms are considered, then $s_\xi^2 = 100.5$, and the respective F-ratios are 2451.7, 494.6, 3.7 and 0.9. The regression equation for this particular trend surface model is, with the affiliated t-statistics,

$$
\begin{aligned}
\hat{Y} &= 0.175 \sum_{j=1}^{360} c_j Y_j + \underset{(9.0)}{23.870} + [\underset{(-2.7)}{-4.611X_1} - \underset{(-0.2)}{0.384X_2}] \\
&\quad + [\underset{(1.4)}{2.551X_1^2} - \underset{(-0.4)}{0.949X_1X_2} - \underset{(-0.2)}{0.448X_2^2}].
\end{aligned}
\tag{10.4}
$$

Again the only significant absolute coordinate term has to do with the X_1 axis measure. Differences between equations (10.3) and (10.4) have to do with the inclusion of cubic, quartic, quintic and sextic terms in (10.3), as opposed to the lagged term in (10.4). Clearly equation (10.4) is much simpler. In addition, both parameter estimates and t-statistics are noticeably different for these two equations.

In equation (10.4) only the simple autoregressive, the mean and the linear terms are significant. The estimate of ϱ is remarkably stable over the set of regression equations. And presumably the need for a very high order trend surface model with OLS estimation results from an attempt to capture the local autoregressive structure with a surface having many hills and valleys in it. Consequently, the standard error estimate has been reduced by 34%, and only a first-order trend surface model has been found to be appropriate. These findings illustrate the concerns spatial analysts need to have about drawing proper inferences, and characterizes much of the thrust of spatial statistics to date.

A concern also is emerging that has to do with further developments of results already derived. Mardia and Marshall (1984) have outlined in detail the properties of consistency and asymptotic normality for spatially autocorrelated errors models. These results have been referred to in previous chapters of this text.

Lele and Ord (1986a) have described some asymptotic results for conditional least squares estimation in the presence of spatial autocorrelation. The main advantage of this method is that rather than requiring knowledge of the conditional distribution, it only requires knowledge of the conditional mean. They have shown that these sorts of estimators are consistent and asymptotically normal. They also found that while point estimates are obtained with both this method and that of pseudo-likelihood, an obvious way of developing confidence intervals is not yet known. Further, for the case of Gaussian random variables, they found that the conditional least squares and the pseudo-likelihood methods are equivalent, although this former method is more flexible and simi-parametric in nature.

Lele and Ord (1986b) also have explored optimality properties of Besag's pseudo-likelihood method, which has been suggested as an effective procedure in those cases where the likelihood function is intractable. They outline the conditions under which the pseudo-likelihood function represents the best linear combination of the coding likelihood equations, and also demonstrate the asymptotic equivalence between such estimators and the conceptually simpler weighted average of coding estimators. In general it is not the best linear combination. It is best though when a weakly stationary spatial process without boundary points satisfies certain necessary and sufficient conditions.

Therefore, much of the skeleton of spatial statistical theory has been constructed already. But considerable virgin territory remains untouched; presumably exploration of this territory will lead to the fleshing out of this body of knowledge. To this end it is hoped that this book has made a contribution.

10.3. REFERENCES

Lele, S., and J. Ord, 1986a, Conditional least squares estimation for spatial processes: some asymptotic results, *Technical Reports and Reprints* No. 65, Department of Statistics, The Pennsylvania State University, University Park.

Lele, S., and J. Ord, 1986b, Besag's pseudo-likelihood: some optimality results, *Technical Reports and Reprints* No. 66, Department of Statistics, The Pennsylvania State University, University Park.

Mardia, K., and R. Marshall, 1984, Maximum likelihood estimation of models for residual covariance in spatial regression, *Biometrika*, Vol. 71: 135–146.

Unwin, D., 1975, *An Introduction to Trend Surface Analysis*, Norwich, England: Geo Abstracts.

Index

ADVANCED STUDIES IN THEORETICAL AND APPLIED ECONOMETRICS
VOLUME 12

---

1. Paelinck J.H.P. (ed.): Qualitative and Quantitative Mathematical Economics, 1982. ISBN 90 247 2623 9.
2. Ancot J.P. (ed.): Analysing the Structure of Economic Models, 1984. ISBN 90 247 2894 0.
3. Hughes Hallett A.J. (ed.): Applied Decision Analysis and Economic Behaviour, 1984. ISBN 90 247 2968 8.
4. Sengupta J.K.: Information and Efficiency in Economic Decision, 1985. ISBN 90 247 3072 4.
5. Artus P. and Guvenen O., in collaboration with Gagey F. (eds.): International Macroeconomic Modelling for Policy Decisions, 1986. ISBN 90 247 3201 8.
6. Vilares M.J.: Structural Change in Macroeconomic Models, 1986. ISBN 90 247 3277 8.
7. Carraro C. and Sartore D. (eds.): Developments of Control Theory for Economic Analysis, 1987. ISBN 90 247 3345 6.
8. Broer D.P. (ed.): Neoclassical Theory and Empirical Models of Aggregate Firm Behaviour, 1987. ISBN 90 247 3412 6.
9. Italianer A. (ed.): Theory and Practice of International Trade Linkage Models, 1986. ISBN 90 247 3407 X.
10. Kendrick D.A.: Feedback, 1988. ISBN 90 247 3593 9 (HB). ISBN 90 247 3650 1 (PB).
11. Sengupta J.K. and Kadekodi G.K. (eds.): Econometrics of Planning and Efficiency, 1988. ISBN 90 247 3602 1.
12. Griffith D.A.: Advanced Spatial Statistics, 1988. ISBN 90 247 3627 7.